Advances in
MARINE BIOLOGY

VOLUME 43

Advances in MARINE BIOLOGY

Series Editors

A. J. SOUTHWARD

Marine Biological Association, The Laboratory, Citadel Hill, Plymouth, PL1 2PB, UK

P. A. TYLER

School of Ocean and Earth Science, University of Southampton, Southampton Oceanography Centre, European Way, Southampton, SO14 3ZH, UK

C. M. YOUNG

Marine Sciences Division, Harbor Branch Oceanographic Institution, 5600 US Highway, 1N, Fort Pierce, Florida 34946, USA

and

L. A. FUIMAN

Marine Science Institute, University of Texas at Austin, 750 Channel View Drive, Port Aransas, Texas 78373, USA

ACADEMIC PRESS

An imprint of Elsevier Science

Amsterdam – Boston – London – New York – Oxford – Paris – San Diego
San Francisco – Singapore – Sydney – Tokyo

Academic Press
An Imprint of Elsevier Science
84 Theobald's Road, London WC1X 8RR, UK
http://www.academicpress.com

Academic Press
An Imprint of Elsevier Science
525 B Street, Suite 1900, San Diego, California 92101-4495, USA
http://www.academicpress.com

ISBN 0-12-026143-X

A catalogue record for this book is available from the British Library

Typeset by Keyset Composition, Colchester, Essex, UK
Printed and bound in Great Britain by MPG, Bodmin, Cornwall, UK

02 03 04 05 06 07 MP 9 8 7 6 5 4 3 2 1

CONTRIBUTORS TO VOLUME 43

A. S. Brierley, *Gatty Marine Laboratory, School of Biology, University of St Andrews, St Andrews, Fife, KY16 8LB, UK*

J. D. Hedley, *Tropical Coastal Management Studies, Department of Marine Sciences and Coastal Management, Ridley Building, University of Newcastle, NE1 7RU, UK*

E. Ramirez Llodra, *School of Ocean and Earth Science, University of Southampton, Southampton Oceanography Centre, European Way, Southampton, SO14 3ZH, UK*

P. J. Mumby, *Tropical Coastal Management Studies, Department of Marine Sciences and Coastal Management, Ridley Building, University of Newcastle, NE1 7RU, UK*

K. Rohde, *Division of Zoology, School of Biological Sciences, University of New England, Armidale, NSW 2351, Australia*

D. N. Thomas, *School of Ocean Sciences, University of Wales-Bangor, Menai Bridge, Anglesey, LL59 5EY, UK*

CONTENTS

Ecology and Biogeography of Marine Parasites

Klaus Rohde

Fecundity and Life-history Strategies in Marine Invertebrates

Eva Ramirez Llodra

Ecology of Southern Ocean Pack Ice

Andrew S. Brierley and David N. Thomas

Biological and Remote Sensing Perspectives of Pigmentation in Coral Reef Organisms

John D. Hedley and Peter J. Mumby

Colour Plate Section appears between pages 276 and 277.

Series Contents for Last Ten Years*

*The full list of contents for volumes 1–37 can be found in volume 38.

Ecology and Biogeography of Marine Parasites

Klaus Rohde

*School of Biological Sciences, University of New England, Armidale
NSW 2351, Australia
FAX: +61 2 6773 3814 email: krohde@metz.une.edu.au*

*This work is dedicated to John Sprent, the eminent
Australian parasitologist.*

ADVANCES IN MARINE BIOLOGY VOL. 43
ISBN 0-12-026143-X

A review is given of (mainly recent) work on the biodiversity, ecology, biogeography and practical importance of marine parasites. Problems in estimating species numbers have been thoroughly discussed for free-living species, and the main points of these discussions are reviewed here. Even rough estimates of the richness of most parasite groups in the oceans are premature for the following reasons: species numbers of host groups, in particular in the deep sea and the meiofauna, are not known; most host groups have been examined only insufficiently for parasites or not at all; even in some of the best known groups, latitudinal, longitudinal and depth gradients in species richness are only poorly understood or not known at all; effects of hosts on parasite morphology and geographical variation have been studied only in a few cases; there are few studies using techniques of molecular biology to distinguish sibling species. Estimates of species richness in the best known groups, trematodes, monogeneans and copepods of marine fishes, are given. Parasites are found in almost all taxa of eukaryotes, but most parasitic species are concentrated in a few taxa.

Important aspects of the ecology of marine parasites are discussed. It is emphasized that host specificity and host ranges should be distinguished, and an index that permits calculation of host specificity is discussed. The same index can be applied to measure site specificity. Central problems in ecology are the importance of interspecific competition and whether equilibrium or non-equilibrium conditions prevail. Marine parasites are among the few groups of organisms that have been extensively examined in this regard. A holistic approach, i.e. application of many methods, has unambiguously shown that metazoan ecto- (and probably endo-) parasites of marine fish live in largely non-saturated niche space under non-equilibrium conditions, i.e. they live in assemblages rather than in communities structured by competition. Nestedness occurs in such assemblages, but it can be explained by characteristics of the species themselves. There is little agreement on which other factors are involved in "structuring" parasite assemblages. Few studies on metapopulations of marine parasites have been made. A new approach, that of fuzzy chaos modelling, is discussed. It is likely that marine parasites are commonly found in metapopulations consisting of many subpopulations, and they are ideally suited to test the predictions of fuzzy chaos. Some recent studies on functional ecology and morphology – especially with regard to host, site and mate finding – are discussed, and attention is drawn to the amazing variety of sensory receptors in some marine parasites. Effects of parasites on hosts, and some studies on the evolution and speciation of marine parasites are discussed as well.

A detailed overview of biogeographical studies is given, with respect to latitudinal gradients in species diversity, reproductive strategies and host ranges/specificity. Studies of marine parasites have contributed significantly to giving a non-equilibrium explanation for latitudinal diversity gradients. Recent studies on longitudinal and depth gradients are discussed, as well as parasites in brackish water, parasites as indicators of zoogeographical regions and barriers, and parasites as biological tags.

The practical importance of marine parasites in mariculture, as monitors of pollution, agents of human disease, the use of parasites for controlling introduced marine pests, and some related aspects, are also discussed.

1. INTRODUCTION

The aim of this article is to give an overview of the most important aspects of the diversity, ecology and zoogeography of marine parasites. Emphasis is on recent work, although some older, particularly biogeographical, studies are referred to as well, if they have not been reviewed before or if they are necessary for a coherent account. The reader is referred to the monograph *Ecology of Marine Parasites* by Rohde (2nd edn, 1993) for a more detailed discussion of older work. Attention is drawn to the lack of studies on certain aspects of marine parasitology and recently-developed methods for such studies. Most of the references used are from parasito-logical and ecological journals, and this review should therefore be of particular value to marine biologists.

2. DIVERSITY OF MARINE PARASITES

Problems in estimating species richness have been extensively discussed for free-living organisms, and estimates of species numbers of parasites depend on our knowledge of the richness of their hosts. Therefore, free-living species are discussed first.

2.1. Problems in estimating species richness in free-living organisms

There is more than one species concept. As recently pointed out by Sluys and Hazevoet (1999), species concepts can be based on interbreeding or common descent, and application of these concepts can give conflicting results. For metazoan animals and most protozoans, the biological species concept based on interbreeding (a species comprises all individuals that can interbreed freely in nature) is usually satisfactory, although intermediate forms may exist ("Rassenkreise", geographical species). For microorganisms, different concepts may be necessary. Although interbreeding for most species has not been tested, it is made likely by morphological (and molecular) similarity that is greater within than between species. However, it is necessary to keep in mind that there will always be a subjective component in decisions regarding species distinctions, and there will always be ambiguous cases of species identity, which may lead to over- or under-estimates of species numbers.

Grassle and Maciolek (1992) sampled along a 176 km transect on the continental slope of the north-east Atlantic Ocean, recording 798 species of macroinvertebrates. They extrapolated from these data and concluded that the global richness of deep-sea soft sentiments is in the order of 10^7 species. Koslow et al. (1997) tested their method by extrapolating global deep-sea richness of fish from a survey along the continental slope of western Australia, arriving at a number of 60,000 species. Only about 2650 deep-sea fish species have been described to date and, adding the number of species estimated as not yet described, there should not be more than 3000–4000 species. The authors concluded that the method of Grassle and Maciolek should not be used. Major errors may be introduced because of our ignorance of habitat specificity and geographical ranges of species.

May (1990) discussed estimates of global species richness and methods used for arriving at these estimates. He begins his account with the statement that the number of species on earth "is currently uncertain to within a factor of 10 or more". Even the number of described species is uncertain, but may be around 1.8 million (Stork, 1988). Methods using projections from past trends in discovering and describing species suffer from the fact that past studies are dominated by those conducted in cool temperate environments. "The tropical insect fauna may have very different patterns of diversity". May referred to the differences between tropical and temperate diversity patterns, but it is important to note that – in the oceans – gradients with depth also have to be considered and relationships are not always linear. For example, Boucher and Lambshead (1995) found a non-linear relationship between depth and species richness for marine nematodes, the bathyal and abyssal being richest. Further, according to May, importantly, we know little about the specificity of animals (e.g. of insects for certain tree species), and estimates based on recording insect species (comprising the majority of animal species) from particular plant species and multiplying them by the number of plant species are therefore invalid. Although May considered the number of insect parasitoids – estimated by Hawkins (1990) to comprise about five species per phytophagous insect species in tropical and temperate regions – the number of endoparasites was not considered. How many species of protozoan, nematode, etc. species are found, for example, in the millions of tropical insect species? No studies have been conducted, at least on parasites of insects from tropical tree canopies. Furthermore, according to May, the animal kingdom comprises more than 70% of all recorded species, but the proportion of microorganisms may be much greater than estimated at present. All these points apply to marine species, including marine parasites, and are discussed further below.

May (1990) also discussed the empirical rule, according to which "for each tenfold reduction in length (1000-fold reduction in body weight) there are 100 times the number of species", pointing out that the rule breaks down at body lengths below 1 cm, which, however, may be due to our incomplete knowledge of small terrestrial animals. This rule is discussed further below with respect to parasites.

May and Nee (1995) drew attention to the "species alias problem", i.e. the proportion of synonyms. Insects have been most thoroughly studied in this respect: synonymy rates were commonly about 20%, but exceeded 50% in some groups. Importantly, rates may be even higher, as studies are continuing and it is unlikely that all synonyms have already been discovered. Taking the approximate proportion of synonyms into account, the number of recorded species worldwide, according to these authors, should be downgraded to about one million.

Genuine intermediate cases between species and subspecies exist, a problem that has been well studied in terrestrial animals. Examples are "Rassenkreise" (geographical species), in which most adjacent populations interbreed but populations at the ends of the species' range that are in secondary contact do not.

Hodkinson and Casson (1990) estimated the number of tropical insect species by determining the fraction of species in a well-known group that had been recorded earlier, and applying this proportion to the total fauna. For example, they found that 37% of species in a Sulawesi National Park was already known before their survey, and arrived at an estimate of a total number of 2.7 million insect species. However, such generalizations are almost certainly wrong (see below for marine parasites).

Molecular studies attempting to distinguish marine species are few. Such studies may reveal the existence of morphologically identical or very similar sibling species, or they may show that "species" distinguished on the basis of slight morphological diffferences may in fact be single species. Etter *et al.* (1999) compared "local" populations of four deep-sea mollusc species not separated by "major topographic features" and from sites only tens to hundreds of kilometres apart and from narrow depth ranges. They found that their genetic divergence is similar to that found between recognized coastal marine and aquatic mollusc species, suggesting that sibling species may exist and deep-sea diversity may be much higher than conventionally estimated on purely morphological grounds (see also Creasey and Rogers, 1999). Another example may be that of *Scomber japonicus*. On the basis of morphological data, populations in the Atlantic and Indo-Pacific were long considered as belonging to the same species. Recently, Scoles *et al.* (1998) and Collette (1999), on the basis of restriction site analysis with 12 restriction enzymes and cyto-chrome *b* sequences, suggested that Atlantic and Indo-Pacific populations of *Scomber japonicus* "may need to be recognized as separate species". However, samples from the Indian Ocean were not examined, and Baker and Collette (1998) have reported that populations long considered to be *S. japonicus* in the northern Indian Ocean and the Red Sea are in fact *S. australasicus*. So, on present knowledge, the former species is disjunct, i.e. populations of this species between the western Pacific and the Atlantic do not exist, which makes a decision difficult on whether a single species or two disjunct conspecific populations with a considerable degree of molecular divergence should be recognized (for parasites of this species see below).

All these difficulties have led to drastically different estimates of the number of free-living animal species in various marine groups and habitats. Hessler and Sanders (1967) were the first to draw attention to the amazing diversity in the deep sea, and Grassle and Maciolek (1992) estimated the deep-sea richness as exceeding 1 million and perhaps reaching close to 10 million (for a further discussion and further references see Solow, 1995; Gage, 1996). According to Lambshead (1993), about 160,000 marine species have been described, and he estimated that the total for macro-fauna could reach 10 million species (1×10^7), with the total meiofauna "an order of magnitude higher", using methods similar to those of Grassle. Briggs (1994), on the other hand, gave much lower estimates on the basis of advice received from specialists on the various taxa. He estimated that terrestrial species diversity is about 12 million, plus or minus 1 million, but that marine diversity appears to be less than 200,000, and this in spite of the much larger area of the seas and the greater evolutionary age. He quotes Thorson (1971) who gave an estimate of a marine living space 300 times larger than that on land; Stanley (1989), according to whom life in the sea appeared about 800 million years ago; and Seldon and

Edwards (1990), according to whom life on land appeared "somewhat more than 400 million year ago". Briggs attributed the discrepancy between land and the sea to the size of the primary producers: on land they are mainly trees and shrubs, whereas in the sea they are mainly unicellular organisms that cannot provide physical support to other plants and metazoan animals. His estimates include that for the Platyhelminthes, given as comprising approximately 5000 species. No estimate is given for the Copepoda, apparently because they are considered to be relatively poor in species. Both the very high and very low estimates for marine species richness are without much basis. Briggs' estimates do not include the very species-rich parasitic Platyhelminthes and Copepoda, nor do they include meiofauna. The estimates of Grassle and Maciolek (1992) for the deep-sea fauna are based on few data, and that of Lambshead (1993) includes an estimate for meiofauna based on few surveys. The only place where micro- and meiofauna have been thoroughly, although by no means exhaustively, studied is a sandy beach on the island of Sylt in the North Sea, where 652 species have been recorded to date, 25 times as many as macrofaunal species. However, many species – particularly of flagellates, foraminiferans, ciliates and nematodes but also of platyhelminths – remain to be described (estimated total of undescribed species about 200) (Armonies and Reise, 2000). The species richness may be above average because the beach is on a gradient from very sheltered to very exposed. Estimates of global meio/microfaunal richness are premature because we know little about the geographical ranges of species and latitudinal gradients in species richness. Nevertheless, the global diversity of meiofauna must be enormous, considering the counts on Sylt just mentioned, and considering the observation that "ubiquital" (widespread) meiobenthic species seem to be rare. According to Faubel (personal communication), of the 259 species of meiobenthic turbellarians collected by him from sandy exposed beaches along the Australian east coast, only two species were found both in tropical northern Queensland (Townsville) and south of temperate Sydney. As regards estimates for deep-sea richness, attempts to describe deep-sea meiofaunal turbellarians have failed because of the fragility of the species, which disintegrate when brought to the surface (Faubel, personal communication). For deep-sea nematodes, Lambshead et al. (1994), pointed out that only 10 studies of deep-sea nematode diversity at the species level have been made, and "a rough calculation, therefore, suggests that nematode diversity from less than a square metre of sea bed has been investigated in a deep-sea environment that covers slightly more than half the Earth's surface".

In summary, even rough estimates of species numbers of free-living marine animals are premature, particularly because of our poor knowledge of deep-sea and meiofaunal groups.

2.2. Problems in estimating species richness in marine parasites

All the difficulties discussed above for estimating species richness in free-living animals also apply to parasites, and there are some additional points that apply only to parasites. Parasites are smaller than their hosts and this is one of the reasons why parasites are less known than hosts. Contributing factors to our poor knowledge of marine parasites are that few studies of the geographical variation and of host effects on parasite morphology have been made. Furthermore, the scarcity of taxonomic studies using techniques of molecular biology, and our poor knowledge of species numbers in many host groups, particularly invertebrates such as the deep-sea fauna, meiofauna and tropical invertebrates (see above) (and to a lesser degree vertebrates), makes even rough estimates of species numbers difficult. The following section gives some examples that illustrate the difficulties.

2.2.1. *Lack of surveys*

Many (and probably most) free-living marine animals have not been examined for parasites. Thus, occasional observations have shown that benthic meiofaunal turbellarians may harbour mesozoans and other microscopic parasites, but there is no published quantitative study of meiofaunal parasites. There is not a single quantitative survey of parasites of deep-sea invertebrates (such as nematodes or molluscs). Also, it must be taken into consideration that even free-living marine invertebrates are very incompletely known, as discussed above.

2.2.2. *Body-size relations*

A rule of thumb, according to which animal species of a certain size should be 100 times as diverse as species 10 times larger, was referred to earlier. If correct, this would imply a number of more than 2,500,000 species of metazoan parasites and many millions of species of micro-organisms infecting the 25,000 species of fishes. This is highly unlikely.

2.2.3. *Synonymy*

For some marine parasites, the synonymy rate is even higher than for the insects mentioned above. In recent revisions of species of Monogenea infecting the heads and gills of Scombridae and belonging to two families (Gotocotylidae and Thoracocotylidae), Hayward and Rohde (1999a, b, c) and Rohde and Hayward (1999a, b) described eight new and eight old species considered as valid, and synonymized about 41 species (old and new synonyms, 72% of the total). However, it would be quite wrong to generalize this finding by claiming, for instance, that the thousands of Monogenea described to date should be reduced by 71%. The reason is that parasites of scombrids have been thoroughly examined by many authors and from many seas, because they are large and of commercial importance. The vast majority of marine fish species have never been examined for parasites or they have been examined poorly; most parasites have never been described under any name and not many synonyms are likely to exist.

2.2.4. *Geographical variation*

Species descriptions of parasites are often based on small morphological differences. In the Monogenea, such differences may be the size and shape of copulatory and attachment sclerites. Rohde and Watson (1985a, b) examined geographical variation of such sclerites in monogeneans of the genera *Kuhnia* and *Pseudokuhnia* on the gills of three species of *Scomber* from various regions in the Atlantic and Indo-Pacific Oceans, and found considerable geographical variation. Populations from widely-separated seas differ so strongly that taxonomists would not hesitate to describe them as different species, if they were not shown to be conspecific by intermediate forms (Table 1).

2.2.5. *Intermediate forms*

The foregoing text has drawn attention to molecular differences between Atlantic and Indo-Pacific populations of *Scomber japonicus*, which may justify distinction of two species. Parasite data may support this suggestion. The monogenean *Grubea cochlear* infects *Scomber scombrus* and *S. japonicus* in the Atlantic Ocean, and *G. australis* infects *S. australasicus* and probably *S. japonicus* in the Indo-Pacific. Morphological differences between the two species are distinct and unambiguous (Rohde, 1987b). Populations of another monogenean from the same hosts, *Pseudokuhnia minor* from the Atlantic and Indo-Pacific, were described as belonging to one species, but there are clear differences between them in the size of attachment and copulatory sclerites, although they are less distinct than

Table 1 Lengths of hamuli (μm) of *Kuhnia scombri* from different seas.

Lengths of hamuli (μm)	91–95	96–100	101–105	106–110	111–115	116–120	121–125	126–130	131–135	136–140
West Australia[a]	8	11	3							
NSW[a]		5	30	51	16	2				
Tasmania[a]		1	2	7	5	1	1			
New Zealand[a]				3	24	65	80	19	1	
Japan[b]				4	4	2				
Philippines[a]	1									
Ecuador[b]							3	13	11	
South Africa[b]					2	2				1
North Sea[c]				3	16	20	12	5	1	
Mediterranean[b,c]	1	5	10	9	1					
N. America[c] (W. Atlantic)		1	1	1	8	15	14	1		

Hosts: [a]*Scomber australasicus*, [b]*S. japonicus*, [c]*S. scombrus*.
Note: differences are not due to host size.

those between *Grubea* spp. (Rohde and Watson, 1985a) (Figures 1 and 2). A decision as to whether two species should be recognized is difficult because intermediate forms from the Indian Ocean may be found in the future and we may simply have a case of geographical variation of populations all belonging to the same species. There had been doubt about the specific status of the parasitic copepods *Lernaeocera branchialis* and *L. lusci*, but genetic analysis confirmed that they are indeed different species (Tirard *et al.*, 1993).

2.2.6. *Effects of latitudinal gradients*

A comparison of the data in Tables 2 and 3 shows that the proportion of Monogenea (and possibly of Copepoda) relative to trematodes (and possibly other metazoan endoparasites of marine fishes) is much greater at low than at high latitudes. Projections of global data based on exclusively tropical or high latitude surveys are therefore suspect. Nevertheless, the vast majority of marine animals live in warm waters. For example, at least three-quarters of marine coastal teleost species are found between latitudes 40°N and 40°S (Figure 72 in Rohde, 1978a). Therefore, even if latitudinal gradients have not been studied for many groups, extensive surveys at low latitudes will at least permit reasonable estimates of global richness. However, unfortunately, most surveys have been conducted in cold-temperate northern waters.

Table 2 Numbers of species of gill Monogenea of coastal teleost fish at different latitudes (after various sources).

Locality	Number of fish species examined	Number of fish examined	Number of species of gill Monogenea	Relative species richness
Pacific Ocean				
Bering Sea	17	213	22	≥ 1.1
Pacific coast of Canada	34	944	34–38	1.0–1.1
New Zealand	17	351	27	1.6
Northern NSW	45	1859	83	1.8
Great Barrier Reef	45	547	89	2.0
Atlantic Ocean				
White Sea	25	1232	9	0.36
Barents Sea	39	967	12	0.31
Argentina	7	422	5	0.7
Gulf of Mexico	37	412	46	1.2
Brazil	17	414	23–24	1.4

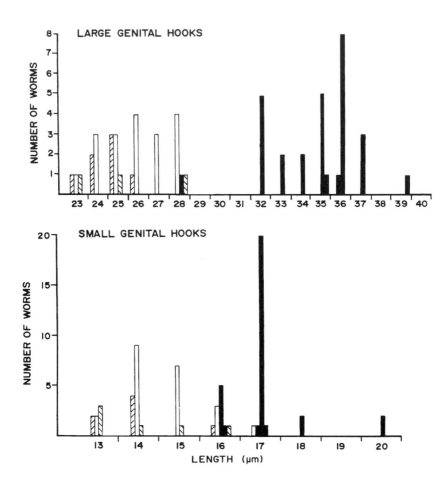

Figure 1 Pseudokuhnia minor. Length of large and small genital hooks in worms from the Atlantic and western Pacific Oceans. ■, three populations from the western Pacific: New South Wales, *Scomber australasicus*; Amoy, China, *Scomber australasicus*; and Philippines, *Scomber australasicus*. ▨, □, ▧, three populations from the Atlantic: Atlantic South Africa, *Scomber japonicus*; Atlantic Spain, *Scomber japonicus*; and Mediterranean, *Scomber japonicus*. Modified from: Rohde, K. and Watson, N. (1985a). Morphology and geographical variation of *Pseudokuhnia minor* n.g., n. comb. (Monogenea Polyopisthocotylea). *International Journal for Parasitology* **15**, 557–567, with the permission of the copyright holders, Elsevier Science.

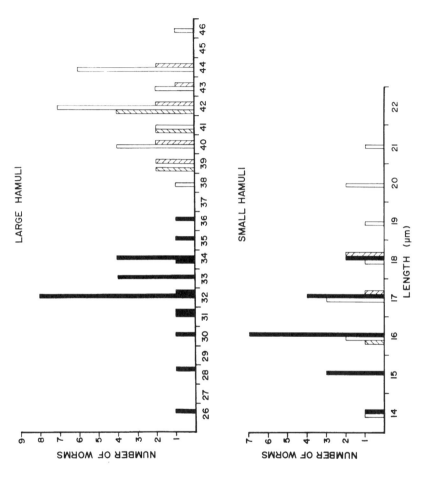

Figure 2 Pseudokuhnia minor. Length of large and small hamuli (attachment sclerites) in worms from the Atlantic and western Pacific Oceans. For explanation of histograms and sources see legend to Figure 1.

Table 3   Species richness of digenean trematodes at different latitudes.[a]

Locality	Number of teleost species examined	Number of fish examined	Number of Digenea species	Relative species richness
Barents Sea 70°N	42	972	43	1.0
White Sea 65°N	29	1354	30	1.0
Moreton Bay, S. Qld 27°30'S	103	1037	98	1.0
Heron Island 23°27'S	214	1352	236	1.1

[a]Reprinted from Rohde, K. and Heap, M. (1998). Latitudinal differences in species and community richness and in community structure of metazoan endo- and ectoparasites of marine teleost fish. *International Journal for Parasitology* **28**, 461–474, with the permission of the copyright holders, Elsevier Science.
Relative species richness = number of parasite species found/number of host species examined.

2.2.7. *Effects of depth gradients*

We have seen above that deep-sea diversity of free-living animals is little known. A further complicating factor for estimating global richness is the existence of gradients in richness with depth in a group that has been relatively well studied, i.e. helminth parasites of teleosts, best documented for monogenean ectoparasites. Rohde (1988) showed that relative species diversity of gill Monogenea is about five times greater in surface than in deep-water fish off the coast of south-eastern Australia (for details see Rohde, 1993 and section on depth gradients below).

2.2.8. *Effects of longitudinal gradients*

Little is known about differences in species diversity along a gradient from east to west, but the few studies available suggest that south-east Asian waters are the richest, with decreased richness in the eastern Pacific and Atlantic Oceans (Rohde and Hayward, 2000). Such gradients also have to be considered when estimating global species richness of parasites.

2.2.9. *Effects of hosts*

Similar parasites infecting different but often closely related host species are frequently described as different species, solely on the basis of slight differences in parasite morphology. However, some studies have shown that even parasites infecting the same host species but individuals of different size may vary significantly, for instance in the size of attachment sclerites of Monogenea (Figure 3), suggesting that there may be significant effects by hosts belonging to different species as well. Experimental studies (infection experiments) would be one approach to decide whether such species are indeed conspecific or not, but such studies have not been made. Likewise, techniques of molecular biology have not been applied to comparisons of similar parasite species infecting different marine host species. Future studies along those lines may well show that many supposedly "good" species are in fact synonyms of others, but they may also show that many morphologically identical or similar species belong to different (sibling) species.

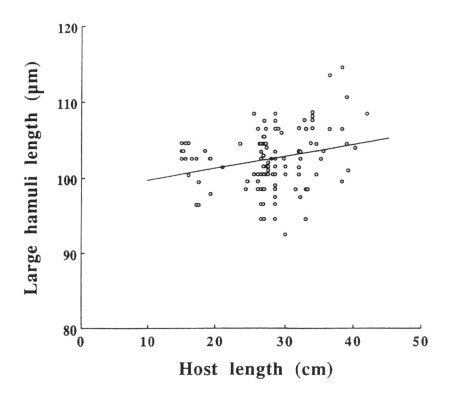

Figure 3. Relationship between length of host (*Scomber australasicus*) and length of large hamuli of *Kuhnia scombri*. Reproduced from: Perera, K. M. L. (1992). The effect of host size on large hamuli length of *Kuhnia scombri* (Monogenea: Polyopisthocotylea) from Eden, New South Wales, Australia. *International Journal for Parasitology* **22**, 123–124, with the permission of the copyright holders, Elsevier Science.

2.3. Estimates of richness in some taxa of marine fish parasites

Parasites of marine fish have been more extensively studied than parasites of any other marine host group. Therefore, in spite of the points mentioned in the last sections, comprehensive surveys permit at least reasonable estimates of species richness in some parasite groups. Rohde (1976, 1977b) examined 381 fish of 74 species at Heron Island, southern Great Barrier Reef, and 169 fish of 54 species at Lizard Island, northern Great Barrier Reef, and found 98 and 55 species of Monogenea, respectively, almost all from the gills. The skin, which also harbours monogeneans, was examined in only a few fish, and single individual fish were examined in 31 and 23 fish species, respectively. Single individuals of monogeneans were found of 21 and 11 species, respectively, indicating that larger surveys would, with certainty, recover additional parasite species. Also, a large proportion of the fish belonged to the very small damsel fishes (Pomacentridae). Monogenea represent only a small, though significant, proportion of all parasite species of marine fish, as indicated by several large surveys in other seas. Rohde (1976) therefore suggested that the total number of parasite species in the southern region of the Great Barrier Reef, where about 1000 fish species occur, must be at least 20,000. In the northern part of the Great Barrier Reef species richness of fish, and therefore of parasites, is even greater. These estimates have to be revised in view of the finding that monogeneans are relatively more diverse in warm than in cold seas (see Table 2 and below). According to Whittington (1998), there should be "almost 25,000 monogenean species on Earth", "if each host species is host to different species". Since there are about 3500 species of fish in Australian waters, there should be 3500 species of Monogenea, of which only about 300 species have been described. This estimate is without much basis; data available suggest that there is more than one species of Monogenea per fish species at low latitudes, whereas there are much fewer at high latitudes (Table 2).

Cribb et al. (1994) compared species of fishes from reef and inshore waters in Australia and found 236 species of trematodes on 218 species of Great Barrier Reef fishes and 98 species of trematodes on 103 species of inshore fishes (Moreton Bay). The authors estimated, without support from quantitative data, that the 1300 fish species on the Great Barrier may harbour about 2270 species of Digenea. According to Cribb (1998), there may be about 6000 species of Digenea on the 5500 species of Australian (terrestrial, freshwater and marine) vertebrate species.

The above estimates have to be revised by taking latitudinal gradients (Tables 2 and 3) and depth gradients (see below: Depth gradients) into account. The total number of fish species worldwide is estimated to be 24,618, of which 9966 are virtually confined to fresh water (Nelson,

1994). Of the 14,652 marine fish species, about 2650 described and a further 1000 species estimated as not yet described, occur in the deep sea (Koslow *et al.*, 1997), i.e. approximately 12,000 species are shallow-water species. Of the latter, each species, on average, harbours one species of trematode at all latitudes (Table 3), and probably about two species of gill and skin monogeneans (Table 2), giving a total number of about 12,000 species of trematodes and 24,000 species of monogeneans. Estimates for Monogenea are based on data in Table 2, which include only gill mono-geneans, and the fact that most fish species occur in warm Pacific waters. The 3000–4000 deep-sea fish harbour only about one adult trematode species per two host species and less than one adult gill monogenean species per two host species (Rohde, 1988), i.e. a total of about 1500 trematode and about 1500 monogenean species (adding some skin mono-geneans). Most species at different latitudes are different, although there is some overlap. However, this overlap is probably more than compensated for by the number of species not yet discovered on the host species examined. We arrive, then, at a global estimate of about 39,000 species of marine digeneans and monogeneans. Nematodes and cestodes are much less diverse than the above groups, although elasmobranchs harbour many cestode species, which make up for the very few digenean trematodes found in these hosts. There are probably not more than a few thousand species of these taxa. Copepods are as diverse as monogeneans or even more so, as indicated by dissections of many fish in Australian waters (unpublished observations), data on scombrid fishes (Tables 4 and 5), and the records from 182 species of Mediterranean fishes, which carried 226 species of parasitic copepods (Raibaut *et al.*, 1998). It is likely that the global number of copepods infecting marine fish is similar to that for monogeneans or greater, i.e. about 22,000–25,000. Isopods are common but not very diverse, probably comprising a few hundred or few thousand species. Myxozoa and various protozoans, in particular Microsporidia, are also common, but estimates of global richness are not possible. A total estimate of about 100,000 species for all parasites of marine fishes seems to be reasonable.

2.4. Concentration of parasites in some clades

Recent studies of the phylogeny of organisms have led to a reappraisal of relationships among taxa. Parasitic species are found in almost all large taxa of eukaryotes (Figure 4). Within smaller groups, parasitic species tend to be clustered in certain taxa. Thus, some invertebrate phyla consist entirely of parasites, although many others contain some parasites as

Table 4 Number of species of the scombrids *Scomberomorus* and *Grammatorcynus* and their copepod and monogenean parasites in different seas.[a]

	All species				Endemic species			
Ocean	Number of fish species	Number of copepod species	Number of monogenean species	Relative species richness	Number of fish species	Number of copepod species	Number of monogenean species	Relative species richness
Indo/West Pacific combined	13	14	14	2.15	12	8	13	1.75
West Pacific	12	14	14	2.33	7	2	7	1.28
Indian	6	12	7	3.17	1	0	1	1.00
East Pacific	2	6	3	4.50	2	0	2	1.00
East Atlantic	2	7	1	4.00	1	1	0	1.00
West Atlantic	4	9	3	3.00	4	5	0	1.25

[a]Reprinted from Rohde, K. (1999). Latitudinal gradients in species diversity and Rapoport's rule revisited: a review of recent work, and what can parasites teach us about the causes of the gradients? *Ecography* **22**, 593–613, with permission from the editor of *Ecography*.

Table 5 Parasite species richness, length and geographical characteristics of species of *Scomberomorus*.[a]

Host species	Geographical area	Number of species of Monogenea	Number of species of Copepoda	Total number of species	Maximum fork length (mm)	Geographical extent (km)	Midpoint latitude	Latitude range
Scomberomorus commerson	Indian, W. Pacific and East Atlantic	10	9	19	220	14,400	5 S	80°
Scomberomorus regalis	West Atlantic	3	7	10	84	1400	23 N	35°
Scomberomorus queenslandicus	West Pacific	6	4	10	100	1125	22 S	27°
Scomberomorus guttatus	Indian and W. Pacific	6	4	10	76	3000	12 N	45°
Scomberomorus maculatus	West Atlantic	4	4	8	77	875	32 N	27°
Scomberomorus tritor	East Atlantic	4	4	8	98	2150	15 N	60°
Scomberomorus niphonius	West Pacific	4	4	8	100	375	38 N	15°
Scomberomorus cavalla	West and Central Atlantic	4	4	8	173	4875	8 N	65°
Scomberomorus brasiliensis	West Atlantic	3	4	7	125	3350	8 S	55°
Scomberomorus concolor	East Pacific	3	3	6	76	195	27 N	15°
Scomberomorus sierra	East Pacific	3	3	6	97	2750	3 N	55°
Scomberomorus koreanus	Indian and W. Pacific	4	2	6	150	2600	16 N	43°
Scomberomorus semifasciatus	West Pacific	3	3	6	120	800	20 S	20°
Scomberomorus sinensis	West Pacific	3	2	5	200	1280	24 S	32°
Scomberomorus lineolatus	Indian and W. Pacific	2	3	5	80	1575	7 N	33°
Scomberomorus plurilineatus	Indian	3	1	4	120	900	20 S	30°
Scomberomorus multiradiatus	West Pacific	3	1	4	35	6	0°	7°
Scomberomorus munroi	West Pacific	4	0?	4?	100	1350	22 S	27°

[a]Reprinted from Rohde, K. (1999). Latitudinal gradients in species diversity and Rapoport's rule revisited: a review of recent work, and what can parasites teach us about the causes of the gradients? *Ecography* 22, 593–613, with permission from the editor of *Ecography*.

Note: Using linear correlations, no significant association was found between parasite species richness and fork length. A significant association was found between parasite species richness and geographical extent ($p = 0.01$, after Bonferoni adjustment) as well as latitudinal range. There were no significant associations when the species with a high leverage (*Scomberomorus commerson*) was omitted.

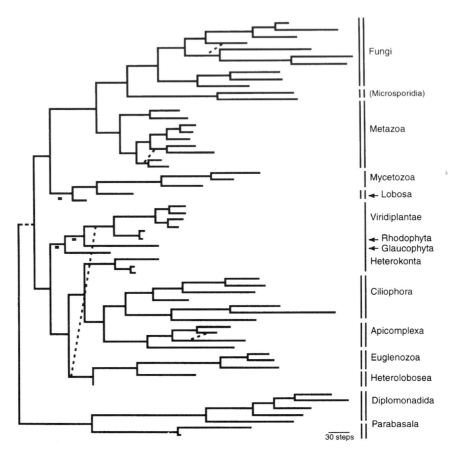

Figure 4 Phylogenetic tree of eukaryotes based on combined protein sequences. Taxa with parasitic species indicated by a double line. Note that almost all taxa include parasitic species, and some parasitic taxa have an early phylogenetic origin. (Based on but strongly modified from Baldauf *et al.*, 2000.)

well as free-living forms (Figure 5). Within the Platyhelminthes, a phylum comprising a large number of parasitic species and many important ones among them, the vast majority of parasites are found in the entirely parasitic Neodermata, and in some turbellarian taxa closely related to each other (the Fecampiida, *Urastoma, Ichthyophaga*), but the latter may in fact be monophyletic with the Neodermata (Figure 6). Nevertheless, parasitic species are found in other platyhelminth taxa as well. According to Jennings (1997), 200 species belonging to 35 families of Turbellaria are known to be parasites or commensals, mainly of invertebrates. Certain species, for example, live in the tube feet or the digestive tract of echinoderms, others infect the gills of fish. However,

Figure 5 Phylogenetic tree of invertebrates and chordates based on 18S rDNA sequences. Taxa with substantial proportion of parasitic species in bold and indicated by a cross. Note: several other taxa (not in bold) have parasitic species, e.g. Porifera, Chordata, Annelida, Mollusca. (Based on but strongly modified from Littlewood *et al.*, 1998.)

compared with the estimated number of 31,000 described species of Neodermata (Gibson, Natural History Museum London, personal communication), they represent only a very small minority of parasitic flatworms.

In summary, then, parasitism has evolved many times, but the majority of parasites are concentrated in a few phyla and in smaller taxa within them. Parasitism, including parasitism in the oceans, is almost as old as life itself.

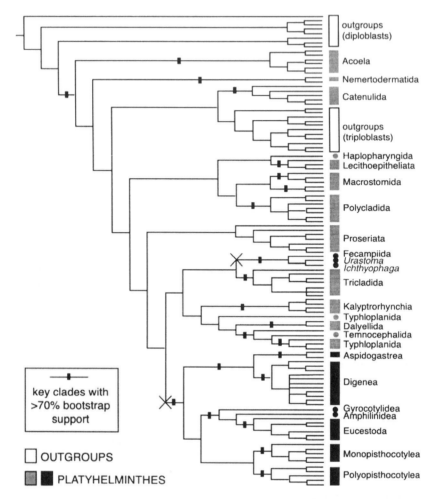

Figure 6 Phylogenetic tree of Platyhelminthes based on complete 18S rDNA sequences and morphology (total evidence). Platyhelminth taxa that are entirely parasitic are indicated by black columns circles, outgroups by white columns circles, and platyhelminth taxa largely consisting of free-living species by stippled columns. Crosses indicate common ancestor of main parasitic taxa. (Based on but strongly modified from Littlewood *et al.*, 1999.)

3. ECOLOGY OF MARINE PARASITES

3.1. Host ranges and host specificity

All parasites are restricted to certain host species, although the degree of host restriction varies greatly. However, even those species that have a wide host range may prefer some hosts over others. Rohde (1980d, 1982, 1984a) therefore distinguished host range and host specificity. Host range refers to the number of host species infected by a parasite species, irrespective of how frequently or how heavily the various species are infected. Host specificity takes prevalence and/or intensity of infection into account. To measure host specificity, Rohde (1980d, 1993, 1994a) proposed a host specificity index as follows:

$$S_i(\text{intensity}) = \frac{\sum(x_{ij}/n_j h_{ij})}{\sum(x_{ij}/n_j)}$$

where S_i = host specificity of ith parasite species, x_{ij} = number of parasite individuals of ith species in jth host species, n_j = number of host individuals in jth species examined, h_{ij} = rank of host species j based on intensity of infection x_{ij}/n_j (species with greatest intensity has rank 1). Numerical values for the indices vary between 0 and 1: the higher the degree of host specificity, the closer to 1. A parasite species infecting many host species, but only few of them heavily, has a greater host specificity than one that infects fewer species, but most of them equally. Poulin (1998) pointed out that the range of values found by application of the index depends on the number of host species. It increases with the number of host species utilized by a parasite. This should be taken into account when interpreting results.

In all groups of marine parasites, some species are strictly host-specific, others have a wide host range. However, most Monogenea have a very high degree of host specificity, whereas most Digenea are less specific. Host ranges, in the Digenea, depend on latitude (see below).

3.2. Site restriction

All parasites prefer certain microhabitats (sites) over others, although the degree of site specificity varies. Rohde's index for host specificity can be used to measure site specificity as well, by simply replacing number of hosts by number of sites. Most accurately, the indices can be used when sites can be quantified, for instance by subdividing the gill habitat into smaller sections of more or less equal size (Rohde, 1981). Among marine parasites, many Monogenea and many didymozoid (trematode) tissue parasites have particularly narrow microhabitats. An example is given in Figure 7. The Atlantic mackerel, *Scomber scombrus,* at Helgoland, North Sea, harbours two species of monogeneans, *Kuhnia sprostonae* on the pseudobranchs, and *K. scombri* at the base of the filaments of the main gills. The same fish

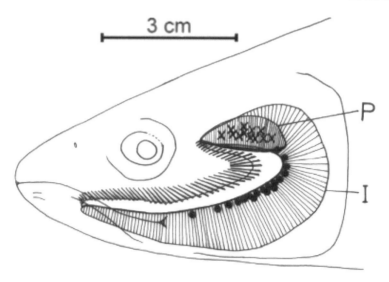

Figure 7 Pseudobranch (P) and first gill (I) of *Scomber scombrus* showing distribution of two species of Monogenea, *Kuhnia sprostonae* (×) and *K. scombri* (●).

species in other areas, and *S. australasicus* in the Indo-Pacific and *S. japonicus* in the Atlantic and Indo-Pacific are infected with further species: *Kuhnia scombercolias* and *Pseudokuhnia minor* on the middle parts of the gill filaments, *Grubea australis* (in the Indo-Pacific) and *G. cochlear* (in the Atlantic) on the most posterior and anterior parts of the gills.

Rohde (1994a) has discussed proximate and ultimate causes of niche restriction, including site restriction, and provided evidence that facilitation of mating is an important ultimate cause of such restriction ("mating hypothesis" of niche restriction; see also Rohde, in press). An example of proximate causation was given by Guitiérrez and Martorelli (1999), who showed that computer models, based on water currents and gill area combined, could explain the actual distribution of some monogeneans on freshwater fish better than either of the models alone.

3.3. Equilibrium or non-equilibrium? Saturation of niches with species and the importance of intra- and inter-specific competition

One of the central problems in ecology is the question of whether habitats are generally saturated with species or not; whether equilibrium or non-equilibrium conditions prevail. Closely connected to this question is the one concerning the importance of competition. If niche space is indeed densely packed, as often assumed, species are more likely to compete with each other than if niche space is largely empty.

Evidence for saturation given by various authors (e.g. Krebs, 1997) is provided by dense species packing leading to frequent interspecific competition as shown by complete or partial competitive exclusion or habitat shifts in the presence of other species, character displacement and particularly differences in the size of feeding organs of species using similar food resources, and an asymptotic relationship between local and regional species richness (Cornell and Lawton, 1992). Rohde and collaborators (Rohde, 1977a, 1979, 1980a, b, 1982, 1984b, 1989b, 1990, 1991a, b, 1993, 1994a, 1998a, 1999, 2001a, b, in press; Rohde et al., 1994, 1995, 1998; Morand et al., 1999) have used parasites of marine teleost fish to examine this question, and there is probably no other group of parasites that has been examined so thoroughly in this respect, by a variety of methods. Some of these methods have sometimes given contradictory results in other groups of organisms, and Rohde (in press) therefore emphasized the necessity of using an holistic approach, i.e. considering all evidence jointly.

The most direct evidence for a high degree of non-saturation is the demonstration that many habitats are indeed empty, and that species are not densely packed (e.g. Rohde, 1979). For parasites, this can be done by comparing species numbers of parasites infecting different host species. For example, Rohde (1998b, earlier references therein) compiled the number of species of metazoan ectoparasites on the heads and gills of 5666 teleost fish belonging to 112 species from his earlier studies (Figure 8). The maximum number of species found was 27, but the vast majority had fewer than six, and many had none or one. A similarly great variance is found for abundance. Maximum abundance was over 3000, but most fish species had fewer than five. Even if only fish of similar size and from similar habitats are compared, very great differences are found. The conclusion is inescapable that many more parasite species could be accommodated, provided that evolutionary time to fill these niches has been great enough. Indeed, if 27 parasite species is considered to be the maximum a species can support (an assumption for which there is no evidence), about 84% of all niches must be considered to be empty. Sasal et al. (1999a) studied communities of digenean endoparasites of 11 species of sparid and 7 species of labrid fishes in the Mediterranean and found a lack of niche saturation: "there was little inter- and/or intraspecific competition or there were enough available space and resources within the host".

Zander et al. (1999a) contradict the concept of "empty niche" because they "understand the niche as a dynamic system that can be either narrow due to the existence and influence of competitors or wide due to their

A

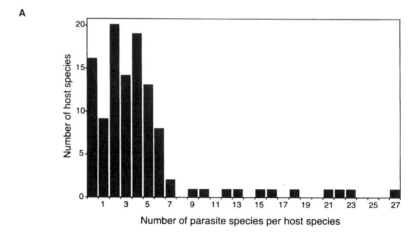

Number of parasite species per host species

B

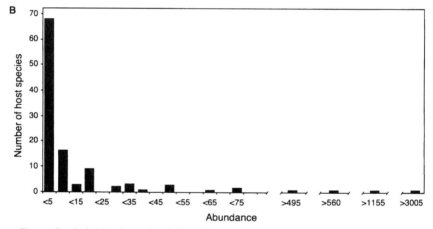

Figure 8 (A) Number of species of metazoan ectoparasites on the heads and gills per species of marine teleost fish (5666 fish of 112 species). Note: maximum number of parasite species per host species 27, most fish species with fewer than five. (B) Abundance (mean number of parasite individuals of all species) per host species. Note: maximum abundance more than 3000, most species with fewer than five. Modified from: Rohde, K. (1998b). Latitudinal gradients in species diversity. Area matters, but how much? *Oikos* **82**, 184–190, with permission from the editors of *Oikos*.

absence". This clearly misses the point: the argument for "empty niches" is not ecological but evolutionary. "Empty niche" simply implies that there is no reason to assume that identical or similar habitats (hosts) cannot support similar numbers of species (parasites) occupying them, provided there has been sufficient evolutionary time. In many fish species, gills are completely empty or only one small microhabitat is occupied; the niche is not wider, and it does not expand in the absence of competitors (see also Rohde, 1981).

Many authors have used reduction of niche overlap as evidence for the evolutionary importance of interspecific competition. An important method used to study interactions between species is the application of similarity indices, but most indices are symmetrical, and assume that each species affects the other equally. However, this is seldom the case. One species may be a stronger competitor than the other. Rohde and Hobbs (1986, 1999, further references therein) therefore proposed an asymmetrical percent similarity index, based on Reekonen's percent similarity index:

$$O_{A,B} = \frac{100A}{N_A} \sum_{i=1}^{k} \min(Q_{iA}, Q_{iB})$$

$$O_{B,A} = \frac{100B}{N_B} \sum_{i=1}^{k} \min(Q_{iA}, Q_{iB}),$$

where $O_{A,B}$ = overlap of A with B, A = number of individuals of species A in those k microhabitats in which B also occurs; N_A = total number of individuals of species A in all microhabitats; Q_{iA} and Q_{iB} = quotient of the number of individuals of species A and B respectively in microhabitat i and the total number of individuals of each species in the k microhabitats in which they co-occur. $O_{B,A}$ = overlap of B with A, etc.

The index was used to clarify whether interspecific competition or reinforcement of reproductive barriers is responsible for segregation in parasites on the gills of marine fishes, by comparing niche overlap of congeners and non-congeners using the same resources (space for attachment and food). Wilcoxon's two sample test performed on the asymmetric indices showed that congeners overlapped less than non-congeners, suggesting that reinforcement and not competition is responsible for the segregation. The results were not significant when symmetric indices were used. These results are supported by the finding that congeneric species of Monogenea with identical copulatory organs always inhabit different microhabitats on the gills, with little or no overlap, whereas congeneric Monogenea with dissimilar copulatory organs may inhabit the same microhabitats (Figure 9). This strongly suggests that reinforcement of reproductive barriers not interspecific competition is responsible for niche segregation in those cases in which it occurs.

Microhabitats (for example of monogeneans on the gills) sometimes change in the presence of other species, but often there are no or slight effects (e.g. Rohde, 1991b). Bagge and Valtonen (1996), in experiments

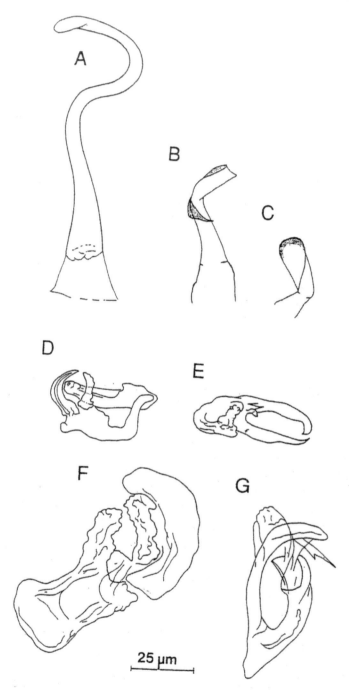

25 µm

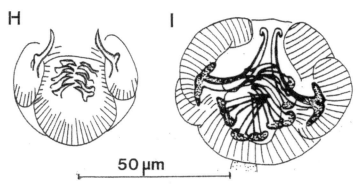

Figure 9 (This page and facing page) Copulatory organs of monopistho-cotylean monogeneans infecting the gills of *Lethrinus miniatus* on the Great Barrier Reef, Australia (A–G), and of polyopisthocotylean monogeneans infecting the gills of *Scomber* spp. (H, I). A–C, *Haliotrema* spp. in overlapping microhabitats. D, F, G, *Calydiscoides* spp. in overlapping microhabitats. E, *Protolamellodiscus* sp. on the pseudobranch. H, three species of *Kuhnia* and two species of *Grubea* spatially segregated in different microhabitats or in different geographical areas. I, *Pseudokuhnia minor* overlapping with four species of *Kuhnia* and *Grubea*. A–G reproduced from: Rohde, K., Hayward, C., Heap, M. and Gosper, D. (1994). A tropical assemblage of ectoparasites: gill and head parasites of *Lethrinus miniatus* (Teleostei, Lethrinidae). *International Journal for Parasitology* **24**, 1031–1053, with the permission of the copyright holders, Elsevier Science.

testing the effect of pollution on ectoparasites of freshwater roach, found that the two most common species of *Dactylogyrus* preferred the same microhabitats on the gills and showed no competition, which was confirmed by increased overlapping indices with increasing abundances between the species. Geets *et al.* (1997) found for all species of gill parasites that niche breadth was independent of the presence of other species, but increased with their own abundance in three of them, suggesting that interspecific effects are less important than intraspecific factors. They provide indirect evidence for the mating hypothesis of niche restriction (see above) for two gill monogeneans (highly aggregated distribution over the gill filaments). Microhabitat restriction is also found on host species that harbour only a single parasite species in a particular habitat, suggesting that microhabitat selection is genetically programmed and is not affected by competing species now, nor has it been affected by them in the evolutionary past. In a recent study, Lo and Morand (2001) found that infracommunities on a fish species in French Polynesia were too poor to "induce processes of interspecific competition".

Differences in the size of feeding organs of species is often given as evidence for interspecific competition. However, Rohde (1991b and references therein) has shown that such differences also occur in mono-geneans using the same resources, space for attachment and food on the same host. In other words, such differences may be fortuitous. Some authors have used an asymptotic relationship between local and regional species richness as evidence for competition. Such relationships have indeed been found when comparing infracommunities of parasites (local richness) with component communities (regional richness). However, computer simulations (Rohde, 1998a) have shown that these relationships may be a consequence of differential likelihoods of species appearing in a community because of different colonization rates and life spans (Figure 10). It is not necessary to invoke interspecific competition to explain the

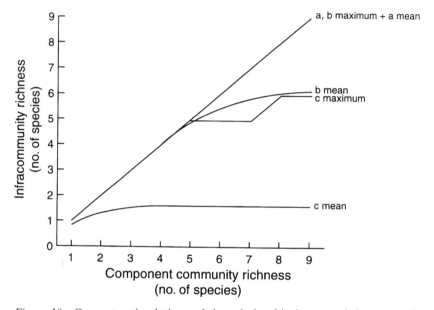

Figure 10 Computer simulations of the relationship between infracommunity and component community richness. The assumption is that each infracommunity can be recruited from any species in the component community, in any order, but species have different likelihoods of appearing in an infracommunity because of different transmission rates and intrinsic life spans. The richness of communities varies between one and nine species; 1000 iterations. (a) All species have a 100% likelihood of appearing in the infracommunity; (b) seven species have an 80% likelihood, and two have a 30% likelihood of appearing in the infracommunity; (c) two species have a 40% likelihood, and seven have a 10% likelihood of appearing. Note: asymptotic relationships for the means and maxima of all infracommunities except a maximum and mean and b maximum. For access to programme see http://www-personal.une.edu.au/~krohde/ (1) Infra- vs component community richness programme (Macintosh only).

relationship. Reduced interspecific aggregation relative to intraspecific aggregation also shows that interspecific competition is not important (Morand *et al.*, 1999).

Nestedness of parasite communities, i.e. the fact that species-poor assemblages represent non-random subsets of progressively richer assemblages, is not evidence for or against the significance of interspecific competition, because it may be caused by other factors, such as differential colonization rates of juvenile and adult hosts. In contrast, the lack of nestedness shows that "communities" are in fact assemblages not structured by competition or any other factors. The studies of Worthen and Rohde (1996) and Rohde *et al.* (1998) have shown that nestedness is not common in marine endo- or ectoparasites. Morand *et al.* (in press) have shown that nestedness, where it occurs, is not the result of inter-specific competition but of epidemiological processes, i.e. characteristics of the species themselves, and communities of marine fish parasites may be "anti-nested" rather than nested (Poulin and Guegan, 2000). Furthermore, positive associations (more common co-occurrence of species than expected on a random basis) are generally much more common than negative ones, and this also suggests that interspecific competition is not important (Rohde, 1994a). Hyperparasites of marine parasites have been demonstrated in a few cases (e.g. the monogenean *Udonella* on parasitic copepods), but such associations are rare, which suggests that a large number of niches are still to be filled.

Most recently, Rohde (2001b) has used a method proposed by Ritchie and Olff (1999) to test for species packing. In their search for "common principles that predict well known responses of biological diversity to different factors", such as the number of available niches in space, productivity, area, body size of species and habitat frag-mentation, these authors used spatial scaling laws based on fractal geometry to derive a rule for the minimum similarity in the size of species that share resources. The rule predicts that the body-size ratio of species of adjacent size should decline with increasing size of organisms, and it also predicts a unimodal distribution skewed to the left, when species numbers are plotted against the size of species. Ritchie and Olff showed that Serengeti (East African) mammalian herbivores and Minnesota savanna plants conform to these patterns. Both these groups can be expected to compete for resources (plants in the case of the former, light and nutrients in the case of the latter), because of their great vagility and dispersal abilities, respectively. Rohde (2001b) used large data sets of endo- and ectoparasites of marine fishes, and showed that the rule does not apply to them (examples in Figure 11). The method is particularly useful, because it permits a distinction between extant and evolutionary effects of interspecific

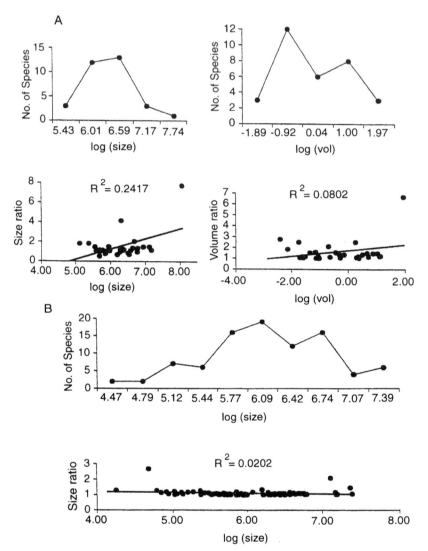

Figure 11 (A) Metazoan endoparasites of the marine teleost fish *Acanthopagrus australis.* Number of species of different sizes (maximum width × maximum length) and volumes (maximum width × maximum length × maximum depth) and size ratios of species of adjacent sizes. Note: 11 trematodes, 10 nematodes, 5 cestodes, 4 acanthocephalans. All except two species are parasites of the digestive tract. Outlier a very large tissue nematode. (B) Metazoan parasites in the digestive tract of marine fishes (1808 fish of 47 species). Note that the parasites do not conform to the packing rules of Ritchie and Olff (1999). Modified from: Rohde, K. (2001b). Spatial scaling laws may not apply to most animal species. *Oikos* **93**, 499–504, with permission from the editor of *Oikos.*

competition. Communities may well be "interactive" and show inter-actions between species, but such interactions are without evolutionary consequences when the packing rule does not apply. The method applied to many plant and animal groups may be used to test the prediction of Rohde (1980b) that large and vagile animal species, as well as species occurring in large populations, generally live in saturated niche space leading to significant interspecific competition, whereas small and sedentary species and species occurring in small populations live in non-saturated niche space and show little interspecific competition.

3.4. Structure of parasite communities

Some recent reviews and studies of marine parasite communities are by Kuris and Lafferty (1994), Lafferty *et al.* (1994), Zander (1998a, b), Sasal *et al.* (1999a), Alves and Luque (2001) and Kleeman (2001). Various factors have been implicated in determining community structure, in addition to species interactions discussed in the previous section. These factors are discussed below.

Many recent studies have used published and unpublished data and re-analysed them controlling for host phylogeny (e.g. Sasal and Morand, 1988). A widely used method is that of phylogenetically independent contrasts. As pointed out by Poulin (1995), there is no general agreement on the phylogeny of any of the major vertebrate groups, and analyses should be repeated when the "exact" phylogenetic relationships are known. Poulin (1995) examined the effects of body size, diet, habitat, latitude and the mean number of parasites per host on the parasite community richness of parasites in various vertebrate hosts. After correction for host phylogeny, he found that only host body size can have an important effect on parasite community richness, but this was not the case for fish ectoparasites (however, it "came close"). There was no evidence that change in habitat had an effect. Other authors have found other factors correlated with species richness of parasites. For example, Morand *et al.* (2000), after controlling for host phylogeny, found that only the percentage of plankton in the diet and conspecific host density were positively correlated with species richness of endoparasites in 21 species of Chaetodontidae in New Caledonia. Poulin (2000) did not find a correlation between fish length and intensity of infection in 76 host–parasites species associations after correction for sample size, although a positive correla-tion was found for larval digeneans, cestodes and gnathiid isopods. Guitiérrez (2001) found no relationship between length or weight of fishes and the number of monogeneans or species richness of monogeneans in a freshwater fish species in the Río de la Plata, Argentina. This casts doubt on the generality of the finding that larger fish harbour more diverse parasite communities, and although the author dealt with a freshwater fish, marine fish should at least be re-examined to test for the pattern. Groenewold *et al.* (1996), on the basis of the examination of parasites of four species of fish in the Wadden Sea, southern North Sea, concluded that the diet of the hosts appears to be the main factor determining the parasite community structure. Heath's (1987) study of endoparasites of deep-sea fishes, which included an ecological analysis, showed that diet and habitat were related to species richness. Larger fish had more parasites. Benthic species had more species than pelagic fish. The analysis was made without correction for host phylogeny. Ectoparasite abundance was positively correlated with host length of three of seven coral reef fishes (Grutter, 1994), and both intra- and interspecific variation in numbers of isopods correlated well with host size, after controlling for host phylogeny and sampling effort (Grutter and Poulin, 1998a).

According to Marcogliese (2001), productivity and diversity are higher in marine than in freshwater systems, and the structure of the food web is more complex, leading to differences in parasite richness and diversity between freshwater and marine systems. Total parasite infracommunity species richness for marine fish is given as 5.6 ± 3.2, that for freshwater fish as 2.8 ± 2.1. Intestinal infracommunities, however, do not differ significantly. Caution is necessary in accepting this generalization. Richness of ectoparasite communities differs significantly between latitudes (see below), and therefore comparisons should be made only between communities at the same latitude.

Poulin (1999b) stressed that parasites play "many roles at many levels" in shaping animal communities. Concerning effects on host communities, they may have different effects on different host species infected by them, they can adversely affect a key species in the host community, or they can affect the host phenotypes and thereby the importance of particular hosts in the community. As regards effects on parasite communities, they may compete with each other, and they can change a host phenotype, making it more or less suitable for other parasite species. These effects, according to Poulin, have been shown to occur, but simultaneous effects by parasites on parasite and host communities have never been considered. He uses an example from an intertidal community to show that "subtle, indirect effects of a parasite species on non-host species" have important effects on the whole community. He studied the cockle *Austrovenus stutchburyi* in areas of New Zealand where oystercatchers *Haematopus* spp. are probably its only predators. The cockle is host to metacercariae of the trematode *Curcuteria australis*. In heavily infected cockles, not only is foot tissue replaced by the larvae, but the size of the foot relative to that of the shell is reduced, leading to a reduced ability of the cockle to bury in the sand and increasing the likelihood (by a factor of seven) of being eaten by the oystercatchers, which are final hosts of the trematode. Effects on cockle communities are difficult to estimate but are probably not large. However, because of infection, more cockles are exposed at the surface and can be colonized by epibionts, i.e. limpets, sea anemones, barnacles and polychaetes. As cockle shells often represent the only hard substratum for epibionts, the parasite indirectly acts as a true "ecosystem engineer", increasing biodiversity. Also, surface cockles infected with *Curcuteria* are five times more commonly infected with another trematode species, possibly because the siphons of the cockles are extended higher above the surface permitting easier infection.

Sasal *et al.* (1999b) examined the question of whether a more predictable environment permits development of more specific adaptations in parasites. The authors distinguish "specialists" restricted to a single host species, and "generalists" using at least two host species. Larger host size (which is correlated with numerous traits, such us longevity) is thought to provide a more predictable environment. It was found that specialists among monogeneans of 48 marine fishes (2547 individuals) do indeed infect larger hosts. Larger hosts also had larger parasites, explained by mechanical problems encountered in the hosts' gill chambers. For generalists there was a negative correlation between parasite body size and prevalence of infection, but the correlation disappeared after correction for host phylogeny (see also Sasal *et al.*, 1997).

Aggregation of hosts may conceivably have an effect on parasite communities. Wikelski (1999) studied Galapagos marine iguanas that were parasitized by mobile (*Ornithodoros*) and contagiously transmitted ticks (*Amblyomma*). Iguanas sleeping alone had a much greater load of mobile ticks, but grouping did not affect infection with contagious ticks. Reduction in infection with mobile ticks, rather than thermoregulation, appears to be the selective advantage of nocturnal aggregations. The author estimates that iguanas sleeping singly would have a 5.4% lower annual energy budget caused by tissue removal by ticks.

Predation is an important factor affecting community structure of free-living animals. For example, it may increase diversity in a community by reducing numbers of superior competitors. Removal of parasites by cleaners may conceivably have a similar effect. The thorough and exten-sive work of Grutter and collaborators has shown that cleaner fish are indeed effective in removing parasites (Grutter, 1996, 1997a, b, 1999b, 2000; Grutter and Poulin, 1998a, Grutter and Hendrikz, 1999). Juveniles of gnathiid isopods infecting many fish species are the main food item of the cleaner wrasse *Labroides miniatus* on the Great Barrier Reef. Infection occurs during the day and at night (Grutter and Hendrikz, 1999, Grutter,

2000). Larger gnathiids are preferred food for the cleaners (Grutter, 1997a). Removal of all cleaners from eight reefs over 6 months did not lead to any effects on the fish, nor did it affect the parasite load (Grutter, 1996, 1997b).

Poulin (1999c) examined the relationship between size of parasites and their abundance, as measured by prevalence and intensity of infection. Among endoparasitic helminths of fish, size of parasites was positively correlated with prevalence and negatively with intensity of infection. Among copepod parasites of fish, size was positively correlated with prevalence as well as intensity of infection. These findings suggest that the size of parasites may play a role in determining abundance and distribution, including aggregation, of parasites. Poulin and Morand (2000b) tested the hypothesis that larger nematodes (from many non-marine hosts) are more highly aggregated than smaller ones, but did not find unambiguous evidence (after correction for host phylogeny) to support it.

In communities of many animals, locally abundant (core) and locally rare (satellite) species are most common, leading to a bimodal distribution (core-satellite hypothesis; Hanski, 1982). Zander et al. (1999b) found that the most important parasites of fish in the south-western Baltic Sea were some generalists and some specialists, supporting the core-satellite species concept. Morand et al. (in press) found the same for a large number of metazoan ectoparasites of marine fish.

3.5. Metapopulations and chaos

Populations often consist of a number of local populations that are largely separated from each other, although migration between them may occur from time to time. Chesson (2001) has given a recent account of meta-population ecology. Populations composed of local populations are referred to as metapopulations (here the term subpopulation is used for local populations, and population is used as a neutral term referring both to meta- and subpopulations). A metacommunity is found when both prey and predators (and/or parasites) have the same metapopulation structure. For example, applied to marine parasites, a metapopulation of a parasite species infecting a particular fish species would comprise all the sub-populations of the parasite infecting subpopulations of that fish species segregated over relatively long periods. If cleaner fish feeding on these parasites show the same segregation into subpopulations, they as well as the fish and their parasites would constitute a metacommunity.

Consequences of metapopulation structure for marine parasite communities have rarely been considered. For example, the fact that different host populations differ in the composition of their parasite assemblages was shown by Poulin and Morand (1999), who demonstrated this for freshwater fish by a multivariate approach based on the permutation of matrices. The degree of difference was shown to be correlated with geographic distance. Rigby *et al.* (1997) examined large-scale patterns of species diversity of *Epinephelus merra*, a serranid reef fish, in the South Pacific and French Polynesia. They found distinct differences in parasite communities between archipelagos. Grutter (1994) found little variation in ectoparasite composition and abundance among local sites, even if they varied physically. Nevertheless, abundance with a particular monogenean infecting *Hemigymnus melapterus* differed between the reef flat and the reef slope (Grutter, 1998).

Bush and Kennedy (1994) discussed the likelihood of parasites becoming extinct due to host fragmentation induced by man in the absence of host extinction. Examining a number of non-marine examples, they concluded that extinctions at the levels of infrapopulations and infracommunities are common, but without serious effects on the survival of the parasites. At higher levels, extinctions are possible if meta- or "suprapopulations" (the latter even more inclusive than metapopulations) become extinct. However, suprapopulations are highly complex and they are unlikely to become extinct due to host fragmentation, because reinvasion may occur from surviving fragments (rescue effect). Overall, the authors consider it unlikely that parasite species become extinct due to host fragmentation, contradicting Sprent (1992).

Chesson (2001) discussed implications of metapopulation structure for single-species dynamics, as well as for multispecies dynamics involving predators and parasitoids or interspecific competition. In single-species systems, he distinguished the density-independent case, in which there is an exponential increase of subpopulation density over time, the case of contest competition (due to competition for a strictly limited resource) in which an equilibrium of subpopulation density is reached after an initial increase, and the case of scramble competition (for example due to great mortality at high densities) leading to chaotic fluctuations in subpopulation density and often to extinction. However, in the dynamics of metapopulations, both contest and scramble competition lead to equilibria, but at a higher level for the latter, if there is much spatial variation in subpopulation densities between subpopulations. Neither Chesson nor other authors have considered the effect of different reproductive rates on the dynamics of metapopulations. Rohde and Rohde (2001) have developed computer simulations which show that averaging the sizes of subpopulations with different reproductive rates leads to a decrease in the extent of chaotic fluctuations predicted by the classic chaotic bifurcation diagrams (Figure 12), the degree of chaos becoming less with increasing number of subpopulations and an increasing range of reproductive rates. This decrease is much more complex than predicted by the simple considerations of Chesson. Most significantly, in these "fuzzy chaos" bifurcation diagrams, more than one population size still occurs at certain reproductive rates (Figure 13). Furthermore, even for populations with low reproductive rates, more than one population size may occur when the heterogeneity of metapopulations is great. Fuzzy chaos modelling of multi-species systems shows that predictions on the behaviour are even more difficult and certainly much more difficult than predicted by Chesson. The full richness of the possibilities can be explored only by making multidimensional animations using programmes based on the Lotka-Volterra equations for species interactions. These programmes describe the competition between metapopulations comprising largely separate subpopulations with different reproductive rates. They simulate popula-tions in interacting two or three species systems and return the time at which the first species becomes extinct. This is used to generate graphs which represent the time of extinction on one of the axes and the species

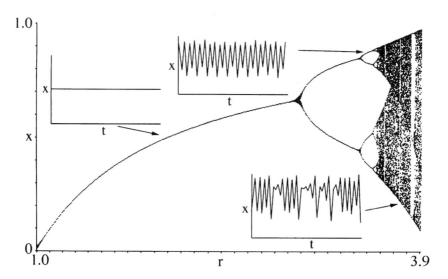

Figure 12 Bifurcation diagram for a single subpopulation (single value of *r*). Population sizes (*x*) plotted against reproductive rates (intrinsic rates of population growth) (*r*). Note the first bifurcation just below *r* = 3, further bifurcations at *r* > 3 and chaos at *r* > 3.57. Insets: population sizes (*x*) plotted against generations (*t*). A single value of *x* when *r* < 3, four alternating values (in successive generations) when *r* = 3 < *r* < 3.57 and chaotic fluctuations when *r* > 3.57.

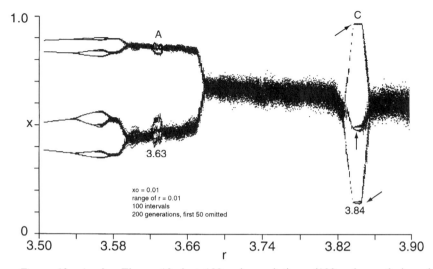

Figure 13 As for Figure 12, but 100 subpopulations (100 values of *r*) and heterogeneity of metapopulation, i.e. range of $r = 10^{-2}$. Note the vertical contraction of the chaotic field and the two windows (A and C) with bifurcations (arrows) in C. Only *r* values from 3.50 and 3.90 are shown.

that became extinct by the colour of the graph. If no species becomes extinct within a designated number of generations, the graph at the corresponding point is left blank. Complex examples can be examined only by using the programmes to generate multidimensional animations in colour. Two simple examples are discussed below, for which black and white diagrams are used.

In Figures 14 and 15, two species (*x* and *z*) are compared, each consisting of 10 subpopulations. In Figure 14 the reproductive rate (the intrinsic rate of population growth) of species *z* is 3.7, that of species *x* is 1.6. The species with the lower reproductive rate becomes extinct only at very great values of alpha and beta (the competition factors of the two species), and the species with a much greater reproductive rate becomes extinct when heterogeneity of the metapopulations, as indicated by the ranges of reproductive rates, is great and competition between species of intermediate strength. In Figure 15, species *x* has a reproductive rate of 2.8, and species *z* has a reproductive rate of 3.7. Species *x* never becomes extinct, and only when competition is low does no extinction of either species occur.

Even the simple examples dicussed here show that competitive effects may be rather unexpected: the species with the lower reproductive rate often is competitively superior.

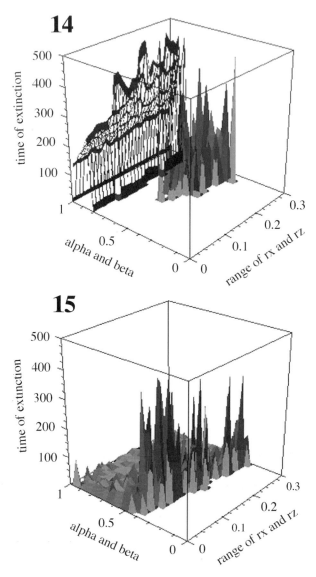

Figures 14 and 15 Ten subpopulations. The ranges of the reproductive rates of species x and z, r_x and r_z (indicative of the heterogeneity of the meta-populations) are plotted against alpha and beta (the competition factors of both species), and the time of extinction in number of generations. The initial population sizes of species x and z, x_0 and $z_0 = 0.5$. The reproductive rate of species z, $r_z = 3.7$. In Figure 14, $r_x = 1.6$, in Figure 15, $r_x = 2.8$. Extinction of species x is indicated by hollow, of species z by grey, no extinction by white. For programmes (by Peter Rohde) see "Fuzzy Chaos- 2 and 3 species competition programs" at http://www-personal.une.edu.au/~krohde.

For three-species systems, the outcome of competition is even more complex and predictions are even more difficult.

Most parasite species are likely to have metapopulations consisting of many subpopulations, although few studies have been made. For example, gyrodactylid monogeneans multiply on host individuals and can easily be segregated sufficiently to permit development of subpopulations on different host individuals. Long-term studies of such populations are ideal models for testing the predictions of fuzzy chaos modelling.

3.6. Functional ecology and morphology

Older papers and monographs on parasite adaptations give many details on morphological adaptations for attachment to hosts and sites within hosts. Here, I wish to mention only that recent electron microscopic studies have shown that even in closely related species, attachment organs that appear to be very similar under the light microscope may in fact not be homologous; evidence for the strong selection pressure that has led to the repeated development of such structures. Thus, the ventral attachment organ of the aspidogastrean trematode *Rugogaster hydrolagi*, and those of digenean trematodes and various monogeneans differ considerably in details of their ultrastructure (Rohde and Watson, 1992b, 1995b).

A considerable amount of work has been done on the behaviour of the infective stages of various parasites leading to host infection. Some of the earlier work has been reviewed by Rohde (1993). Buchmann and Nielsen (1999) have shown that sera and mucus from several unrelated teleost fishes effectively attract the infective stage of the ciliophoran *Ichthyophthirius multifiliis*. The attractive agents were present in host immunoglobulin and proteins. Monogeneans have been particularly well studied and a review of this work has been given by Whittington *et al.* (2001). The reader is referred to this review for more detailed information. In a recent study, Rothsey and Rohde (2001) have, for the first time, demonstrated a reaction of copepod larvae and oncomiracidia of several marine monogeneans to magnetic stimuli. Such stimuli may contribute to host finding. It is not known which cells or tissues react to magnetism, but an electron microscopic study of the so-called terminal globule of poly-opisthocotylean Monogenea has shown the presence of large crystal-like inclusions in vacuoles of posterior cells, which may be responsible (Figures 16 and 17). In monopisthocotylean Monogenea, terminal globules have not been found so far, and the only species tested did not react to magnetic fields, lending some support to the assumption that the terminal globules are indeed involved in magnetic orientation.

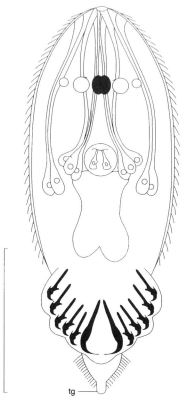

Figure 16 Oncomiracidium of a polyopisthocotylean monogenean. Note the terminal globule (tg). (Scale bar 50 µm.) (Strongly modified and redrawn from Whittington *et al.*, 2001.)

As discussed above, all parasites show preferences for certain host species and microhabitats within or on hosts. Work on host and micro-habitat finding was reviewed by Rohde (1994a). Altogether, little is known about these processes in marine parasites, and the same refers to mate finding. Parasites are generally very small relative to hosts and micro-habitats, and mechanisms for mate finding are therefore essential, not only for bisexual species but for hermaphroditic ones as well, because cross-fertilization has been shown to occur in several hermaphroditic species and may be of considerable evolutionary importance. Rohde (in press) has reviewed mate finding in parasites. The very narrow niches (strict host specificity and narrow microhabitats) may, at least partly, be the result of selection to ensure cross-fertilization ("mating hypothesis" of niche restriction, see above). The reader is referred to this review for more detailed information.

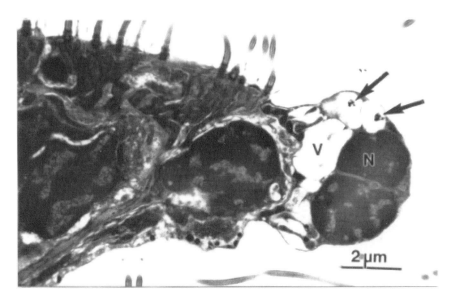

Figure 17 Posterior globule of the oncomiracidium of *Zeuxapta seriolae* (Monogenea Polyopistocotylea). Note: crystal-like inclusions (arrows) in large vacuoles (V) and nuclei (N).

We have little information about the sensory organs or receptors involved in host, microhabitat and mate finding. However, many parasites have remarkably complex arrays of sensory receptors. A comparison of the infective juveniles of two closely related species of aspidogastrean trematodes that differ in the complexity of their life cycles has shown distinct differences in the number of types of sensory receptors, lending suppport to the view that they play a role in host finding. *Multicotyle purvisi* actively swims upwards in the water column where the likelihood of being ingested by the intermediate host, a snail, is greatest (Rohde, 1972, further references therein). It has a pair of eye spots and, in addition, 12 types of ciliated or non-ciliated sensory receptors. *Lobatostoma manteri* is ingested by its snail host that feeds on the eggs containing the juveniles, i.e. there is no hatching (Rohde, 1973). The juvenile lacks eyes and has only nine types of sensory receptors (Rohde, 1989a, 1994b; Rohde and Watson, 1989, 1992a). Juvenile and adult *L. manteri* have a total of 20,000–40,000 receptors, truly remarkable for a parasite without free-living stages. It is likely that these receptors are important for life in the host (Rohde, 1989a).

The few species of Monogenea that have been examined also have remarkably complex sensory receptors – eyes as well as ciliated and non-ciliated receptors. Rohde and Watson (1995a, 1996) examined the anterior sensory receptors of two species of monogeneans belonging to two families by electron microscopy, and found five and seven types of receptors, respectively. Most were non- or uniciliate, one type had several cilia, and in one species, aggregation of receptors in a "taste organ" was shown. Their function is not known, but it is likely that they play a role in feeding, mating, and/or microhabitat finding.

The antennules of the parasitic copepod *Lepeophtheirus salmonis* carry many receptors (Gresty *et al.*, 1993). Ablation of the tips of the antennules reduces mating success, but pair formation still occurs (Hull *et al.*, 1998).

3.7. Effects of parasites on hosts

The behaviour of parasitized animals, terrestrial, fresh water and marine, has been discussed by Moore (1995). A marine example is the snail *Hyanassa obsoleta*, which, when infected with the trematode *Gynaecotyla adunca*, spent more time in the high intertidal during nocturnal low tides. This leads to release of larval trematodes near the next intermediate host, a semi-terrestrial crustacean. McCarthy *et al.* (2000) gave another example. The trematode *Microphallus piriformis* uses periwinkles, *Littorina saxatilis*, as intermediate host and silver gulls, *Larus argentatus*, as final host. It lacks free-swimming cercariae and gulls become infected by eating snails. Laboratory and field experiments showed that snails with infective stages moved further upwards than uninfected snails or snails with "immature" parasites, increasing the chances of infection.

Moore (1995, also Poulin and Thomas, 1999) points out that behavioural changes may be adaptations of hosts to minimize effects by parasites, or they may be adaptations that favour the parasite, for example, by ensuring transmission. In some cases, a shift between these two possibilities may occur depending on environmental conditions. Also, changes may be without any adaptive value. Evidence for induced changes rests on comparison of naturally infected and uninfected hosts and aberrant phenotypes may be the cause of infection and not vice versa. Only experiments can provide definitive evidence (Poulin and Thomas, 1999).

Poulin (1999b) stressed that parasites play "many roles at many levels" in shaping animal communities. Concerning effects on host communities, they may have different effects on different host species infected by them, they can adversely affect a key species in the host community, or they can affect the host phenotypes and thereby the importance of particular hosts in the community.

Poulin and Thomas (1999) reviewed work that shows a normal frequency distribution of values of various traits in uninfected hosts. Parasitic infection can lead to a shift in the mean value of the trait (in either direction) and increase the variance, depending on the intensity of infection. If prevalence of infection is high, the distribution may be displaced but remain normal, if prevalence is greater, the distribution may become skewed. If parasite infection leads to a large shift in phenotype, and if prevalence of infection is below 100%, the distribution can become bimodal. They cite the work of Lauckner (1984) on parasite effects on North Sea snails. Lauckner (1984) gave a thorough account of the effect of larval trematodes on the shell sizes of marine snails: uninfected snails showed a unimodal and normal distribution in shell sizes, the overall population (including uninfected and infected snails), depending on the locality, was either multimodal or highly skewed.

Poulin and Thomas further point out that selection can become "myopic" when environmental effects on phenotypes are strong. This means that only those phenotypes are affected by selection that are present under certain conditions. The authors discuss various possibilities of how parasitism can affect host speciation. One important phenomenon that has been demonstrated is a change in sex ratio by parasites. For example, parasitic Cirripedia (Rhizocephala) are well-known to femininize and sterilize infected male crabs (e.g. Hoeg, 1995).

Pathological effects of parasites on many hosts are extensively discussed in Kinne (1980–1990) and in various monographs. They will not be discussed here. However, it should be emphasized that hosts in their natural environment are often not affected or only slightly affected by parasites, even at high infection intensities. For example, gnathiid isopods are among the most important parasites of coral reef fish (e.g. Grutter and Poulin, 1998b; Grutter, 1999a; Grutter and Hendrikz, 1999; Grutter et al., 2000), but parasitism across an order of magnitude of parasite numbers had no effect on plasma levels, cortisol or glucose of *Hemigymnus melapterus*, i.e. even high intensities of infection do not appear to be stressful to the fish in their natural environment (Grutter and Pankhurst, 2000).

3.8. Evolution and speciation

Important evolutionary problems concern the factors that determine parasite diversification, the occurrence and importance of sympatric speciation, the commonness of co-evolution of parasites and hosts, and the origin of marine parasites. Some recent studies are discussed below.

Poulin and Morand (2000a) gave a stimulating review of parasite diversity, suggesting that sympatric speciation may be important in some groups of parasites. Little is known about which parasite and host traits favour diversification, although small parasites appear to be more speciose. Other characters, such as a high basic reproductive rate, may also be important, but no evidence in support of this assumption is available. Temperature may be important for parasites of marine fish, as suggested by the studies of Rohde (1992, 1998b; see below: latitudinal gradients). With regard to sympatric speciation, Poulin (1999a), discussing multiple congeners in hosts (including 15 species of the monogenean *Dactylogyrus* in freshwater fish), concluded that "perhaps the most parsimonious explanation is to view these multiple congeners as the product of speciation and diversification within the host", i.e. as the product of synxenic (= in the same host) speciation. The only study attempting an evaluation of the occurrence of synxenic speciation in parasites by molecular techniques, was that by Littlewood *et al.* (1997). They showed that speciation in poly-stomes infecting different sites in the same turtle species must have been allo- and not synxenic, because species infecting the same site in different host species (even if they are from different continents) are more closely related than species infecting different sites in the same host species.

Paterson and Poulin (1999) found that chondracanthid copepods have significantly co-evolved with their host fish, and few instances of host-switching occurred. Gusev (1995) emphasized that Monogenea have co-evolved with their hosts, usually diverging faster than the hosts.

The monogeneans are an excellent model for studying some evolu-tionary mechanisms. They comprise two (possibly paraphyletic) groups, the Polyopisthocotylea and Monoopisthocotylea. Congeneric species of the former, infecting the gills of the same host species, are usually segregated in different parts of the gills, whereas congeneric species of the latter often co-occur in the same microhabitat. The reason for this difference appears to be the enormous variability of copulatory sclerites in the mono- but not the polyopisthocotyleans. Congeneric Mono-pisthocotyles are reproductively segregated even if they co-occur. Poly-opisthocotylea achieve reproductive segregation by spatial segregation (Rohde, in press, further references therein). The difference may also explain possible differences in speciation rates between the two groups. I suggest (so far without evidence) that one or few mutations may lead to morphological differences in copulatory sclerites, resulting in fast and per-haps synxenic speciation in the Monopisthocotylea. This would explain the very large number of congeneric monopisthocotyleans infecting one host species (see above: 15 species on one freshwater host species). Studies on insects have indeed shown that fast sympatric speciation by few mutations is possible (references in Rohde, 1993, page 213).

Parasites of pinnipeds are mainly of terrestrial origin, but some are of marine origin (Menier, 2000). For the marine origin of some freshwater parasites (e.g. in Lake Baikal) see Rohde (1993).

4. THE DISTRIBUTION OF MARINE PARASITES

4.1. Latitudinal gradients in species richness

An increase in species numbers from high to low latitudes is well documented for many terrestrial plant and animal groups, although there are many exceptions (for reviews see Rohde, 1992, 1999). Such a gradient also exists for marine organisms. Clarke (1992) claimed that there is no convincing evidence for a general latitudinal trend in the sea comparable to that on land. But it is known to exist, for example, in marine teleosts (Rohde, 1982, 1993). Among marine parasites, the gradient has been well documented for metazoan ecto- and endoparasites of marine fishes. However, there is an important difference between ecto- and endoparasites. In ectoparasites there is a relatively greater increase in species numbers than of host species, whereas in endoparasites, relative species richness (number of parasite species per host species) is more or less the same at all latitudes. The increase in species richness of endoparasites towards the equator is entirely (or almost entirely) due to the increase in species numbers of hosts (Figure 18). These differences remain after correction for phylogeny, using the method of phylogenetically independent contrasts (Figure 19). The same difference is shown by two major groups of ecto- and endoparasites, the monogeneans and digeneans. Whereas relative species richness of the former increases towards the equator (Table 2), it is more or less the same at all latitudes for the latter (Table 3). As a consequence of these differences, infracommunity richness of ectoparasites is greater in the tropics, but does not differ significantly between latitudes for endoparasites (Figure 20). Species diversity of Monogenea is greater in the Indo-Pacific than in the Atlantic Ocean (Rohde, 1980c, 1986) (Table 2).

Many hypotheses attempting to explain latitudinal gradients in species richness have been proposed (reviews in Rohde, 1992, 1999; Willig, 2001). Most of these are either circular or insufficiently supported by evidence, and most of them are equilibrium explanations, based on the assumption that habitats are saturated with species and that species richness is somehow restricted by limiting factors. In other words, tropical habitats can accommodate more species because their limits are higher. Rohde (1992, 1999) provided evidence that equilibrium has not been reached by

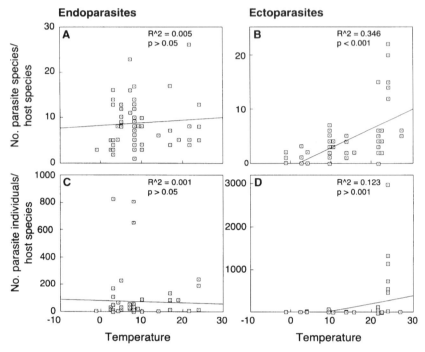

Figure 18 Metazoan parasites of the digestive tract ("endoparasites") of 55 marine teleost species and of the head and gills ("ectoparasites") of 108 marine teleost species at different latitudes (mean sea surface temperatures). Note: more or less the same richness and abundances at all latitudes for endoparasites, a distinct increase of richness and abundances towards the equator for ectoparasites. Modified from: Rohde, K. and Heap, M. (1998). Latitudinal differences in species and community richness and in community structure of metazoan endo- and ectoparasites of marine teleost fish. *International Journal for Parasitology* **28**, 461–474, with the permission of the copyright holders, Elsevier Science.

most taxa and that many more species can be accommodated. He concluded that the gradients can best be explained by a gradient in "effective evolutionary time" modulated by several other factors. Effective evolutionary time is defined as the evolutionary time under which communities have existed under relatively constant conditions, and evolutionary speed. Evolutionary speed is determined by mutation rates, generation times and the speed of physiological processes leading to faster selection. There is some evidence (although more is needed) that these three parameters are correlated with temperature, evolutionary speed being greatest in the tropics. The result, in largely empty niche space, must be a faster accumulation of species at low latitudes, even if

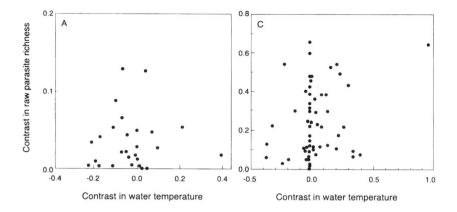

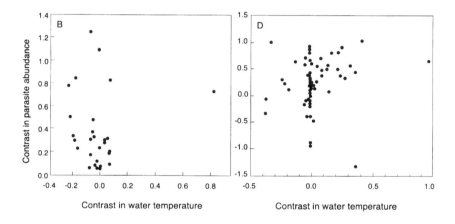

Figure 19 Metazoan parasites of the digestive tract of 55 marine teleost species (A, B) and of the heads and gills of 111 marine teleost species (C, D). A, C, contrasts in raw species richness plotted against corresponding contrasts in water temperature. B, D, contrasts in abundance plotted against corresponding contrasts in water temperature. Note: increase of richness and abundance from high to low latitudes for ecto- but not for endoparasites. If outlier is removed in B, abundance is greater at high latitudes. A and B reproduced from: Rohde, K. and Heap, M. (1998). Latitudinal differences in species and community richness and in community structure of metazoan endo- and ectoparasites of marine teleost fish. *International Journal for Parasitology* **28**, 461–474. With permission from Elsevier Science. C and D reproduced from: Poulin, R. and Rohde, K. (1997). Comparing the richness of metazoan ectoparasite communities of marine fishes: controlling for host phylogeny. *Oecologia* **110**, 278–283, with permission from Springer Verlag.

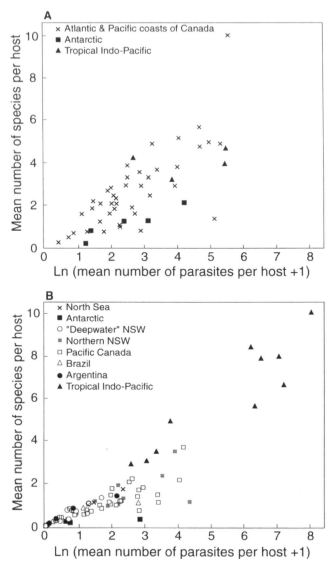

Figure 20 Community richness (as expressed by species richness and abundance) of metazoan endoparasites of the digestive tract of 51 teleost species (A) and of metazoan ectoparasites on the heads and gills of 105 teleost species (B). Note much greater community richness of ecto- but not of endoparasites in tropical waters. Modified from: Rohde, K. and Heap, M. (1998). Latitudinal differences in species and community richness and in community structure of metazoan endo- and ectoparasites of marine teleost fish. *International Journal for Parasitology* **28**, 461–474, with the permission of the copyright holders, Elsevier Science.

extinction rates are higher there as well. An intriguing consequence of this hypothesis is that, in principle, habitats that are very species-poor today, such as the Arctic tundra or cold high latitude seas, may perhaps be able to support as many species as the tropics, provided we wait long enough, although the possibility cannot be excluded that different limits to species richness in different habitats do exist, even if they have not been reached at this point in evolutionary time.

A recent study on bird diversity in South America (Rahbeck and Graves, 2001) suggests that topographic heterogeneity and climate (precipitation, cloud cover, etc.) are the primary determinants of species richness. As pointed out by Rohde (1980b), the assumptions of largely empty niche space with all its consequences do not apply to large animals with great vagility, such as birds and mammals (and they do not apply to small species with great vagility occurring in large populations). Gradients in species richness for such groups may therefore be different and have different explanations. Furthermore, spatial heterogeneity and some of the factors comprising "climate" except for temperature are of much less importance for marine animals and cannot serve as primary explanations for diversity gradients.

4.2. Latitudinal gradients in reproductive strategies

Many marine benthic invertebrates in warm waters tend to produce large numbers of small pelagic planktotrophic larvae, whereas high latitude species tend to produce fewer and larger offspring, often by viviparity or ovoviviparity, and development is in egg capsules or by brooding. Thorson (1957) was the first to draw attention to this phenomenon and it is therefore referred to as Thorson's rule. Rohde (1985) demonstrated a similar phenomenon for gill Monogenea. Whereas species infecting teleosts in warm or temperate waters generally produce eggs from which free-swimming larvae hatch, most species infecting high latitude fish (particularly in the northern hemisphere) belong to the Gyrodactylidae, a family comprising small species that reproduce by viviparity. Juveniles remain on the same fish, and other fish become infected by contact transfer. The gradient is distinct both in the Pacific and the Atlantic Oceans (Figure 21). Rohde (1985, 1999) suggested that the most likely explanation for this phenomenon in the Monogenea is that small larvae cannot locate and infect suitable hosts in the vast spaces of the ocean at temperatures where physiological processes are greatly slowed down, and that the number of pelagic larvae that can be produced at such low temperatures is too small to guarantee infection by host searching. Selection has favoured a strategy that guarantees a suitable habitat by

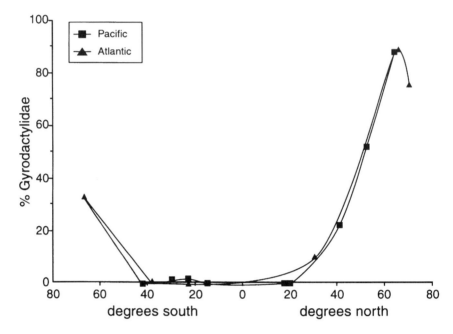

Figure 21 Proportion of gill Gyrodactylidae at various latitudes. (Data from Rohde, 1985, 1999, after various sources.)

remaining on the same host individual. Since the trend in parasites so closely resembles that found in benthic invertebrates, Rohde (1985) suggested a similar explanation for the latter.

4.3. Latitudinal gradients in host ranges and host specificity

For the definition of host range and host specificity see the section on host specificity (p. 22). Comparative studies of monogenean ectoparasites and digenean endoparasites of marine teleosts from various latitudes have shown that host ranges are very narrow for monogeneans at all latitudes. In contrast, host ranges for digeneans are much greater at high than at low latitudes (Rohde, 1978b, 1980d) (Figure 22). Host specificity is similarly great for both groups at all latitudes (Rohde, 1980d). Apparently, even in cold waters, digeneans infect only some host species heavily. However, a word of caution is necessary: only one large survey at high latitudes permits calculation both of host ranges and host specificity for Digenea. More studies are necessary, and other parasite groups should be assessed for latitudinal gradients in host ranges and specificity.

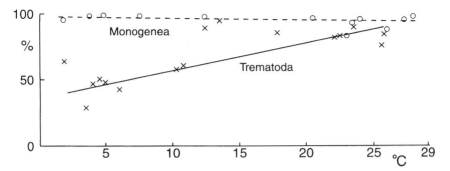

Figure 22 Host ranges of Monogenea (○) and Digenea (×) of marine teleosts at various mean surface water temperatures. Ordinate: host ranges as percentage of species found in one or two host species. Modified from: Rohde, K. (1978b). Latitudinal differences in host specificity of marine Monogenea and Digenea. *Marine Biology* **47**, 125–134, with permission from Springer Verlag.

4.4. Longitudinal gradients and geographical ranges

Even less known than latitudinal gradients are differences in species diversity in a west-easterly direction. Studies of parasites of rabbit fish, *Siganus* spp. from the coasts of eastern Australia and East Africa have shown a much greater species richness of ectoparasites, but not of endo-parasites, in Australian waters. Kleeman (2001) found 19 species of metazoan ectoparasites on *Siganus doliatus* on the Great Barrier Reef, whereas Martens and Moens (1995) and Geets *et al.* (1997) found only seven or eight species on *S. sutor* from the western Indian Ocean. Endo-parasite richness did not show significant differences: seven vs five species. This agrees with the finding of Rohde and Hayward (2000) on ecto-parasites of scombrid fishes, that the tropical/subtropical western Pacific is the primary centre of diversity, at least for scombrid fishes and their copepod and monogenean ectoparasites. These authors also found a secondary centre in the tropical/subtropical western Atlantic. Coastal scombrid fishes and their parasites, apparently spread between the Indo-West Pacific and Atlantic Oceans via the Tethys Sea, and between the western Atlantic and the eastern Pacific, before the Central American landbridge was established. These migrations led to marked differences in species diversity between the oceans. Fish species found in low diversity seas have relatively more parasite species than those found in high diversity seas (Table 4). When considering only endemic fish and parasite species, however, the Indo-West Pacific and the West Atlantic, i.e. the primary and secondary centres of diversity, have parasite–host ratios of

1.75 and 1.25, respectively, vs a ratio of 1.00 for all other seas (Table 4). As pointed out by Rohde (1999), these differences strongly suggest that historical processes are responsible for the differences and that more species could be accommodated at least in the seas with few species.

The lack of longitudinal differences in the diversity of endoparasites, mainly digeneans, corresponds to a similar lack of latitudinal differences for these parasites.

The effects of geographical ranges of hosts on species richness of parasites have been little studied. Data in Table 5 suggest that fish species with the widest geographical distribution have the greatest number of parasite species, but significance of the results is lost after Bonferroni correction for multiple comparisons. Much work has been done on Rapoport's rule, which claims that latitudinal ranges of tropical species of plants and animals are generally narrower than those of high latitude species (for a recent review see Rohde, 1999). Marine parasites have not been examined in this respect.

4.5. Depth gradients

Noble (1973) reviewed earlier work on deep-sea parasites. Rohde (1993) also discussed them. Heath (1987) has made a large survey of (mainly pelagic) deep-sea fish (45 species, 1308 specimens) for metazoan endoparasites off the south-eastern Australian coast, which has never been discussed in the literature. Thirty-eight (84.4%) of the species and 499 (38.2%) of the specimens were infected. Digenea were the most common group: 28 (62.2%) of the species and 280 (21.4%) of the specimens were infected. Next were the Cestoda, with 24 (53.3%) of the species and 307 (23.5%) of the specimens infected. Also, 24 (53.3%) of the species and 307 (23.5%) of the specimens were infected with Nematoda, and 8 (17.8%) of the species and 29 (2.2%) of the specimens were infected with Acanthocephala. A total of 73,775 parasite specimens was collected. The Digenea included 22 subadult/adult and 2 larval species, the Acanthocephala 3 adult and 2 larval species, the Cestoda 1 adult and 19 larval species, and the Nematoda 6 adult and 1 larval species. In general, the parasite fauna was poor relative to inshore shallow-water and offshore deep-water benthic faunas.

The findings on low diversity, particularly of pelagic deep-water fish, agree with earlier findings by various authors (Manter, 1967; Collard, 1968, 1970; Noble and Collard, 1970; Noble, 1973; Armstrong, 1974; Campbell et al., 1980; Campbell, 1983, further references therein; review in Rohde, 1993). However, Heath's results differ from previous studies in that digeneans were more abundant and important in parasite communities of deep-sea fish than nematodes (Heath, 1987). Also, although adult helminths have been reported by others, they were surprisingly diverse among the digeneans, nematodes and acanthocephalans in Heath's study.

4.6. Parasites in brackish waters

Zander (1998a, b) reviewed parasitism in the largely brackish Baltic Sea. The parasite fauna is poor relative to truly marine habitats, specificity is low, new hosts have been acquired, the number of hosts in life cycles is reduced, and there are adaptations to brackish water hosts. Valtonen *et al.* (2001) examined 11 marine and 22 freshwater fish species from the northeastern Baltic Sea, which is the most oligohaline area of the Baltic Sea. Of the 63 parasite species found, only 8 were marine. Marine fish species harboured marine and freshwater parasites, and freshwater fish harboured marine and freshwater parasites as well. Parasites showed an antinested pattern. Kesting and Zander (2000) found that the locality with the lowest salinity (6–8) in the south-western Baltic Sea had the smallest number of parasite species of fishes, molluscs and crustaceans.

4.7. Parasites as indicators of zoogeographical regions, oceanic barriers and ancient dispersal routes

There are many studies that use parasites as biological monitors for tracing host migrations and distinguishing host populations (see below). In contrast, very few studies using parasites as indicators of large-scale zoogeographical patterns and their history have been made. Von Ihering was the first to make such studies, and the method is therefore referred to as the Von Ihering method. Byrnes (1987) and Byrnes and Rohde (1992) used the distribution of four species of the teleost *Acanthopagrus* and their 25 species of Copepoda, 15 species of Monogenea, two of Branchiura, five of Isopoda, and one of Hirudinea to examine the validity of the zoogeographical regions in Australian coastal waters for parasites. Based on the study of various marine invertebrates, five such regions have been distinguished (Figure 23); data clearly show that parasites do not fit into them (Figure 23). Species recovered repeatedly generally extend over more than one region.

A recent study by Hayward (1997) used Indo-Pacific whiting (family Sillaginidae) and their ectoparasites to distinguish two provinces with great parasite species richness on the shelves of Australia and Asia

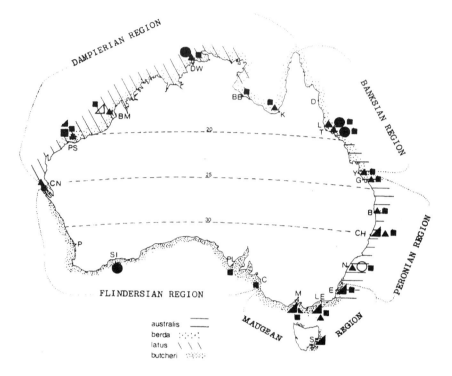

Figure 23 Australian coastal zoogeographical regions and the distribution of four species of the teleost *Acanthopagrus* (*australis, berda, latus* and *butcheri*) and some of their copepod ectoparasites. Different copepod species are indicated by different symbols. Note: copepod species recovered repeatedly not restricted to single regions. (Data from Byrnes, 1987.)

(Figures 24, 25). The Australian province has 15 endemics including five monogeneans, nine copepods and one leech, and the Asian province has 14 endemics including five monogeneans and 12 copepods. Most whiting, apparently, cannot migrate across large deep-water regions, such as those found between the Asian and Australian shelves, and consequently only four parasite species are shared by both provinces. According to Hayward, three Australian parasite species also appear to be encroaching on the southern periphery of the Asian shelf at present, and at least one less-recent invasion of Asian sillaginids into Australian waters would explain the occurrence of six pairs of congeneric copepods with one species in each province. Furthermore, one species seems to have spread relatively recently from the Arabian Sea to East African shores.

Rohde and Hayward (2000), based on large surveys of 26 species of the genera *Scomberomorus*, *Grammatorcynus*, *Scomber* and *Rastrelliger*

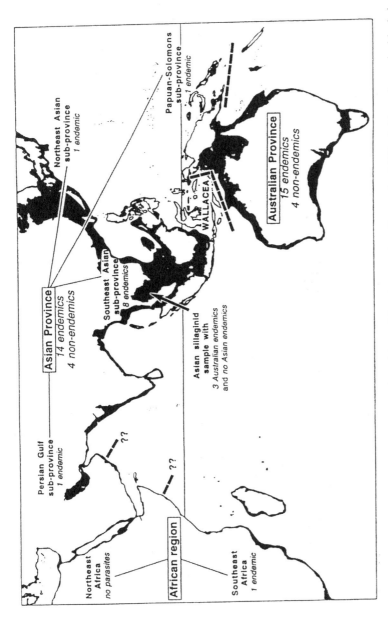

Figure 24 Ectoparasite provinces and proposed position of barriers to the dispersal of Sillaginidae. Areas in black indicate the lower limits of the changing coastline over the past 18,000 years. Reproduced from: Hayward, C. J. (1997). Distribution of external parasites indicates boundaries to dispersal of sillaginid fishes in the Indo-West Pacific. *Marine Freshwater Research* **48**, 391–400, with permission from CSIRO.

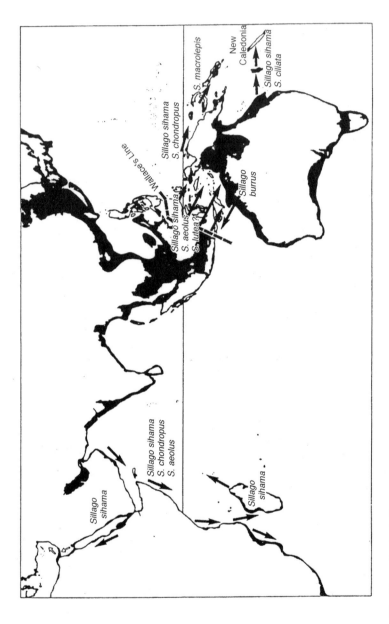

Figure 25 Hypothetical dispersal routes of Sillaginidae resulting in the occurrence of four widespread ectoparasites and six sympatric congeners since the mid-Miocene. Unfilled arrows show invasion routes of sillaginids into the eastern Mediterranean since the opening of the Suez Canal. Areas in black indicate the lower limits of the changing coastline over the past 18,000 years. Reprinted from: Hayward, C. J. (1997). Distribution of external parasites indicates boundaries to dispersal of sillaginid fishes in the Indo-West Pacific. *Marine Freshwater Research* **48**, 391–400, with permission from CSIRO.

(Scombridae) and their 32 species of copepod and 25 species of mono-genean ectoparasites, tested the hypothesis that the East Pacific Barrier (the large extent of open water in the eastern Pacific) and not the New World Land Barrier is responsible for the most pronounced break in the circumtropical warm water fauna of the continental shelves. They found unequivocal evidence that largely coastal species of fish and their parasites have not dispersed across the eastern Pacific but across the New World Land Barrier before the central American land bridge was formed: the few species shared between the western and eastern Pacific are circumtropical and may have dispersed between the western Atlantic and the eastern Pacific (Figures 26, 27). In contrast, more pelagic species (*Scomber*) have dispersed across the eastern Pacific, although at the genus level, dispersal has exclusively been between the western Atlantic and eastern Pacific. The findings on *Scomberomorus* agree with those based on molecular (nuclear and mitochondrial DNA) and morphological data of the fish: the *Scomberomorus regalis* group of Spanish mackerel, comprising species in the western Atlantic and eastern Pacific, was found to be monophyletic (Banford *et al.*, 1999; Collette, 1999).

4.8. Parasites as biological markers for distinguishing host populations and studying host migrations

Many authors have used parasites to distinguish host populations and/or study their migrations (examples in Rohde, 1993). A brief review was given by Moser (1991). MacKenzie and Abaunza (1998) provided a concise guide to the criteria necessary for the successful use of parasites as biological markers for stock discrimination of marine fish. They list key literature and discuss the general principles of the method with respect to what parasites to select, statistical analysis of whole parasite assemblages, methods for collecting and preserving parasites, and their identification. Mosquera *et al.* (2000) presented a simple mathematical model to evaluate the potential use of parasites in distinguishing fish populations. Pascual and Hochberg (1996) briefly discussed the use of marine parasites as biological tags, with emphasis on cephalopod hosts. Some of the many recent studies that use parasites as biological markers for fish populations are by Moser and Hsieh (1992), MacKenzie and Longshaw (1995), Larsen *et al.* (1997) and Comiskey and MacKenzie (2000).

The following section discusses a few examples to demonstrate the use-fulness and limitations of the method. A prerequisite for such studies is that only parasites that survive for long periods in the host are used. Larval trematodes, cestodes and nematodes in the tissues fulfil these requirements best. In a classical study, Margolis (1963) used a larval cestode and a larval

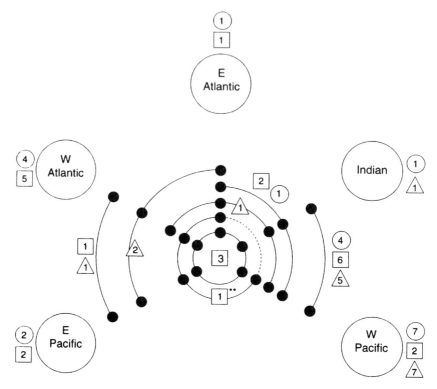

Figure 26   Geographical distribution of 18 species of *Scomberomorus* (Spanish mackerel), two species of *Grammatorcynus*, 23 species of parasitic Copepoda and 17 species of Monogenea. Species numbers restricted to certain seas are given beside the large circles that indicate the different seas, numbers of species shared by different seas are given on or near the lines that indicate sharing. ●, fish; △, Monogenea; □, Copepoda. Indian Ocean defined does not include the coastal zone along the west coasts of Malaysia, Indonesia and Australia. Note that the only species shared by the western and eastern Pacific have a circumglobal distribution, whereas all other seas share species, indicating that the eastern Pacific has been the most effective barrier to dispersal. Modified from: Rohde, K. and Hayward, C. J. (2000). Oceanic barriers as indicated by scombrid fishes and their parasites. *International Journal for Parasitology* **30**, 579–583, with the permission of the copyright holders, Elsevier Science.

nematode as indicators of the geographical origin of sockeye salmon, *Oncorhynchus nerka*, distinguishing two stocks originating in North America and Kamchatka, respectively. The nematode is acquired in Kamchatka fresh water and the cestode in North American fresh water, leaving no doubt about the origin of salmon infected with a particular parasite.

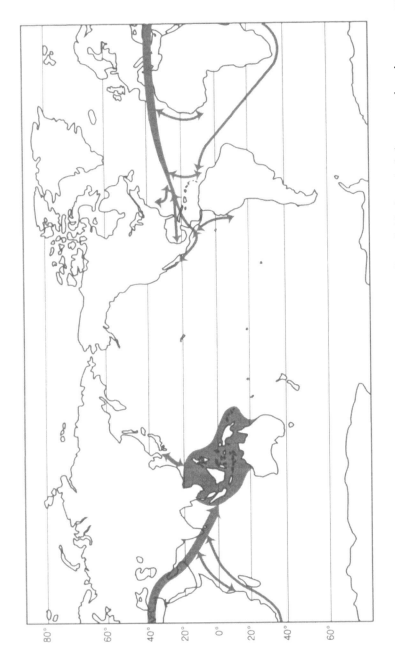

Figure 27 Inter-relationships of *Scomberomorus* and *Grammatorcynus* (Scombridae) and their copepod and monogenean ectoparasites. Note: primary centre of species richness in the western Pacific; the East Pacific Barrier has been a 100% effective barrier to dispersal. Modified from: Rohde, K. and Hayward, C. J. (2000). Oceanic barriers as indicated by scombrid fishes and their parasites. *International Journal for Parasitology* **30**, 579–583, with the permission of the copyright holders, Elsevier Science.

Lester *et al.* (1988) examined 1251 orange roughy (*Hoplostethus atlanticus*) from eight areas off southern Australia and three areas off New Zealand (Figure 28). Fish were divided into three length groups and canonical multivariate analysis of data on three species of larval nematodes and two species of larval cestodes distinguished five Australian and three New Zealand stocks, distinctly differing in their parasite faunas (Figure 29). These findings indicate that orange roughy is a sedentary species with little movement between the management zones. However, interestingly, isoenzyme studies on the same fish in New Zealand led to somewhat different conclusions (Smith, 1986). There was little evidence for separation between samples from different areas in New Zealand including those examined by Lester *et al.* Rohde (1993), partly following Lester *et al.* (1988), suggested several reasons for the contradictory results: (1) separation of populations may be relatively recent with insufficient time to develop genetic differences, although parasites are different; (2) gene flow between populations may be so small that recognition of different stocks is difficult; (3) eggs or larvae of fish disperse before they become infected by parasites, which would make them unsuitable for distinguishing fish stocks. All this means that great caution is necessary; application of several methods should be attempted and – if results differ – conclusions should be made with great care.

Rohde (1987a, 1991c) suggested a novel method for distinguishing host populations. He used geographical variation in sclerite size of Monogenea to distinguish two populations of the blue (slimy) mackerel, *Scomber australasicus*, in New Zealand and New South Wales, Australia. A prerequisite of this method is that worms are from fish of approximately the same size (because sclerite length may be affected by host size, see Figure 3). Fixation of the worms (freezing, formalin fixation) has no effect. A disadvantage of the method is that the lifespan of most monogeneans is not known, and conclusions concerning the length of separation of populations are not possible. Further research is necessary.

5. PRACTICAL IMPORTANCE OF MARINE PARASITES

Marine parasites have been shown to be of very great importance in mariculture. Pollution may make marine animals, in particular those in aquaculture, more susceptible to infection; pollution may lead to an increase or decrease in parasite loads and such effects may be useful in monitoring environmental conditions; marine parasites may be infective to man, and they can be used to control introduced marine pests. All these aspects are discussed in the following pages. The use of parasites as biological tags was discussed in the previous section.

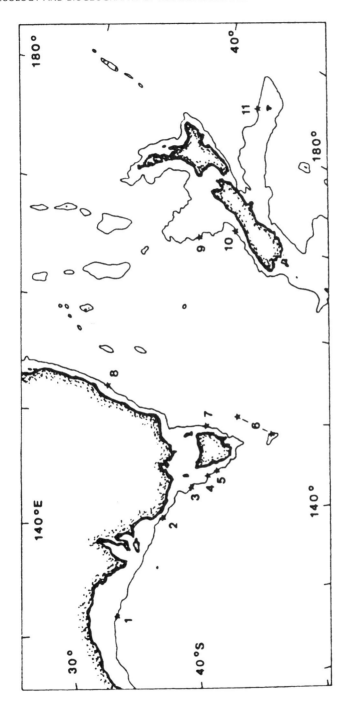

Figure 28 Origins of 11 samples of orange roughy, *Hoplostethus atlanticus* used for the analyses. The 1000 m isobath is shown. Modified from: Lester, R. J. G., Sewell, K. B., Barnes, A. and Evans, K. (1988). Stock discrimination of orange roughy, *Hoplostethus atlanticus*, by parasite analysis. *Marine Biology* **99**, 137–143, with permission from Springer-Verlag.

(a) small

(b) medium

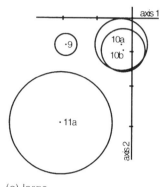

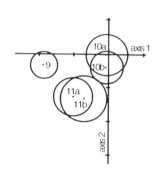

(c) large

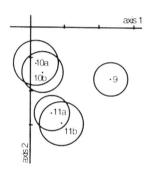

Figure 29 *Hoplostethus atlanticus*. Results of multivariate analysis on small, medium and large fish from New Zealand (areas 9–11 in Figure 28) using eight parasites, with 99% confidence range. Note: three separate populations for all three size classes. Modified from Lester, R. J. G., Sewell, K. B., Barnes, A. and Evans, K. (1988). Stock discrimination of orange roughy, *Hoplostethus atlanticus*, by parasite analysis. *Marine Biology* **99**, 137–143, with permission from Springer-Verlag.

5.1. Importance of parasites in mariculture

Earlier discussions were reviewed in Rohde (1993), and only some recent studies are discussed here.

Hecht and Endemann (1998) have reviewed aquaculture in Africa. In 1994, production amounted to 76,600 metric tonnes, more than half in countries along the Mediterranean, about 46% in sub-Saharan Africa. Altogether, aquaculture production is still low and "Africa is considered the sleeping giant of aquaculture". Disease has been little studied in sub-Saharan Africa, but seems to be relatively unimportant, apparently because of the low intensity of aquaculture. Some cases of viral infections have been reported, but most disease is caused by bacteria, protozoans, various helminths and crustaceans. The authors stress that introduction of fish, in particular of ornamental fish as well as hybrid and genetically modified fish, represents the greatest risk, as shown by the great range of parasites introduced into Israel with uncontrolled import of live fish.

According to Ogawa (1996) and Ogawa and Yokoyama (1998), parasites are more difficult to control in Japanese marine than in freshwater aquaculture, because of the use of net cage systems. More than 30 species of marine fish are cultured, with a wide range of parasite infections, but few life cycles are known (Ogawa and Yokoyama, 1998). Monogenea are of great importance, and their eggs – many possessing long filaments and laid in clusters – become entangled in floating net cages, increasing the risk of infection. Parasites with direct life cycles can easily reach high prevalences and intensities of infection, as demonstrated by the occurrence of a leech in aquacultured orange-spotted grouper, *Epinephelus coioides*, in the Philippines. Prevalence reached 83%, with large numbers of leeches attached to various organs and tissues responsible for frayed fins, haemorrhages and skin swellings; 50 ppm formalin baths were an effective treatment (Cruz-Lacierda *et al.*, 2000).

Mortality of tilapia in the highly saline Salton Sea, California, due to the dinoflagellate *Amyloodinium ocellatum*, was reported by Kuperman and Matey (1999). *Sparus auratus*, the gilthead sea bream, cultured along the Spanish coast, was found to be infected with the myxozoan *Polysporoplasma sparis*, causing serious damage to the trunk kidney. The infection was restricted to semi-intensive cultures and it did not occur in open and intensive closed systems (Palenzuela *et al.*, 1999). Rigos *et al.* (2001) reported considerable mortality caused by *Cryptocaryon irritans* of some but not all marine fish broodstock held under intensive culture conditions. Extended, rapid and repeated exposure to hyposalinity was effective, but chemical treatment was not effective in controlling the disease. Larval *Eubothrium* sp. significantly reduced the weight and haematocrit level of Atlantic salmon, *Salmo salar*, kept in tanks, even at low intensities of infection (Saksvik *et al.*, 2001).

5.2. Effects of pollution on parasites in mariculture

According to Soniat (1996), temperature and salinity and particularly their interactions have important effects on infection of the eastern oyster, *Crassostrea virginica*, with *Perkinsus marinus* along the Gulf of Mexico. However, most of the variation in infection is not explained by these factors. Pollution, as well as oyster density and various other factors, may be important. Burreson and Calvo (1996) found little evidence for the assumption that pollution was and is responsible for the remarkable increase of infection of *Crassostrea virginica* with *Perkinsus marinus* in Chesapeake Bay since 1985, although there may be subtle effects of poor water quality or toxicants. Chu (1996) set up experiments to test for the effects of temperature, salinity, number of infective cells and pollution on infection of *Crassostrea virginica* with *Perkinsus marinus*. Temperature, cell dose and salinity, in that order, were found to be the most important factors. Chu and Hale (1994) showed experimentally that pollutants enhance pre-existing infections with this parasite, and also the susceptibility of the oysters. On the other hand, Daros *et al.* (1998) tested two populations of a marine clam for infections with *Perkinsus* sp. in the Lagoon of Venice, Italy, one from a clean and one from a polluted area. They found a higher prevalence of infection in the clean site, but nevertheless clams from the clean site had greater digestive cell height and longer lysosomal latency time (see also Dusek *et al.*, 1998 for a recent study of effects of pollution on freshwater parasites).

5.3. Use of parasites for pollution monitoring

Costs of monitoring water pollution with conventional methods may be very high, and parasites of both freshwater and marine fish and of other hosts may provide a cheaper alternative (e.g. Anonymous, 1998; Landsberg *et al.*, 1998). Lafferty (1997) has discussed how parasites can be used to test environmental quality, and MacKenzie (1999) has given reasons why parasites are useful pollution indicators: there are more parasitic than free-living species, many parasites have several life cycle stages, and many parasites have infective stages that are particularly sensitive to environmental change. Several recent studies deal with the effects of pollution on parasites of marine fish (e.g. Siddall *et al.*, 1994; Marcogliese *et al.*, 1998), mussels (e.g. Powell *et al.*, 1999: effect on petroleum seep mussels) and snails (e.g. Siddall *et al.*, 1993). Of particular importance is the finding that some parasites accumulate heavy metals orders of magnitude higher than host tissue or the environment. Such parasites include the Acanthocephala and fish cestodes (e.g. Sures *et al.*, 1999). (For a detailed study of freshwater parasites see, for example, Sures *et al.*, 1994; Zimmermann *et al.*, 1999 found lead accumulation in an acanthocephalan but not a nematode of eel, *Anguilla anguilla*.)

Diamant *et al.* (1999) pointed out that parasites with complex life cycles involving several host species can survive only in habitats where conditions are right for all hosts, whereas parasites with direct life cycles involving only a single host survive and predominate in impoverished environments. Proceeding from these assumptions, they tested environmental quality by determining the ratio between parasite species with complex and those with simple life cycles, as well as species diversity, of macroparasite communities of the rabbit fish *Siganus rivulatus* from several localities in the Red Sea and the eastern Mediterranean. They used macroparasites because quantification is more accurate for them. The authors found higher ratios (relatively more species with complex life cycles) and greater species diversity in coral reef habitats that were ecologically stable, than in sandy habitats or habitats affected by mariculture. In communities near mariculture farms diversity was lowest, and even gill monogeneans with direct life cycles were reduced. Communities in the eastern Mediterranean had no parasites with complex life cycles at all, interpreted by the authors as evidence for extreme environmental deterioration. Testing the activity of cytochrome P-450-dependent mono-oxygenase EROD and lysosomal stability confirmed the conclusion: EROD activity in the Mediterranean was twice, and lysosomal membrane stability, half that in the Red Sea.

The effect of pollution on the ratio of parasites with complex and simple life cycles depends on the mobility of the host species. Siddall *et al.* (1994) found that the long rough dab, *Hippoglossoides platessoides*, in spite of the absence of intermediate hosts of certain parasite species at sites exposed to sewage sludge dumping, did not have a reduced number of parasites with complex life cycles, explained by the high mobility of the fish. A direct effect of pollution on larval stages of trematodes, i.e. parasites with complex life cycles, was demonstrated by Siddall *et al.* (1993). Infection with metacercariae in the common whelk, *Buccinum undatum*, decreased significantly towards a sewage sludge dump site in the Firth of Clyde, Scotland (prevalence of infection reduced from 19.7% at 13 km from the dumping area to 2% at the periphery of the dumping area).

Lafferty and Kuris (1999) discussed the effects that environmental stressors can have on the impact parasites have on hosts. (1) Pollution may increase host susceptibility to parasites; (2) pollution may increase the abundance of intermediate hosts and vectors; (3) pollution may decrease parasitism by causing higher mortality in infected hosts, or by affecting parasites or infected intermediate hosts or vectors negatively; (4) habitat alterations can lead to increased or decreased parasitism; (5) fisheries can affect host populations more negatively when already under stress; or it may lower host population density below a critical density necessary for survival of a parasite population; (6) introduction of species may introduce parasites dangerous to the less resistant native population, or they are at a competitive advantage over the native species because they are introduced without their parasites.

Experimental evidence for the effect of pollution on ectoparasites of fish is mainly available from freshwater habitats. For example, Bagge and Valtonen (1996) kept roach in cages in an unpolluted lake and samples were examined for ectoparasites. Some of the fish were then transferred to a lake influenced by pulp mill effluent. Prevalence of infection of one monogenean species was increased, and the abundance of three monogenean species was significantly increased in the fish in the polluted lake.

Broeg *et al.* (1999) used integrated biological effect monitoring of flounder, *Platichthys flesus*, to measure pollution at four locations in the German Bight, North Sea. Parasites, a molecular test measuring the integrity of lysosomal membranes in hepatocytes and the neutral lipids content in the liver, and measurements of macrophage aggregate activity in the liver were used, as well as a chemical analysis of standard organochlorine and heavy metal residues in the tissue. The protozoan *Trichodina* sp. proved to be the best indicator species for pollution, with the greatest intensity of infection at the most polluted site. Species diversity was significantly lower in the polluted estuary of the Elbe.

Gomez-Bautista *et al.* (2000) found mussels (*Mytilus galloprovincialis*) and cockles (*Cerastoderma edule*) in north-western Spain infected with oocysts of *Cryptosporidium parvum* infective to neonatal mice. Infected molluscs were found only near the mouths of rivers with many grazing ruminants, i.e. they are indicators of pollution. The same parasite was also found in marine mussels, *Mytilus edulis*, in the Sligo area, Ireland (Chalmers *et al.*, 1997).

According to Zander (1998a, b), eutrophication increases favour both common and rare parasites. Infracommunity richness and intensities of infection were increased in gobiid fishes of the south-western Baltic Sea, and this is explained by increased parasite density in intermediate hosts. However, in extremely eutrophic areas both species richness and intensities of infection were decreased. The author points out that these findings need confirmation by including other hosts.

Kesting and Zander (2000) found an impoverishment of the parasite fauna of small or juvenile fish, crustaceans and snails at four localities in the south-western Baltic Sea, demonstrated over a period of 18 years. It is suggested that this may be due to increasing eutrophication. The spectrum of parasites was reduced to those species with direct life cycles, and to species that prefer planktonic or suprabenthic (not benthic) hosts. The locality with the lowest salinity (6–8) had the smallest number of species.

5.4. Human disease caused by marine parasites

References on earlier work can be found in Rohde (1993). Marine parasites that infect humans exist but their overall importance is much less than that of non-marine parasites. Nevertheless, some marine helminths pose a significant risk to man. These include anisakids (Nematoda) acquired by eating raw or undercooked marine fish and invertebrates, and *Trichinella*. Forbes (2000) has given a recent discussion of *Trichinella* in the sea. Polar bears, walruses and less frequently seals (and once a Beluga whale) carry the infection, which was identified as *T. nativa*. Larvae from marine animals are cold-tolerant and infective to man. The infection is so common in circumpolar Arctic regions that safety programmes to prevent infection have been introduced in some Arctic communities.

Gomez-Bautista *et al.* (2000) reported that mussels (*Mytilus galloprovincialis*) and cockles (*Cerastoderma edule*) in north-western Spain harboured oocysts of *Cryptosporidium parvum* infective to neonatal mice (see above). The parasite has become increasingly important, particularly in immunodepressed patients, and the above observations show that marine bivalves can serve as reservoirs for human infection.

5.5. Introduction of parasites into new habitats

A case of great economic losses due to introduction of a parasite into an area previously free of it, well documented by Russian authors, is the monogenean *Nitzschia sturionis*, which was introduced into the Aral Sea with sturgeon from the Caspian Sea and devastated sturgeon fisheries there for many years (for details see the brief review in Rohde, 1993). A recently examined example is that of *Haplosporidium nelsoni*, a major disease agent of the eastern oyster, *Crassostrea virginica* on the mid-Atlantic coast of the USA since 1957. Molecular work by Burreson *et al.* (2000) has shown that a parasite of the Pacific oyster, *C. gigas*, is that species, and that the parasite was probably introduced into Californian waters and into the Atlantic with *C. gigas* from Japan.

Hecht and Endemann (1998), in their review of parasites in African aquaculture (see above) have emphasized that the great variety of disease and parasites in Israel aquaculture was caused by uncontrolled introductions. Ogawa (1996), in his review of marine parasites with reference to Japanese fisheries and mariculture, has pointed out that the international trade with live fish and shellfish has introduced previously unknown parasites into Japan. He urges establishment of an efficient quarantine system. Seng (1997), in his review of parasite problems in south-east Asian aquaculture, also stressed that large-scale international movement of fingerlings or juveniles, in addition to the rapid expansion and concentration of fish farms, has led to severe problems due to parasites, in particular monogeneans.

Mass mortalities of abalone in California were reported in 1985, and it was suggested that an infective agent was responsible. However, no such agent was found (Lafferty and Kuris, 1993). In 1993, a polychaete (apparently introduced) was brought to the attention of biologists. Some abalone cultures in California were heavily infected, leading to large economic losses. Other gastropods, but not bivalves, were infected as well. Attempts to find a Californian predator on the polychaete were not successful (Kuris and Culver, 1999).

Hines *et al.* (1997) made a detailed study of introduced and native populations of the rhizocephalan *Loxothylacus panopei*, which castrates xanthid crabs. Prevalence of infection ranged from 0 to 83%, with a mean smaller than 1. Epidemic outbreaks were recorded over a 15-year period at the Rhode River sub-estuary of Chesapeake Bay, following slow spreading of the parasite. A maximum prevalence of 72% was recorded.

5.6. Use of marine parasites for biological control of introduced marine pests

Lafferty and Kuris (1996) discussed guidelines for using biological control agents in the marine environment, based on experience gathered in terrestrial and freshwater habitats.

The European green crab, *Carcinus maenas*, has spread into many habitats on the Atlantic and Pacific coasts of the USA, into South Africa and southern Australia. Lafferty and Kuris (1996) proposed a strategy for biological control of the crab. Torchin *et al.* (1996) gave further details on this strategy. They tested a nemertean, *Carcinonemertes epialti*, a predator on eggs of the native Californian crab, *Hemigrapsus oregonensis*, for its host specificity in the laboratory, and found that it feeds on eggs of the introduced crab as well. Introduced crabs in the field also had reproducing and egg-feeding nemerteans, suggesting that the worm may be a useful control agent. A knowledge of the biology of potential control agents is important for such programmes: Kuris (1993) has discussed life cycles of various nemerteans, an important basis on which to develop such knowledge.

Control of *Carcinus maenas* by parasitic barnacles, i.e. rhizocephalans, also seems feasible. Thresher *et al.* (2000) conducted field and laboratory experiments to test host selection and specificity in the rhizocephalan *Sacculina carcini,* and host specificity in the related species *Heterosaccus lunatus*. They found that host specificity in the former species depends on the interaction between the parasite and host physiology. Therefore, safety trials are difficult, probably more difficult than for many terrestrial parasites. Before using barnacles for biological control, detailed studies of the behaviour of the infective planktonic larvae and physiological interactions with the host are necessary, and this may also apply to other marine parasites with planktonic larvae.

ACKNOWLEDGEMENTS

I am particularly indebted to Peter Rohde for writing the programmes used to generate Figures 10 and 12–15. Dr A. Faubel, University of Hamburg, kindly advised me on meiofauna. Peter Rohde, Craig Lawlor and Louise Streeting helped with redrawing many of the figures, and Libby Moodie kindly helped with Figure 4.

REFERENCES

Alves, D. R. and Luque, J. L. (2001). Community ecology of the metazoan parasites of white croaker, *Micropogonias furnieri* (Osteichthyes: Sciaenidae), from the coastal zone of the State of Rio de Janeiro, Brazil. *Memorias do Instituto Oswaldo Cruz* **96**, 145–153.

Anonymous. (1998). Parasite indicators for water pollution analysis. *South African Journal of Science* **94**, 95–96.

Armonies, W. and Reise, K. (2000). Faunal diversity across a sandy shore. *Marine Ecology Progress Series* **196**, 49–57.

Armstrong, H. W. (1974). A Study of the Helminth Parasites of the Family Macrouridae from the Gulf of Mexico and Caribbean Sea: Their Systematics, Ecology and Zoogeographical Implications. PhD thesis, Texas A&M University.

Bagge, A. M. and Valtonen, E. T. (1996). Experimental study on the influence of paper and pulp mill effluent on the gill parasite communities of roach (*Rutilus rutilus*). *Parasitology* **112**, 499–508.

Baker, E. A. and Collette, B. B. (1998). Mackerel from the northern Indian Ocean and the Red Sea are *Scomber australasicus*, not *Scomber japonicus*. *Ichthyological Research* **45**, 29–33.

Baldauf, S. L., Roger, A. J., Wenk-Siefert, I. and Doolittle, W. F. (2000). A kingdom-level phylogeny of eukaryotes based on combined protein data. *Science* **290**, 972–977.

Banford, H. M., Bermingham, E., Collette, B. B. and McCafferty, S. S. (1999). Phylogenetic systematics of the *Scomberomorus regalis* (Teleostei: Scombridae) species group: molecules, morphology and biogeography of Spanish mackerels. *Copeia* **1999**, 596–613.

Boucher, G. and Lambshead, P. J. D. (1995). Ecological biodiversity of marine nematodes in samples from temperate, tropical, and deep-sea regions. *Conservation Biology* **9**, 1594–1604.

Briggs, J. C. (1994). Species diversity: land and sea compared. *Systematic Biology* **43**, 130–135.

Broeg, K., Zander, S., Diamant, A., Korting, W., Kruner, G., Paperna, I. and von Westernage, H. (1999). The use of fish metabolic, pathological and parasitological indices in pollution monitoring – 1. North Sea. *Helgoland Marine Research* **53**, 171–194.

Buchmann, K. and Nielsen, M. E. (1999). Chemoattraction of *Ichthyophthirius multifiliis* (Ciliophora) theronts to host molecules. *International Journal for Parasitology* **29**, 1415–1423.

Burreson, E. M. and Calvo, L. M. R. (1996). Epizootiology of *Perkinsus marinus* disease of oysters in Chesapeake Bay, with emphasis on data since 1985. *Journal of Shellfish Research* **15**, 17–34.

Burreson, E. M., Stokes, N. A. and Friedman, C. S. (2000). Increased virulence in an introduced pathogen: *Haplosporidium nelsoni* (MSX) in the eastern oyster *Crassostrea virginica*. *Journal of Aquatic Animal Health* **12**, 1–8.

Bush, A. O. and Kennedy, C. R. (1994). Host fragmentation and helminth parasites: hedging your bets against extinction. *International Journal for Parasitology* **24**, 1333–1343.

Byrnes, T. (1987). Caligids (Copepoda: Caligidae) found on the bream (*Acanthopagrus* spp.) of Australia. *Journal of Natural History* **21**, 363–404.

Byrnes, T. and Rohde, K. (1992). Geographical distribution and host specificity of ectoparasites of Australian bream, *Acanthopagrus* spp. (Sparidae). *Folia Parasitologica* **39**, 249–264.

Campbell, R. A. (1983). Parasitism in the deep sea. *In* "The Sea" (G. T. Rowe, ed.), pp. 473–552, John Wiley and Sons, New York.

Campbell, R. A., Haedrich, R. L. and Munroe, T. A. (1980). Parasitism and ecological relationships among deep sea benthic fishes. *Marine Biology* **57**, 301–313.

Chalmers, R. M., Sturdee, A. P., Mellors, P., Nicholson, V., Lawlor, F., Kenny, F. and Timpson, P. (1997). *Cryptosporidium parvum* in environmental samples in the Sligo area, Republic of Ireland – a preliminary report. *Letters in Applied Microbiology* **25**, 380–384.

Chesson. P. (2001). Metapopulations. *In* "Encyclopedia of Biodiversity", pp. 161–176. Academic Press, New York.

Chu, F. L. E. (1996). Laboratory investigations of susceptibility, infectivity, and transmission of *Perkinsus marinus* in oysters. *Journal of Shellfish Research* **15**, 57–66.

Chu, F. L. E. and Hale, R. C. (1994). Relationship between pollution and susceptibility to infectious disease in the eastern oyster, *Crassostrea virginica*. *Marine Environmental Research* **38**, 243–256.

Clarke, A. (1992). Is there a latitudinal diversity cline in the sea? *Tree* **7**, 286–287.

Collard, S. B. (1968). A Study of Parasitism in Mesopelagic Fishes. PhD thesis, University of California, Santa Barbara.

Collard, S. B. (1970). Some aspects of host parasite relationships in mesopelagic fishes. *In* "A Symposium on Diseases of Fishes and Shellfishes" (S. F. Sniezko, ed.), pp. 41–56. American Fisheries Society Special Publication 5, Washington, DC.

Collette, B. B. (1999). Mackerels, molecules, and morphology. *In* "Proceedings of the 5th Indo-Pacific Fish Conference, Noumea" (B. Séret and J.-Y. Sire, eds), pp. 149–164. Societé Francaise Ichthyologie.

Comiskey, P. and MacKenzie, K. (2000). *Corynosoma* spp. may be useful biological tags for saithe in the northern North Sea. *Journal of Fish Biology* **57**, 525–528.

Cornell, H. V. and Lawton, J. H. (1992). Species interactions, local and regional processes, and limits to the richness of ecological communities: a theoretical perspective. *Journal of Animal Ecology* **61**, 1–12.

Creasey, S. S. and Rogers, A. D. (1999). Population genetics of bathyal and abyssal organisms. *Advances in Marine Biology* **35**, 1–151.

Cribb, T. H. (1998). The diversity of the Digenea of Australian animals. *International Journal for Parasitology* **28**, 899–911.

Cribb, T. H., Bray, R. A., Barker, S. C., Adlard, R. D. and Anderson, G. R. (1994). Ecology and diversity of digenean trematodes of reef and inshore fishes of Queensland.. *International Journal for Parasitology* **24**, 851–860.

Cruz-Lacierda, E. F., Toledo, J. D., Tan-Fermin, J. D. and Burreson, E. M. (2000). Marine leech (*Zeylanicobdella arguamensis*) infestation in cultured orange-spotted grouper, *Epinephelus coicoides*. *Aquaculture* **183**, 191–196.

Daros, L, Marin, M. G., Nesto, N. and Ford, S. E. (1998). Preliminary results of a field study on some stress-related parameters in *Tapes philippinarum* naturally infected by protozoan *Perkinsus* sp. *Marine Environmental Research* **46**, 249–252.

Diamant, A., Banet, A., Paperna, I., von Westerhagen, H., Broeg, K., Kruener, G., Koerting, W. and Zander, S. (1999). The use of fish metabolic, pathological and parasitological indices in pollution monitoring – II. The Red Sea and Mediterranean. *Helgoland Marine Research* **53**, 195–208.

Dusek, L., Gelnar, M. and Sebelova, S. (1998). Biodiversity of parasites in a fresh-water environment with respect to pollution – metazoan parasites of chub (*Leuciscus cephalus* L.) as a model for statistical evaluation. *International Journal for Parasitology* **28**, 1555–1571.

Etter, R. J., Rex, M. A., Chase. M. C. and Quattro, J. M. (1999). A genetic dimension to deep-sea diversity. *Deep-Sea Research I* **46**, 1095–1099.

Forbes, L. B. (2000). The occurrence and ecology of *Trichinella* in marine mammals. *Veterinary Parasitology* **93**, 321–334.

Gage, J. D. (1996). Why are there so many species in deep-sea sediments? *Journal of Experimental Marine Biology and Ecology* **200**, 257–286.

Geets, A., Coene, H. and Ollevier, F. (1997). Ectoparasites of the whitespotted rabbitfish, *Siganus sutor* (Valenciennes, 1835) off the Kenyan coast: distribution within the host population and site selection on the gills. *Parasitology* **115**, 69–79.

Gomez-Bautista, M., Ortega-Mora, L. M., Tabares, E., Lopez-Rodas, V. and Costas, E. (2000). Detection of infectious *Cryptosporidium parvum* oocysts in mussels (*Mytilus galloprovincialis*) and cockles (*Cerastoderma edule*). *Applied and Environmental Microbiology* **66**, 1866–1870.

Grassle, J. F. and Maciolek, N. J. (1992). Deep-sea species richness: regional and local diversity estimates from quantitative bottom samples. *American Naturalist* **139**, 313–341.

Gresty, K. A., Boxshall, G. A. and Nagasawa, K. (1993). Antennulary sensors of the infective copepodid larva of the salmon louse, *Lepeophtheirus salmonis* (Copepoda: Caligidae). *In* "Pathogenesis of Wild and Farmed Fish, Sea Lice" (G. A. Boxshall and D. Defaye, eds), pp. 83–98. Ellis Horwood, New York.

Groenewold, S., Berghahn, R. and Zander, C.-D. (1996). Parasite communities of four fish species in the Wadden Sea and the role of fish discarded by the shrimp fisheries in parasite transmission. *Helgoländer Meeresuntersuchungen* **50**, 69–85.

Grutter, A. S. (1994). Spatial and temporal variations of the ectoparasites of seven reef fish species from Lizard Island and Heron Island, Australia. *Marine Ecology Progress Series* **115**, 21–30.

Grutter, A. S. (1996). Experimental demonstration of no effect by the cleaner wrasse *Labroides dimidiatus* (Cuvier and Valenciennes) on the host fish *Pomacentrus moluccensis* (Bleeker). *Journal of Experimental Marine Biology and Ecology* **196**, 285–298.

Grutter, A. S. (1997a). Size-selective predation by the cleaner fish *Labroides dimidiatus*. *Journal of Fish Biology* **50**, 1303–1308.

Grutter, A. S. (1997b). Effect of the removal of cleaner fish on the abundance and species composition of reef fish. *Oecologia* **111**, 137–143.

Grutter, A. S. (1998). Habitat-related differences in the abundance of parasites from a coral reef fish – an indication of the movement patterns of *Hemigymnus melapterus*. *Journal of Fish Biology* **53**, 49–57.

Grutter, A. S. (1999a). Infestation dynamics of gnathiid isopod juveniles parasitic on the coral-reef fish *Hemigymnus melapterus* (labridae). *Marine Biology* **135**, 545–552.

Grutter, A. S. (1999b). Cleaner fish really do clean. *Nature* **398**, 672–673.

Grutter, A. S. (2000). Ontogenetic variation in the diet of the cleaner fish *Labroides dimidiatus* and its ecological consequences. *Marine Ecology Progress Series* **197**, 241–246.

Grutter, A. S. and Hendrikz, J. (1999). Diurnal variation in the abundance of juvenile parasitic gnathiid isopods on coral reef fish: implications for parasite-cleaner fish interactions. *Coral Reefs* **18**, 187–191.

Grutter, A. S. and Pankhurst, N. W. (2000). The effects of capture, handling, confinement and ectoparasite load on plasma levels of cortisol, glucose and lactate in the coral reef fish *Hemigymnus melapterus*. *Journal of Fish Biology* **57**, 391–401.

Grutter, A. S. and Poulin, R. (1998a). Cleaning of coral reef fishes by the wrasse *Labroides dimidiatus* – influence of client body size and phylogeny. *Copeia* **1**, 120–127.

Grutter, A. S. and Poulin, R. (1998b). Intraspecific and interspecific relationships between host size and the abundance of parasitic larval gnathiid isopods on coral reef fishes. *Marine Ecology Progress Series* **164**, 263–271.

Grutter, A. S., Lester, R. J. G. and Greenwood, J. (2000). Emergence rate from the benthos of the parasitic juveniles of gnathiid isopods. *Marine Ecology Progress Series* **207**, 123–127.

Guitiérrez, P. A. (2001). Monogenean community structure on the gills of *Pimelodus albicans* from Río de la Plata (Argentina): a comparative approach. *Parasitology* **122**, 465–470.

Guitiérrez, P. A. and Martorelli, S. R. (1999). Hemibranch preference by freshwater monogeneans a function of gill area, water currents, or both? *Folia Parasitologica* **46**, 263-266.

Gusev, A. V. (1995). Some pathways and factors of monogenean micro-evolution. *Canadian Journal of Fisheries and Aquatic Sciences* **52** (Suppl. 1), 52–56.

Hanski, I. (1982). Dynamics of regional distribution: the core and satellite species hypothesis. *Oikos* **38**, 210–221.

Hawkins, B. A. (1990). Global patterns of parasitoid assemblage size. *Journal of Animal Ecology* **59**, 57–72.

Hayward, C. J. (1997). Distribution of external parasites indicates boundaries to dispersal of sillaginid fishes in the Indo-West Pacific. *Marine Freshwater Research* **48**, 391–400.

Hayward, C. J. and Rohde K. (1999a). Revision of the monogenean family Gotocotylidae (Polyopisthocotylea). *Invertebrate Taxonomy* **13**, 425–460.

Hayward, C. J. and Rohde, K. (1999b). Revision of the monogenean subfamily Thoracocotylinae Price, 1936 (Polyopisthocotylea: Thoracocotylidae) with description of a new species of the genus *Pseudothoracocotyla* Yamaguti, 1963. *Systematic Parasitology* **44**, 157–169.

Hayward, C. J. and Rohde, K. (1999c). Revision of the monogenean subfamily Neothoracocotylinae Lebedev, 1969 (Polyopisthocotylea: Thoracocotylidae). *Systematic Parasitology* **44**, 183–191.

Heath, B. M. (1987). Study of the Endoparasitic Helminths of Deep-Sea Fishes from South Eastern Australia: Taxonomy, Zoogeography and Host-Parasite Ecology. PhD thesis, University of New England, Armidale, Australia.

Hecht, T. and Endemann, F. (1998). The impact of parasites, infections and diseases on the development of aquaculture in sub-Saharan Africa. *Journal of Applied Ichthyology* **14**, 213–221.

Hessler, R. R. and Sanders, H. L. (1967). Faunal diversity in the deep sea. *Deep-Sea Research* **14**, 65–78.

Hines, A. H., Alvarez, F. and Reed, S. A. (1997). Introduced and native populations of a marine parasitic castrator – variation in prevalence of the Rhizocephalan *Loxothylacus panopaei* in xanthid crabs. *Bulletin of Marine Science* **61**, 197–214.

Hodkinson, I. D. and Casson, D. (1990). A lesser predilection for bugs: Hemiptera (Insecta) diversity in tropical rain forests. *Biological Journal of the Linnean Society* **43**, 101–109.

Hoeg, J. T. (1995). The biology and life cycle of the Rhizocephala (Cirripedia). *Journal of the Marine Biological Association of the United Kingdom* **75**, 517–550.

Hull, M. Q., Pike, A. W., Mordue, A. J. and Rae, G. H. (1998). Patterns of pair formation and mating in an ectoparasitic caligid copepod *Lepeophtheirus salmonis* (Kroyer 1837) – implications for its sensory and mating biology. *Philosophical Transactions of the Royal Society of London B* **353**, 753–764.

Jennings, J. B. (1997). Nutritional and respiratory pathways to parasitism exemplified in the Turbellaria. *International Journal for Parasitology* **27**, 679–691.

Kesting, V. and Zander, C. D. (2000). Alteration of the metazoan parasite faunas in the brackish Schlei Fjord (Northern Germany, Baltic Sea). *Internationale Revue der gesamten Hydrobiologie* **85**, 325–340.

Kinne, O. ed. (1980–1990). "Diseases of Marine Animals", vols. 1–4. John Wiley, Chichester, and Biologische Anstalt Helgoland, Hamburg.

Kleeman, S. (2001). The development of the community structure of the ecto- and endoparasites of *Siganus doliatus*, a tropical marine fish. BSc Honours thesis, University of New England, Armidale, Australia.

Koslow, J. A., Williams, A. and Paxton, J. R. (1997). How many demersal fish species in the deep sea – a test of a method to extrapolate from local to global diversity. *Biodiversity & Conservation* **11**, 1523–1532.

Krebs, C. J. (1997). "Ecology: the Experimental Analysis of Distribution and Abundance" 4th edn. Harper Collins, New York.

Kuperman, B. I. and Matey, V. E. (1999). Massive infestation by *Amyloodinium ocellatum* (Dinoflagellidae) of fish in a highly saline lake, Salton Sea, California, U.S.A. *Diseases of Aquatic Organisms* **39**, 65–73.

Kuris, A. M. (1993). Life cycles of nemerteans that are symbiotic egg predators of decapod crustacea – adaptations to host life histories. *Hydrobiologia* **266**, 1–14.

Kuris, A. M. and Culver, C. S. (1999). An introduced sabellid polychaete pest infesting cultured abalones and its potential spread to other California gastropods. *Invertebrate Biology* **118**, 391–403.

Kuris, A. M. and Lafferty, K. D. (1994). Community structure – larval trematodes in snail hosts. *Annual Review of Ecology and Systematics* **25**, 189–217.

Lafferty, K. D. (1997). Environmental parasitology – what can parasites tell us about human impacts on the environment? *Parasitology Today* **13**, 251–255.

Lafferty, K. D. and Kuris, A. M. (1993). Mass mortality of abalone *Haliotis cracherodii* on the California Channel Islands – tests of epidemiological hypotheses. *Marine Ecology Progress Series* **96**, 239–248.

Lafferty, K. D. and Kuris, A. M. (1996). Biological control of marine pests (Review). *Ecology* **77**, 1989–2000.

Lafferty, K. D. and Kuris, A. M. (1999). How environmental stress affects the impacts of parasites. *Limnology and Oceanography* **44**, 925–931.

Lafferty, K. D., Sammon, D. T. and Kuris, A. M. (1994). Analysis of larval trematode communities. *Ecology* **75**, 2275–2285.

Lambshead, P. J. D. (1993). Recent developments in marine benthic biodiversity research. *Océanis* **19**, 5–24.

Lambshead, P. J. D., Elge, B. M., Thistle, E., Eckman, J. E. and Barnett, P. R. O. (1994). A comparison of the biodiversity of deep-sea marine nematodes from three stations in the Rockall Trough, Northeast Atlantic, and one station in the San Diego Trough, Northeast Pacific. *Biodiversity Research* **2**, 95–107.

Landsberg, J. H., Blakesley, B. A., Reese, R. O., McRae, G. and Forstchen, P. R. (1998). Parasites of fish as indicators of environmental stress. *Environmental Monitoring and Assessment* **51**, 211–232.

Larsen, G., Hemmingsen, W., MacKenzie, K. and Lysne, D. A. (1997). A population study of cod, *Gadus morhua* L., in northern Norway using otolith structure and parasite tags. *Fisheries Research* **32**, 13–20.

Lauckner, G. (1984). Impact of trematode parasitism on the fauna of a North Sea tidal flat. *Helgoländer Meeresuntersuchungen* **37**, 185–199.

Lester, R. J. G., Sewell, K. B., Barnes, A. and Evans, K. (1988). Stock discrimination of orange roughy, *Hoplostethus atlanticus*, by parasite analysis. *Marine Biology* **99**, 137–143.

Littlewood, D. T. J., Rohde, K. and Clough, K. A. (1997). Parasite speciation within or between host species? – phylogenetic evidence from site-specific polystome monogeneans. *International Journal for Parasitology* **27**, 1289–1297.

Littlewood, D. T. J., Telford, M. J., Clough, K. A. and Rohde, K. (1998). Gnathostomulida – an enigmatic metazoan phylum from both morphological and molecular perspectives. *Molecular Phylogenetics and Evolution* **9**, 72–79.

Littlewood, D. T. J., Rohde, K., Bray, R. A. and Herniou, E. A. (1999). Phylogeny of the Platyhelminthes and the evolution of parasitism. *Biological Journal of the Linnean Society* **68**, 257–287.

Lo, C. M. and Morand, S. (2001). Gill parasites of *Cephalopolis argus* (Teleostei: Serranidae) from Moorea (French Polynesia): site selection and coexistence. *Folia Parasitologica* **48**, 30–36.

McCarthy, H. O., Fitzpatrick, S. and Irwin, S. W. B. (2000). A transmissible trematode affects the direction and rhythm of movement in a marine gastropod. *Animal Behaviour* **59**, 1161–1166.

MacKenzie, K. (1999). Parasites as pollution indicators in marine ecosystems: a proposed early warning system. *Marine Pollution Bulletin* **38**, 955–959.

MacKenzie, K. and Abaunza, P. (1998). Parasites as biological tags for stock discrimination of marine fish – a guide to procedures and methods. *Fisheries Research* **38**, 45–56.

MacKenzie, K. and Longshaw, M. (1995). Parasites of the hakes *Merluccius australis* and *M. hubbsi* in the waters around the Falkland Islands, southern Chile and Argentina, with an assessment of their potential value as biological tags. *Canadian Journal of Fisheries & Aquatic Sciences* **52**, 213–224.

Manter, H. W. (1967). Some aspects of the geographical distribution of parasites. *Journal of Parasitology* **53**, 3–9.

Marcogliese, D. J. (2001). Pursuing parasites up the food chain: implications of food web structure and function on parasite communities in aquatic systems. *Acta Parasitologica* **46**, 82–93.

Marcogliese, D. J., Nagler, J. J. and Cyr, D. G. (1998). Effects of exposure to contaminated sediments on the parasite fauna of American plaice (*Hippoglossoides platessoides*). *Bulletin of Environmental Contamination and Toxicology* **61**, 88–95.

Margolis, L. (1963). Parasites as indicators of the geographical origin of sockeye salmon, *Oncorhynchus nerka* (Walbaum), occurring in the North Pacific Ocean and adjacent seas. *Bulletin of the International North Pacific Fisheries Commission* **11**, 101–156.

Martens, E. and Moens, J. (1995). The metazoan ecto- and endoparasites of the rabbitfish, *Siganus sutor* (Cuvier and Valenciennes, 1835) of the Kenyan coast. I. *African Journal of Ecology* **33**, 405–416.

May, R. M. (1990). How many species? *Philosophical Transactions of the Royal Society of London B* **330**, 293–304.

May, R. M. and Nee, S. (1995). The species alias problem. *Nature* **378**, 447–448.

Menier, K. (2000). Origins and evolution of parasites in marine mammals. Pinnipeds example (in French). *Revue de Medecine Veterinaire* **151**, 275–280.

Moore, J. (1995). The behaviour of parasitized animals. *Bioscience* **45**, 89–96.

Morand, S., Poulin R., Rohde, K. and Hayward, C. J. (1999). Aggregation and species coexistence of ectoparasites of marine fishes. *International Journal for Parasitology* **29**, 663–672.

Morand, S., Rohde, K. and Hayward, C. J. (in press). Order in parasite communities of marine fish is explained by epidemiological processes. *Parasitology*.

Morand, S., Cribb, T. H., Kulbicki, M., Rigby, M. C., Chauvet, C., Dufour, V., Faliex, E., Galzin, R., Lo, C. M., Lo-Ya, A., Pichelin, S. and Sasal, P. (2000). Endoparasite species richness of New Caledonian butterfly fishes: host density and diet matter. *Parasitology* **121**, 65–73.

Moser, M. (1991). Parasites as biological tags. *Parasitology Today* **7**, 182–185.

Moser, M. and Hsieh, J. (1992). Biological tags for stock separation in Pacific herring *Clupea harengus pallasi* in California. *Journal of Parasitology* **78**, 54–60.

Mosquera, J., Gomez-Gesteira, M. and Perez-Villar, V. (2000). Using parasites as biological tags of fish populations. A dynamical model. *Bulletin of Mathematical Biology* **62**, 87–99.

Nelson, J. S. (1994). "Fishes of the World", 3rd edn. Wiley, New York.

Noble, E. R. (1973). Parasites and fishes in a deep-sea environment. *Advances in Marine Biology* **11**, 121–195.

Noble, E. R. and Collard, S. B. (1970). The parasites of mid water fishes. *In* "A Symposium on Diseases of Fishes and Shell Fishes" (S. F. Sniezko, ed.), pp. 57–68. American Fisheries Society Special Publication 5, Washington, DC.

Ogawa, K. (1996). Marine parasitology with special reference to Japanese fisheries and mariculture. *Veterinary Parasitology* **64**, 95–105.

Ogawa, K. and Yokoyama, H. (1998). Parasitic diseases of cultured marine fish in Japan. *Fish Pathology* **33**, 303–309.

Palenzuela, O., Alvarez-Pellitero, P. and Sitja-Bobadilla, A. (1999). Glomerular disease associated with *Polysporoplasma sparis* (Myxozoa) infections in cultured gilthead sea bream, *Sparus aurata* L. (Pisces: Teleostei). *Parasitology* **118**, 245–256.

Pascual, S. and Hochberg, F. G. (1996). Marine parasites as biological tags of cephalopod hosts. *Parasitology Today* **12**, 324–327.

Paterson, A. M. and Poulin, R. (1999). Have chondracanthid copepods co-speciated with their teleost hosts? *Systematic Parasitology* **44**, 79–85.

Perera, K. M. L. (1992). The effect of host size on large hamuli length of *Kuhnia scombri* (Monogenea: Polyopisthocotylea) from Eden, New South Wales, Australia. *International Journal for Parasitology* **22**, 123–124.

Poulin, R. (1995). Phylogeny, ecology and the richness of parasite communities in vertebrates. *Ecological Monographs* **65**, 283–302.

Poulin, R. (1998) "Evolutionary Ecology of Parasites. From Individuals to Communities". Chapman and Hall, London.

Poulin, R. (1999a). Speciation and diversification of parasite lineages: an analysis of congeneric parasite species in vertebrates. *Evolutionary Ecology* **13**, 455–467.

Poulin, R. (1999b). The functional importance of parasites in animal communities: many roles at many levels? *International Journal for Parasitology* **29**, 903–914.

Poulin, R. (1999c). Body size vs. abundance among parasite species: positive relationship? *Ecography* **22**, 246–250.

Poulin, R. (2000). Variation in the intraspecific relationship between fish length and parasitic infection: biological and statistical causes. *Journal of Fish Biology* **56**, 123–137.

Poulin, R. and Guegan, J. F. (2000). Nestedness, anti-nestedness, and the relationship between prevalence and intensity in ectoparasites assemblages of marine fish: a spatial model of species coexistence. *International Journal for Parasitology* **30**, 1147–1152.

Poulin, R. and Morand, S. (1999). Geographical distance and the similarity among parasite communities of conspecific host populations. *Parasitology* **119**, 369–374.

Poulin, R. and Morand, S. (2000a). The diversity of parasites. *Quarterly Review of Biology* **75**, 277–293.

Poulin, R. and Morand, S. (2000b). Parasite body size and interspecific variation in levels or aggregation among nematodes. *Journal of Parasitology* **86**, 642–647.

Poulin, R. and Rohde, K. (1997). Comparing the richness of metazoan ectoparasite communities of marine fishes: controlling for host phylogeny. *Oecologia* **110**, 278–283.

Poulin, R. and Thomas, F. (1999). Phenotypic variability induced by parasites: extent and evolutionary implications. *Parasitology Today* **15**, 28–32.

Powell, E. N., Barber, R. D., Kennicutt, M. C. and Ford, S. E. (1999). Influence of parasitism in controlling the health, reproduction and PAH body burden of petroleum seep mussels. *Deep-Sea Research I* **46**, 2053–2078.

Rahbeck, C. and Graves, G. R. (2001). Multiscale assessment of patterns of avian species richness. *Proceedings of the National Academy of Science USA* **98**, 4534–4539.

Raibaut, A., Combes, C. and Benoit, F. (1998). Analysis of the parasitic copepod species richness among Mediterranean fish. *Journal of Marine Systems* **15**, 185–206.

Rigby, M. C., Holmes, J. C., Cribb, T. H. and Morand, S. (1997). Patterns of species diversity in the gastrointestinal helminths of a coral reef fish, *Epinephelus merra* (Serranidae), from French Polynesia and the South Pacific Ocean. *Canadian Journal of Zoology* **75**, 1818–1827.

Rigos, G., Pavlidis, M. and Divanach, P. (2001). Host susceptibility to *Cryptocaryon* sp. infection of Mediterranean marine broodfish held under intensive culture conditions: a case report. *Bulletin of the European Association of Fish Pathologists* **21**, 33–36.

Ritchie, M. E. and Olff, H. (1999). Spatial scaling laws yield a synthetic theory of biodiversity. *Nature* **400**, 557–562.

Rohde, K. (1972). The Aspidogastrea, especially *Multicotyle purvisi* Dawes, 1941. *Advances in Parasitology* **10**, 77–151.

Rohde, K. (1973). Structure and development of *Lobatostoma manteri* sp. nov. (Trematoda, Aspidogastrea) from the Great Barrier Reef, Australia. *Parasitology* **66**, 63–83.

Rohde, K. (1976). Species diversity of parasites on the Great Barrier Reef. *Zeitschrift für Parasitenkunde* **50**, 93–94.

Rohde, K. (1977a). A non-competitive mechanism responsible for restricting niches. *Zoologischer Anzeiger* **199**, 164–172.

Rohde, K. (1977b). Species diversity of monogenean gill parasites of fish on the Great Barrier Reef. *In* "Proceedings of the Third International Coral Reef Symposium", pp. 585–591. Miami, Florida.

Rohde, K. (1978a). Latitudinal differences in species diversity and their causes. I. A review of the hypotheses explaining the gradients. *Biologisches Zentralblatt* **97**, 393–403.

Rohde, K. (1978b). Latitudinal differences in host specificity of marine Monogenea and Digenea. *Marine Biology* **47**, 125–134.

Rohde, K. (1979). A critical evaluation of intrinsic and extrinsic factors responsible for niche restriction in parasites. *American Naturalist* **114**, 648–671.

Rohde, K. (1980a). Comparative studies on microhabitat utilization by ecto-parasites of some marine fishes from the North Sea and Papua New Guinea. *Zoologischer Anzeiger* **204**, 27–63.

Rohde, K. (1980b). Warum sind ökologische Nischen begrenzt? Zwischenartlicher Antagonismus oder innerartlicher Zusammenhalt? *Naturwissenschaftliche Rundschau* **33**, 98–102.

Rohde, K. (1980c). Diversity gradients of marine Monogenea in the Atlantic and Pacific Oceans. *Experientia* **36**, 1368–1369.

Rohde, K. (1980d). Host specificity indices of parasites and their application. *Experientia* **36**, 1369–1371.

Rohde, K. (1981). Niche width of parasites in species-rich and species-poor communities. *Experientia* **37**, 359–361.

Rohde, K. (1982). "Ecology of Marine Parasites". University of Queensland Press, St. Lucia, Australia.

Rohde, K (1984a). Ecology of marine parasites. *Helgoländer Meeresuntersuchungen* **37**, 5–33.

Rohde, K. (1984b). Zoogeography of marine parasites. *Helgoländer Meeresuntersuchungen* **37**, 35–52.

Rohde, K. (1985). Increased viviparity of marine parasites at high latitudes. *Hydrobiologia* **127**, 197–201.

Rohde, K. (1986). Differences in species diversity of Monogenea between the Pacific and Atlantic Oceans. *Hydrobiologia* **137**, 21–28.

Rohde, K. (1987a). Different populations of *Scomber australasicus* in New Zealand and southeastern Australia, demonstrated by a simple method using mono-genean sclerites. *Journal of Fish Biology* **30**, 651–657.

Rohde, K. (1987b). *Grubea australis* n. sp. (Monogenea, Polyopisthocotylea) from *Scomber australasicus* in southeastern Australia, and *Grubea cochlear* Diesing, 1858 from *S. scombrus* and *S. japonicus* in the Mediterranean and western Atlantic. *Systematic Parasitology* **9**, 29–38.

Rohde, K. (1988). Gill Monogenea of deepwater and surface fish in southeastern Australia. *Hydrobiologia* **16**, 271–283.

Rohde, K. (1989a). At least eight types of sense receptors in an endoparasitic flatworm: a counter-trend to sacculinization. *Naturwissenschaften* **76**, 383–385.

Rohde, K. (1989b). Simple ecological systems, simple solutions to complex problems? *Evolutionary Theory* **8**, 305–350.

Rohde, K. (1990). Marine parasites: an Australian perspective. *International Journal for Parasitology,* **20**, 565–575.

Rohde, K. (1991a). "Ekologi Parasit Laut" (Malay/Indonesian translation of "Ecology of Marine Parasites"). Dewan Bahasa dan Pustaka, Kuala Lumpur, Malaysia.

Rohde, K. (1991b). Intra- and interspecific interactions in low density populations in resource-rich habitats. *Oikos* **60**, 91–104.

Rohde, K. (1991c). Size differences in hamuli of *Kuhnia scombri* (Monogenea Polyopisthocotylea) from different geographical areas not due to differences in host size. *International Journal for Parasitology* **21**, 113–114.

Rohde, K. (1992). Latitudinal gradients in species diversity: the search for the primary cause. *Oikos* **65**, 514–527.

Rohde, K. (1993). "Ecology of Marine Parasites", 2nd edn. CAB International, Wallingford, Oxfordshire.

Rohde, K. (1994a). Niche restriction in parasites: proximate and ultimate causes. *Parasitology* **109**, S69–S84.

Rohde, K. (1994b). The minor groups of parasitic Platyhelminthes. *Advances in Parasitology* **33**, 145–234.

Rohde, K. (1998a). Is there a fixed number of niches for endoparasites of fish? *International Journal for Parasitology* **28**, 1861–1865.

Rohde, K. (1998b). Latitudinal gradients in species diversity. Area matters, but how much? *Oikos* **82**, 184–190.

Rohde, K. (1999). Latitudinal gradients in species diversity and Rapoport's rule revisited: a review of recent work, and what can parasites teach us about the causes of the gradients? *Ecography* **22**, 593–613. Also published *In* "Ecology 1999 – and tomorrow" (T. Fenchel, ed.), pp. 73–93. Ecology Institute, University of Lund, Sweden.

Rohde, K. (2001a). Parasitism. *In* "Encyclopedia of Biodiversity" (S. Levin, ed.), vol. IV, pp. 463–484. Academic Press, New York.

Rohde, K. (2001b). Spatial scaling laws may not apply to most animal species. *Oikos* **93**, 499–504.

Rohde, K. (in press). Niche restriction and mate finding in vertebrates, *In* "Behavioral Ecology of Parasites" (E. E. Lewis, J. F. Campbell and M. V. K. Sukhdeo, eds). CAB International, Wallingford, Oxon.

Rohde, K. and Hayward, C. J. (1999a). Revision of the monogenean subfamily Priceinae Chauhan, 1953 (Polyopisthocotylea: Thoracocotylidae). *Systematic Parasitology* **44**, 171–182.

Rohde, K. and Hayward, C. J. (1999b). *Scomberomorocotyle munroi* n.g., n.sp. (Scomberomorocotylinae n.subf.), a thoracocotylid monogenean from *Scomberomorus munroi* (Scombridae) off Australia and Papua New Guinea. *Systematic Parasitology* **43**, 1–6.

Rohde, K. and Hayward, C. J. (2000). Oceanic barriers as indicated by scombrid fishes and their parasites. *International Journal for Parasitology* **30**, 579–583.

Rohde, K. and Heap, M. (1998). Latitudinal differences in species and community richness and in community structure of metazoan endo- and ectoparasites of marine teleost fish. *International Journal for Parasitology* **28**, 461–474.

Rohde, K. and Hobbs, R. (1986). Rarity in marine Monogenea. Does an Allee-effect or parasite-induced mortality explain truncated frequency distributions? *Biologisches Zentralblatt* **107**, 327–338.

Rohde, K. and Hobbs, R. P. (1999). An asymmetric percent similarity index. *Oikos* **87**, 601–602.

Rohde, K. and Rohde, P. P. (2001). Fuzzy chaos: reduced chaos in the combined dynamics of several independently chaotic populations. *American Naturalist* **158**, 553–556.

Rohde, K. and Watson, N. (1985a). Morphology and geographical variation of *Pseudokuhnia minor* n.g., n. comb. (Monogenea Polyopisthocotylea). *International Journal for Parasitology* **15**, 557–567.

Rohde, K. and Watson, N. (1985b). Morphology, microhabitats and geographical variation of *Kuhnia* spp. (Monogenea Polyopisthocotylea). *International Journal for Parasitology* **15**, 569–586.

Rohde, K. and Watson, N. (1989). Sense receptors in *Lobatostoma manteri* (Trematoda, Aspidogastrea). *International Journal for Parasitology* **19**, 847–858.

Rohde, K. and Watson, N. A. (1992a). Sense receptors of larval *Lobatostoma manteri* (Trematoda, Aspidogastrea). *International Journal for Parasitology* **22**, 35–42.

Rohde, K. and Watson, N. A. (1992b). Ultrastructure of tegument, ventral sucker and rugae of *Rugogaster hydrolagi* (Trematoda, Aspidogastrea). *International Journal for Parasitology* **22**, 967–974.

Rohde, K. and Watson, N. A. (1995a). Ultrastructure of the buccal complex of *Polylabroides australis* (Monogenea, Polyopisthocotylea, Microcotylidae). *International Journal for Parasitology* **25**, 307–318.

Rohde, K. and Watson, N. A. (1995b). Comparative ultrastructural study of the posterior suckers of four species of symbiotic Platyhelminthes, *Temnocephala* sp. (Temnocephalida), *Udonella caligorum* (Udonellidea), *Anoplodiscus cirrusspiralis* (Monogenea: Monopistocotylea), and *Philophthalmus* sp. (Trematoda: Digenea). *Folia Parasitologica* **43**, 11–28.

Rohde, K. and Watson, N. A. (1996). Ultrastructure of the buccal complex of *Pricea multae* (Monogenea, Polyopisthocotylea, Gastrocotylidae). *Folia Parasitologica* **43**, 117–132.

Rohde, K., Hayward, C., Heap, M. and Gosper, D. (1994). A tropical assemblage of ectoparasites: gill and head parasites of *Lethrinus miniatus* (Teleostei, Lethrinidae). *International Journal for Parasitology* **24**, 1031–1053.

Rohde, K., Hayward, C. and Heap, M. (1995). Aspects of the ecology of metazoan ectoparasites of marine fishes. *International Journal for Parasitology* **25**, 945–970.

Rohde, K., Worthen, W., Heap, M., Hugueny, B. and Guégan, J.-F. (1998). Nestedness in assemblages of metazoan ecto- and endoparasites of marine fish. *International Journal for Parasitology* **28**, 543–549.

Rothsey, S. and Rohde, K. (2001). The response of oncomiracidia to light, and magnetic fields. *In* "Proceedings of the 4th International Symposium on Monogenea", p. 43. University of Queensland, Brisbane.

Saksvik, M., Nilsen, F., Nylund, A. and Berland, B. (2001). Effect of marine *Eubothrium* sp. (Cestoda: Pseudophyllidea) on the growth of Atlantic salmon, *Salmo salar* L. *Journal of Fish Diseases* **24**, 111–119.

Sasal, P. and Morand, S. (1988). Comparative analysis – a tool for studying monogenean ecology and evolution. *International Journal for Parasitology* **28**, 1637–1644.

Sasal, P., Morand, S. and Guegan, J. F. (1997). Determinants of parasite species richness in Mediterranean marine fishes. *Marine Ecology Progress Series* **149**, 61–71.

Sasal, P., Niquil, N. and Bartoli, P. (1999a). Community structure of digenean parasites of sparid and labrid fishes of the Mediterranean sea: a new approach. *Parasitology* **119**, 635–648.

Sasal, P., Trouvé, S., Müller-Graf, C. and Morand, S. (1999b). Specificity and host predictability: a comparative analysis among monogenean parasites of fish. *Journal of Animal Ecology* **68**, 437–444.

Scoles, D. R., Collette, B. B. and Graves, J. E. (1998). Global phylogeography of mackerels of the genus *Scomber*. *Fisheries Bulletin* **96**, 823–842.

Seldon, P. A. and Edwards, D. (1990). Colonisation of the land. *In* "Evolution and the fossil record" (K. Allen and D. Briggs, eds), pp. 122–152. Smithsonian Institution Press, Washington, DC.

Seng, L. T. (1997). Control of parasites in cultured marine finfishes in Southeast Asia – an overview. *International Journal for Parasitology* **27**, 1177–1184.

Siddall, R., Pike, A. W. and McVicar, A. H. (1993). Parasites of *Buccinum undatum* (Mollusca, Prosobranchia) as biological indicators of sewage-sludge dispersal. *Journal of the Marine Biological Association of the United Kingdom* **73**, 931–948.

Siddall, R., Pike, A. W. and McVicar, A. H. (1994). Parasites of flatfish in relation to sewage sludge dumping. *Journal of Fish Biology* **45**, 193–209.

Sluys, R. and Hazevoet, C. J. (1999). Pluralism in species concepts: dividing nature at its diversity joints. *Species Diversity* **4**, 243–256.

Smith, P. J. (1986). Genetic similarity between samples of the orange roughy *Hoplostethus atlanticus* from the Tasman Sea, south-west Pacific Ocean and North-east Atlantic Ocean. *Marine Biology* **81**, 173–180.

Solow, A. R. (1995). Estimating biodiversity: calculating unseen riches. *Oceanus* **38**, 9–10.

Soniat, T. M. (1996). Epizootiology of *Perkinsus marinus* disease of eastern oysters in the Gulf of Mexico. *Journal of Shellfish Research* **15**, 35–43.

Sprent, J. F. A. (1992). Parasites lost. *International Journal for Parasitology* **22**, 139–151.

Stanley, S. M. (1989). "Earth and Life Through Time", 2nd.edn. W. H. Freeman, New York.

Stork, N. E. (1988). Insect diversity: facts, fiction and speculation. *Biological Journal of the Linnean Society* **35**, 321–337.

Sures, B., Taraschewski, H. and Jackwerth, E. (1994). Lead accumulation in *Pomphorhynchus laevis* and its host. *Journal of Parasitology* **80**, 355–357.

Sures, B., Siddall, R. and Taraschewski, H. (1999). Parasites as accumulation indicators of heavy metal pollution. *Parasitology Today* **15**, 16–21.

Thorson, G. (1957). Bottom communities (sublittoral or shallow shelf). *In* "Treatise on Marine Ecology and Paleoecology" (J. W. Hedgpeth, ed.), pp. 461–534. Geological Society of America.

Thorson, G. (1971). "Life in the Sea." McGraw Hill, New York.

Thresher, R. E., Werner, M., Hoeg, J. T., Svane, I., Glenner, H., Murphy, N. E. and Wittwer, C. (2000). Developing the options for managing marine pests: specificity trials on the parasitic castrator, *Sacculina carcini*, against the European crab, *Carcinus maenas*, and related species. *Journal of Experimental Marine Biology and Ecology* **254**, 37–51.

Tirard, C., Berrebi, P., Raibaut, A. and Renaud, F. (1993). Biodiversity and biogeography in heterospecific teleostean (Giadidae)-Copepod (Lernaeocera) associations. *Canadian Journal of Zoology* **71**, 1639–1645.

Torchin, M. E., Lafferty, K. D. and Kuris, A. M. (1996). Infestation of an introduced host, the European green crab, *Carcinus maenas*, by a symbiotic nemertean egg predator, *Carcinonemertes epialti. Journal of Parasitology* **82**, 449–453.

Valtonen, E. T., Pulkinnen, K., Poulin, R. and Julkunen, M. (2001). The structure of parasite component communities in brackish water fishes of the northeastern Baltic Sea. *Parasitology* **122**, 471–481.

Whittington, I. D. (1998). Diversity down under – monogeneans in the antipodes (Australia) with a prediction of monogenean biodiversity worldwide. *International Journal for Parasitology* **28**, 1481–1493.

Whittington, I. D., Chisholm, L. A. and Rohde, K. (2001). The larvae of Monogenea (Platyhelminthes). *Advances in Parasitology* **44**, 139–232.

Wikelski, M. (1999). Influences of parasites and thermoregulation on grouping tendencies in marine iguanas. *Behavioral Ecology* **10**, 22–29.

Willig, M. R. (2001). Latitude, common trends within. *In* "Encyclopedia of Biodiversity", vol. 3 (S. Levin, ed.), pp. 701–714. Academic Press, New York.

Worthen, W. B. and Rohde, K. (1996). Nested subset analysis of colonisation-dominated communities: metazoan ectoparasites of marine fish. *Oikos* **75**, 471–478.

Zander, C. D. (1998a). Ecology of host parasite relationships in the Baltic Sea. *Naturwissenschaften* **85**, 426–436.

Zander, C. D. (1998b). Parasitengemeinschaften bei Grundeln (Gobiidae, Teleostei) der südwestlichen Ostsee. *Verhandlungen der Gesellschaft für Ichthyologie* **1**, 241–252.

Zander, C. D., Reimer, L. W. and Barz, K. (1999a). Parasite communities of the Salzhaff (Nordwest Mecklenburg, Baltic Sea). I. Structure and dynamics of communities of littoral fish, especially small-sized fish. *Parasitology Research* **85**, 356–372.

Zander, C. D., Reimer, L. W., Barz, K, Dietel, G. and Strohbach, U. (1999b). Parasite communities of the Salzhaff (Nordwest Mecklenburg, Baltic Sea). II. Guild communities, with special regard to snails, benthic crustaceans, and small-sized fish. *Parasitology Research* **86**, 359–372.

Zimmerman, S., Sures, B. and Taraschewski, H. (1999). Experimental studies on lead accumulation in the eel-specific endoparasites *Anguillicola crassus* (Nematoda) and *Paratenuisentis ambiguus* (Acanthocephala) as compared with their host, *Anguilla anguilla. Archives of Environmental Contamination and Toxicology* **37**, 190–195.

ADDENDUM

Several contributions that have appeared or have been brought to my notice since this review was submitted need to be considered.

O'Neill *et al.* (1982) modelled populations in which reproductive rates (*r*) varied over time. They found that ecological systems, in certain situations, become more predictable, similar to some of the effects discussed in the section on metapopulations and chaos. However, details of the effects are quite different. Whereas the width of the chaotic band becomes narrower with increasing numbers of subpopulations, and the first bifurcation depends on the range of reproductive values (see section 3.5), varying reproductive rates over time has none of these effects, but tends to lead to a disappearance of the "windows" in the chaotic region (Rohde and Rohde, unpublished).

Ecosystem engineering has attracted much attention in recent years. Ecosystem engineers are organisms that cause physical state changes in biotic or abiotic factors and thus modulate the availability of resources to other species. Thomas *et al.* (1999, further references therein) have discussed the role of parasites as possible ecosystem engineers.

In a fascinating study, Bartoli *et al.* (2000) showed that social rank and behavioural changes in a labrid fish from the western Mediterranean had a distinct effect on infection with a parasite species. The parasite was present only (with one exception) in large males and particularly those that were involved in nest building.

Lo and Morand (2000) studied spatial distribution and coexistence of monogeneans on the gills of two pomacentrid fishes in French Polynesia. They concluded that interspecific competition was not important, and that reinforcement of reproductive barriers may have led to the avoidance of hybridization between congeners, as suggested by Rohde.

Madhavi and Ram (2000) studied the community structure of helminths of the tuna, *Euthynnus affinis*, from the Bay of Bengal. They found a profound effect of host size on community structure: parasite density decreased and diversity increased with increasing host size.

Taraschewski (2000) gave a detailed account of host–parasite interactions in acanthocephalans, using recent morphological (largely ultrastructural) and histochemical descriptions. Consequences for microhabitat selection, host specificity and defences against host reactions are discussed, as well as high capacity for heavy metal uptake, resulting in the usefulness of acanthocephalans as bioindicators.

Ernst *et al.* (2001), contrary to what they say in their paper, provide further support for the conclusion that gill monogeneans of the viviparous Gyrodactylidae are much less common in tropical and subtropical than in high latitude seas (Thorson's rule, see Figure 21). The gills of only two fish species from along the Queensland coast (of a total of 104 species examined) had gyrodactylids. Twenty percent of the fish species had between 16 and 19 gyrodactylid species (apparently including those from the gills, the others from the fins and body surfaces). These data permit extension of Thorson's rule to monogeneans infecting body parts other than the gills, although more surveys from various latitudes are needed. Table 2 shows that the 104 fish species examined by the authors can be expected to harbour approximately 200 species of gill Monogenea alone, and 19 species of Gyrodactylidae would represent about 10% of this total. However, the total is likely to be greater because only gill monogeneans are included in Table 2. At high northern latitudes, at least 70% of the species are likely to be gill, body and fin gyrodactylids.

Some recent attention has been paid to a new category of parasites, "hitchhikers", parasites that exploit host manipulation by other parasites to gain entry into hosts. Mouritsen (2001, further references therein) has shown that distinction between a hitchhiker and the parasite used for hiking may be difficult thus a trematode species, originally thought to be the hitchhiker, turned out to be the species being used by the other trematode.

Gotelli and Rohde (2002), using null model analyses, have given further evidence that most ectoparasites of marine fish live in assemblages largely unstructured by interactions between species. In most analyses, co-occurrence patterns could not be distinguished from those that might arise by random colonization and extinction events. Importantly, they found a continuum of co-occurrence patterns from small bodied taxa with low vagility and/or small populations (such as marine ectoparasites), to large bodied taxa with high vagility and/or large populations, such as birds and mammals. The former live in largely random assemblages, the latter in highly structured communities.

Ectoparasites of marine crustaceans (e.g. Itani, in press; Itani et al., in press) face the problem of surviving ecdyses of their hosts. Fascinating studies, including time-lapse photographs, have shown that some parasite species are simply shed and replaced by new infection, whereas other species survive. The latter may never move during the intermoult stages of the host, but they move either just after the host starts moulting, or somewhat earlier, thus avoiding shedding. Parasites initiating their avoidance movements before moulting need some cue indicating the impending moult, possibly provided by host hormones.

A detailed account of experiments that demonstrate reactions of larval copapods and monogeneans to light, gravity and magnetic fields (section 3.6) is given by Rothsey and Rohde (in press).

ACKNOWLEDGEMENTS

I thank Bob Gardner for drawing my attention to the paper by O'Neill *et al.* (1982), and also Dr Itani for sending preprints of his forthcoming papers and giving permission to discuss them here.

REFERENCES

Bartoli, P., Morand, S., Riutort, J. J. and Combes, C. (2001). Acquisition of parasites correlated with social rank and behavioural changes in a fish species. *Journal of Helminthology* **74**, 289–293.

Ernst, I., Whittington, I. D. and Jones, M. K. (2001). Diversity of gyrodactylids from marine fishes in tropical and subtropical Queensland, Australia. *Folia Parasitologica* **48**, 165–168.

Gotelli, N. J. and Rohde, K. (2002). Co-occurrence of ectoparasites of marine fishes: a null model analysis. *Ecology Letters* **5**, 86–94.

Itani, G. (in press). Two types of symbioses between grapsid crabs and a host thalassinidean shrimp. *Publications of the Seto Marine Biological Laboratory*.

Itani, G., Kato, M. and Shirayama, Y. (in press). Behaviour of the shrimp ectosymbionts *Peregrinamor ohshimai* (Mollusca: Bivalvia) and *Phyllodorus* sp. (Crustacea: Isopoda) through host ecdyses. *Journal of the Maraine Biological Association of the United Kingdom*.

Lo, C. M. and Morand, S. (2000). Spatial distribution and coexistence of monogenean gill parasites inhabiting two damselfishes from Moorea Island in French Polynesia. *Journal of Helminthology* **74**, 329–336.

Madhavi, R. and Ram, S. B. K. (2000). Community structure of helminth parasites of the tuna, *Euthynnus affinis*, from the Visakhapatnam coast, Bay of Bengal. *Journal of Helminthology* **74**, 337–342.

Mouritsen, K. N. (2001). Hitch-hiking parasite: a dark horse may be the real rider. *International Journal for Parasitology* **31**, 1417–1420.

O'Neill, R. V., Gardner, R. H. and Weller, D. E. (1982). Chaotic models as representations of ecological systems. *American Naturalist* **120**, 259–263.

Rothsey, S. and Rohde, K. (in press). The responses of larval copepods and monogeneans to light, gravity and magnetic fields. *Acta Parasitologica*.

Taraschewski, H. (2000). Host-parasite interactions in Acanthocephala: a morphological approach. *Advances in Parasitology* **46**, 1–179.

Thomas, F., Poulin, R., de Meeüs, T., Guégan, J.-L. and Renaud, F. (1999). Parasites and ecosystem engineers: what roles could they play? *Oikos* **84**, 167–171.

Fecundity and Life-history Strategies in Marine Invertebrates

Eva Ramirez Llodra

School of Ocean and Earth Science, Southampton University,
Southampton Oceanography Centre, European Way,
Southampton SO14 3ZH, UK
E-mail: eramirez_llodra@hotmail.com

The reproductive strategies of an organism play a major role in the dynamics of the population and the biogeography and continuity of the species. Numerous processes are involved in reproduction leading to the production of offspring. Although diverse processes are involved in oogenesis (the production of eggs) and spermatogenesis (the production of sperm), the basic patterns of gametogenesis are similar amongst invertebrates, with the proliferation and differentiation of germ cells leading

ADVANCES IN MARINE BIOLOGY VOL. 43
ISBN 0-12-026143-X

to the final production of mature gametes. The production of gametes, especially eggs, is energetically expensive, and therefore strongly sensitive to selective pressures. An organism can ingest and assimilate a limited amount of energy from the environment. The different ways by which energy is allocated to growth and reproduction in order to maximize fitness forms the basis of the differing life-history strategies that have developed in marine invertebrates.

Fecundity is defined as the number of offspring produced by a female in a determined time period. The term fecundity needs to be explicitly defined in each study in order to obtain the maximum information from the data analysed. Because of the variety of egg production patterns found among marine invertebrates, a wide range of methodologies has been developed to quantify fecundity. These include direct egg counts in brooding species, spawning induction in live individuals and histological studies of preserved material. Specific considerations need to be taken into account for colonial organisms, because of their modular organization.

The production of eggs requires an optimal allocation of energy into growth and reproduction for the maximization of parental fitness. Fecundity is central in studies of life-history theory and in the development of life-history models because it is directly related to energy allocation and partitioning. There are important relationships and trade-offs between fecundity and other life-history traits, such as egg size, female size and age, age at first reproduction, reproductive effort and residual reproductive value. These trade-offs, together with morpho-functional constraints and genetic variation determine the evolution of life histories through natural selection.

Fecundity is a highly plastic character within the limits defined by the bioenergetics and life-history strategy of the organism. Egg production is affected mainly by environmental factors such as food quantity and quality, temperature or presence of toxic elements in the habitat. The differences in fecundity found among closely related species from different biogeographical locations reflect, at least in part, the differing environmental conditions of their habitat.

1. INTRODUCTION

Fecundity refers to the number of offspring produced by a female in a certain period of time (see section 2 for definitions). There is a wide variability in the types of gamete production, larval development, metamorphosis and settlement of juveniles among marine invertebrates. These processes have important ecological and biogeographical consequences in population dynamics and for the continuity of the species (Giangrande *et al.*, 1994). A long sequence of reproductive and developmental processes precedes. These include the proliferation, differentiation, growth and maturation of gametes, spawning events, fertilization and embryonic development. All these reproductive phases can be influenced both by phylogeny and by environmental conditions that exert selective pressure on life-history traits (Giese and Pearse, 1974; Stearns, 1992; Giangrande *et al.*, 1994).

To understand fully the functioning of an ecosystem, it is imperative to have a sound knowledge of the reproductive strategies of its components (Giangrande *et al.*, 1994). Sexual reproduction ensures not only the production of offspring to replace the parents and maintain the population, but also genetic recombination, which allows for evolution over geological timescales. Within the different reproductive processes, gametogenesis (oogenesis and spermatogenesis) plays a central role.

The processes involved in oogenesis and spermatogenesis are diverse among invertebrates (Eckelbarger, 1994). For example, (1) oogenesis may be extra-ovarian or intra-ovarian; (2) the developing oocytes may be associated with a variety of cells such as follicle cells, nurse cells, nutritive eggs or other accessory cells; (3) vitellogenesis may be autosynthetic, heterosynthetic or mixed; (4) oogenesis may be synchronous or asynchronous; (5) different amounts of energy may be stored in oocytes leading to differences in egg size and larval development (see section 5.1); and (6) the eggs might be retained in the female, spawned as free cells or enclosed in egg masses of diverse types (see section 3). Spermatogenesis can give rise to ect-aquasperm, aflagellate spermatozoa and many other kinds of sperm. However, the basic patterns of gametogenesis are markedly similar among most invertebrates (Giese and Pearse, 1974). Gametogenesis starts with the proliferation of gonial cells (oogonia and spermatogonia) by mitotic divisions. Oogonia and spermatogonia differentiate into primary oocytes and spermatocytes, respectively. At this stage, the processes of meiosis start. The first meiotic division produces secondary oocytes and spermatocytes.

In spermatogenesis, the second meiotic division occurs soon after the first meiotic division, producing four haploid spermatids. These mature into spermatozoa during the processes of spermiogenesis.

In oogenesis, the first meiotic division is arrested at the diplotene stage of prophase I when the paired chromosomes are dispersed in the enlarged nucleus (named the germinal vesicle in oocytes). This stage is followed by oocyte growth during vitellogenesis, usually assisted by accessory cells. Vitellogenesis is the synthesis and accumulation of ooplasmic reserves (yolk) in the oocytes. This is the longest and most expensive process during oogenesis. The common term "yolk" includes a variety of elements, such as lipids, proteins, carbohydrates, pigments, free sugars, free amino acids, nucleotides and nucleic acids (Krol et al., 1992; Eckelbarger, 1994).

There are three major vitellogenetic pathways in invertebrates: autosynthetic, heterosynthetic and mixed. In species with autosynthetic vitellogenesis, there is an uptake of exogenous low molecular weight precursors and the subsequent synthesis of vitellin by the proteosynthetic organelles of the oocytes. This mechanism results in a slow egg production rate. In species with heterosynthetic vitellogenesis, there is the transport of externally-synthesized yolk proteins into the oocyte, allowing for fast egg production. Finally, the mixed pathway is a combination of the first two (Eckelbarger, 1983, 1994; Jaeckle, 1995). The vitellogenic pathways are determined by ovarian morphology and reflect the species' ancestral history (Eckelbarger, 1994; Eckelbarger and Watling, 1995). There is a strong phylogenetic basis in the gametogenetic process of invertebrates, which constrains the evolution of life-history patterns (see section 5.1.3).

After reaching full size through growth during vitellogenesis, primary oocytes undergo the two meiotic divisions (maturation divisions). Germinal vesicle breakdown can occur before spawning, through hormonal stimulation, or after spawning, either spontaneously or through induction by fertilization. One haploid ovum and three polar bodies are formed from each primary oocyte.

Gametogenesis is strongly dependent on the bioenergetic budget of the species and the optimal allocation of energy between somatic growth and synthesis of reproductive material (Giese and Pearse, 1974; Olive, 1985; Giangrande et al., 1994). The egg is one of the most valuable single cells in the life history of any species. Eggs are highly specialized, very large cells, which, following activation during fertilization, initiate the processes of embryogenesis (Eckelbarger, 1983, 1994). The egg not only gives half of the genetic information and all the mitochondrial DNA to the zygote in sexually reproducing species, but also provides the energetic reserves and structural material for embryo development (Jaeckle, 1995). Furthermore, in species with lecithotrophic larval development, the energy invested in the egg represents the total nutrient reserve that will sustain the larva until metamorphosis occurs.

The ecological consequences of parental investment per offspring have received substantial attention in the last eighty years. Such studies of marine invertebrates aim to understand the evolution of reproductive patterns and the importance of both phylogeny and the environment in the acquirement of adaptive strategies (Orton, 1920; Thorson, 1950; Pianka, 1970; Vance, 1973a, b; Smith and Fretwell, 1974; Bell, 1980; Stearns, 1977, 1980, 1992; Charlesworth, 1980, 1990; Levitan, 1993, 1996; Podolsky and Strathmann, 1996; McHugh and Rouse, 1998). Natural selection tends to optimize the energy allocation that would confer maximal fitness to the individual. Because the production of eggs is an energetically-expensive process, reproductive traits related to egg production (such as egg size and fecundity) play important roles in the evolution of life-history strategies. Fecundity is directly related to life-history traits such as egg size, age at maturity, lifespan and reproductive effort. These life-history traits have complex interactions determining specific trade-offs between them. The evolution of the reproductive traits and their interactions are moulded by natural selection to produce the wide variety of optimal life-history strategies found among marine invertebrates (Stearns, 1992; Giangrande *et al.*, 1994).

In the following sections, fecundity will be defined and the methodologies used to quantify this parameter in species with differing reproductive patterns will be outlined. The role of fecundity in life-history theory will be described and the different models that have been developed in life-history theory will be summarized. An analysis of the relationships of fecundity to other life-history traits such as egg size, female size and age, age at first reproduction and reproductive effort will follow. The limitations of morphological barriers to evolution of fecundity and reproductive output will be exposed, and examples of the plasticity of fecundity related to environmental factors will be presented.

The published work cited here is not intended to be exhaustive, but relevant examples in invertebrate life-history studies are used to illustrate each section.

2. DEFINITIONS

In general terms, fecundity refers to the total number of offspring produced by a female. The basic unit of fecundity measurements is the offspring, of which the different forms are: oocytes, eggs, embryos, larvae and juveniles. Fecundity is expressed as the number of offspring produced over a certain period of time. In order to relate fecundity data to the life history of a species, the term fecundity needs to be explicitly defined. For each specific study, fecundity data should indicate (1) the unit counted (e.g. oocytes, eggs, embryos, larvae); (2) the individual in which the unit is counted (batch of eggs, female, colony); and (3) the timescale (e.g. spawning event, breeding season, year, lifetime).

Most studies refer to fecundity as the number of eggs per female in one breeding season. Scheltema (1994) called this variable "apparent fecundity", because it indicates the number of gametes produced at a particular instant, rather than the rate of gamete production over the lifetime of the organism. However, this quantification of fecundity is widely used because it is the most direct and available way of estimating egg production in most species. In species that produce several broods in one breeding season, the term brood size is used to define the number of eggs produced during one breeding event, while fecundity refers to the total number of eggs produced during the whole breeding season (Fonseca-Larios and Briones-Fourzan, 1998). Lifetime fecundity is of major interest in the life history strategy of a species, because it integrates the variation of the energy budget with age. But lifetime fecundity is a difficult, or often unworkable, variable to quantify for many species for which variables such as lifespan, the relationship between age and fecundity (see section 5.2) and breeding frequency (Scheltema, 1994) are unknown.

Anger and Moreira (1998), working with decapods, distinguished three categories of fecundity: (1) *potential fecundity*, defined as the number of oocytes in the ovary, including developing and mature cells; (2) *realized fecundity*, defined as the number of eggs carried in the pleopods; and (3) *actual fecundity*, defined as the number of hatched larvae, and therefore related to fertilization success and embryo mortality. The terms potential and actual fecundity have also been used in reproductive studies of broadcast spawning invertebrates (mainly echinoderms). In these cases, potential fecundity refers to all oocytes in the ovary; actual fecundity refers to oocytes nearing maximum development, oocyte size reached before spawning (Tyler and Gage, 1983; Tyler and Billett, 1987).

3. QUANTIFICATION OF FECUNDITY

Invertebrates have a wide variety of gamete release patterns spread over a range of possibilities. At one extreme are the broadcast spawners that release a large number of small eggs, while at the other extreme there are brooding species with internal fertilization and direct development. Within these two limits there are a large number of species with intermediate strategies. Broadcast spawning species release their mature gametes into the water column where the eggs are fertilized. The embryos can hatch either into larvae that develop and feed in the plankton (planktotrophic larvae), or into non-feeding larvae that develop in the plankton but are sustained by internal reserves of the egg (lecithotrophic larvae). In some cases, the males are free spawners, while the females retain the eggs in body structures. For example, many groups of the lower Caenogastropoda, such as the Vermetidae, are aphallic and release spermatophores into the sea. These are trapped by the mucous feeding nets of females and fertilization is internal, followed by the production of egg capsules, which are retained in the mantle cavity (Calvo *et al.*, 1998). Other species such as the asteroids *Leptasterias* sp. and *Henricia* sp. (Chia and Walker, 1991; Chia *et al.*, 1993; Levin and Bridges, 1995) have external fertilization but incubate their embryos in external brooding structures. At the other end of the spawning mode spectrum, there are species with internal fertilization by copulation or pseudo-copulation. These species incubate batches of eggs in brooding structures or lay egg masses or egg cases that are attached to the substratum. The embryos may develop into larvae, or there may be direct development, in which there is no larval phase. However, for the purpose of this section, the patterns of spawning have been grouped into two categories: broadcast spawners and brooders.

Many different methods have been used to quantify fecundity. The optimal method to be used in each study will be determined by the reproductive characteristics of the species considered. These characteristics are: (1) spawning pattern (brooding or broadcast spawner); (2) ability to respond to a spawning-inducing chemical; (3) ovary morphology (relative complexity); and (4) the number and size of eggs produced. The following text provides a description of the most widely used methodologies for the quantification of fecundity for brooding and broadcast spawners. The case of colonial species, which may be either broadcast spawners or brooders, will be considered separately, because of the basic differences in quantifying any parameter in modular organisms. Table 1 summarizes the different methodologies for the quantification of fecundity in a variety of taxa.

3.1. Brooding species

In brooding species, the females, or in some cases the males (as in brooding pycnogonids), incubate batches of developing embryos either internally or externally. Internal fertilization in brooding species increases fertilization success, and parental care with its high energetic cost allows for a high level of embryo survivorship. Fecundity in brooding species can be quantified easily and precisely by directly counting the number of embryos developing in the brooding structures.

Many crustaceans, including some decapods, amphipods, isopods and copepods, brood their embryos until the larvae hatch. Embryos are thus easily obtained and can be manipulated in the laboratory for studies on fecundity and related life-history parameters. Species studied come from environments as diverse as shallow water (Hines, 1982, 1986, 1988; Corey, 1987; Corey and Reid, 1991; Reid and Corey, 1991; Clarke, 1993a; Stella *et al.*, 1996), the deep sea (Herring, 1967, 1974a, b; King and Butler, 1985;

Table 1 Summary of methodologies for the quantification of fecundity in marine invertebrates, with relevant examples.

Methodology	Characteristics of the organism	Main examples	References
Direct egg counts			
	Brooding species	Copepods Decapods	Razouls et al., 1991; Laabir et al., 1995 Corey and Reid, 1991; Reid and Corey, 1991
	Species that lay egg masses or egg cases	Molluscs	Chester, 1996; Cheung and Lam, 1999
		Polychaetes	McHugh, 1993
Spawning induction			
Environmental shock (Response to a sharp change in environmental conditions)			
Thermal shock	Response to sharp changes in temperature	Bivalves	Thompson, 1979; Langton et al., 1987; Honkoop and Van der Meer, 1997; Hamel et al., 2001
Osmotic shock	Response to exposure to air	Corals Abalone and other molluscs	Wallace, 1985; Richmond, 1987 Baker, 2001; Baker and Tyler, 2001
Mechanical stimulus	Response to wave action	Polychaete	Mauro, 1975
Chemical induction (Response to the injection or external application of a chemical solution)			
Coelomic injection of 2 ml of 0.55 M KCl	Induces contraction of gonad muscles	Echinoids	Young et al., 1997, 1998
Coelomic injection of 0.5 ml of 10^{-4} M of 1-methyladenine	Induces meiosis and gonad wall contraction	Asteroids	Kanatani, 1975; McEdward and Colter, 1987; Bosch, 1989

Method	Description	Organism	References
Foot injection or external application of 0.4 ml of 2 mM serotonin	Induces germinal vesicle breakdown and/or release of gametes	Bivalves	Ram et al., 1993; Fong et al., 1996; Juneja and Koide, 1996
Injection in coelom of a solution of reproductive pheromone	Induces oocyte maturation and stimulates spawning	Polychaetes	Olive and Bentley, 1980; Bentley et al., 1984; Hardege et al., 1996
External application of hydrogen peroxide	Activates prostaglandin peroxidase synthetase producing active compounds involved in spawning processes	Abalone and other molluscs	Morse, 1977; Beckevar, 1981; Moss et al., 1995
Non-destructive method			
Weight differences in the organism before and after natural spawning	Organisms that need to be kept alive and healthy	Bivalves	Langton et al., 1987; Barber et al., 1988
Body dissections			
Dissection of ovaries (strip spawning)	Species with simple ovarian organization	Bivalves	Nakaoka, 1994
Smear of ovaries on microscope slide	Species with simple ovarian organization	Corals	Parker et al., 1997; Smith and Hughes, 1999
		Asteroids Ophiuroids	George, 1994a, b Sumida et al., 2000
Extraction of coelomic fluid	Species where oocytes develop in the coelom	Polychaetes	McHugh, 1993; Zal et al., 1995; Olive et al., 1997, 1998
Ovarian tissue cleared with Histoclear and oocytes visualized with a transmitted light microscope	Species with small gonads, thin gonad walls or very large mature oocytes	Some holothurians	Tyler and Gage, 1983; Tyler et al., 1985, 1987
Histological methods			
Serial sectioning of the ovary	Complex ovarian organization, small gonads and low fecundity	Ophiuroids	Tyler and Gage, 1980
Volumetric method relating ovary volume to oocyte volume	Complex ovarian organization and densely packed ovaries	Asteroids	Ramirez Llodra et al., in press

Clarke *et al.*, 1991; Hartnoll *et al.*, 1992; Company and Sardà, 1997; Ramirez Llodra, 2000), polar regions (Clarke, 1979, 1987, 1993b; Gorny *et al.*, 1992) and hydrothermal vents at mid-oceanic ridges (Van Dover *et al.*, 1985; Van Dover and Williams, 1991; Ramirez Llodra, 2000; Ramirez Llodra *et al.*, 2000). Fecundity in most decapod studies refers to realized fecundity *sensu* Anger and Moreira (1998) and the unit counted is the developing embryo (see section 2). In decapods, the sperm are packed into spermatophores that are deposited on the female's ventral surface during copulation. The ova are fertilized at the time of spawning, and the developing eggs are incubated on the pleopods (Adiyodi and Subramoniam, 1983; Nelson, 1991). In invertebrates in general, and decapods in particular, there is an important trade-off between the size and number of eggs related to the allocation of limited resources for reproduction (see section 5.1.1). There is a continuum of strategies, from species that produce a small number of large eggs to species that produce a large number of small eggs. Two methods have been used to quantify fecundity in brooding decapods, depending on the size and number of eggs carried on the pleopods.

In species that brood few large eggs, all the eggs can be carefully removed from the abdomen with a spatula and counted directly under a stereomicroscope (Clarke, 1993b). For species that produce a large number of small eggs, an extrapolation of fecundity from the whole egg mass becomes more accurate and less tedious to undertake than direct counts of ova. For this, the total egg mass is weighed, and three subsamples of 100 eggs are weighed. From the average weight of the three 100 egg aliquots and the weight of the total egg mass, the total number of eggs can be estimated (King and Butler, 1985; Clarke, 1993b). This quantification can be obtained by using wet weights, or by drying the egg mass and the subsamples of eggs for dry weights (Anger and Moreira, 1998; Wehrtmann and Andrade, 1998). The use of dry weights gives a more accurate measure of the oogenic material produced because it avoids variability caused by hydration of eggs.

Although fecundity estimates in decapod crustaceans can be very precise, some limitations need to be taken into account. First, the number of eggs carried on the pleopods of a decapod is physically limited by the space available on the female's abdomen. Therefore, larger females carry more eggs, as shown in Figure 1 (King and Butler, 1985; Ivanova and Vassilenko, 1987; Mauchline, 1988; Corey and Reid, 1991; Reid and Corey, 1991; Clarke, 1993b; Bell and Fish, 1996; Stella *et al.*, 1996; Ohtomi, 1997; Thessalou-Legaki and Kiortsis, 1997). If we were to compare fecundity (1) between animals of the same species at different locations or different times, or (2) between different species, the number of eggs needs to be expressed in relation to female size. In these cases, fecundity should be

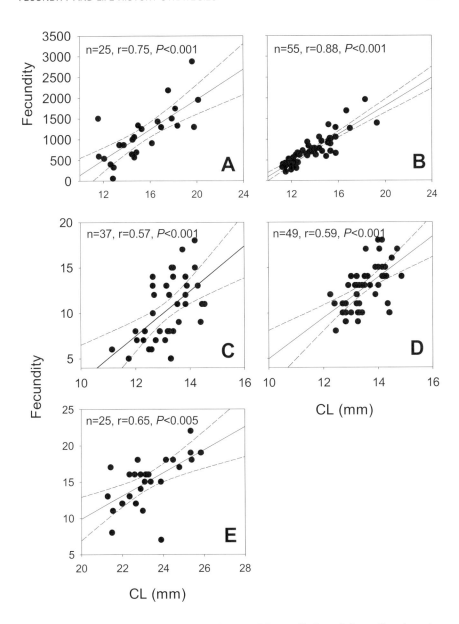

Figure 1 Linear regression and 95% confidence limits of fecundity (number of eggs per female) against carapace length in five species of caridean shrimp from the north-east Atlantic. A, *Acanthephyra purpurea*; B, *Acanthephyra kingsleyi*; C, *Systellaspis debilis* from 40°N; D, *Systellaspis debilis* from 11°N; E, *Parapasiphae sulcatifrons*. CL, carapace length in mm; n, number of females analysed. (Ramirez Llodra, unpublished data.)

quantified as number of eggs per 1 g of female weight or per 1 mm of carapace length. The variation in fecundity caused by the size of the female is thus eliminated and the relative production of eggs can be compared among females from different populations or species (Barnes and Barnes, 1968; Hines, 1982; Somers, 1991; Clarke, 1993b).

Secondly, egg size increases during embryonic development, causing an increase in the egg mass volume. Because the physical space for brooding is limited and not expandable in invertebrates with a hard exoskeleton, some eggs may be extruded from a large brood (Gorny et al., 1992; Ohtomi, 1997; Thessalou-Legaki and Kiortsis, 1997; Wehrtmann and Andrade, 1998). This loss of embryos is difficult to quantify, and error can be minimized by analysing fecundity soon after the fertilized eggs are deposited on the pleopods.

Thirdly, females caught in large trawls, because many can lose eggs during sampling and sorting. In many species, the embryos are not protected but are exposed to the external environment, and in large, well-developed broods from large females embryo loss can be significant (Gorny et al., 1992; Bell and Fish, 1996; Wehrtmann and Andrade, 1998; Ramirez Llodra, personal observations). This causes variability in fecundity data within samples and populations because most data are obtained from large females. Gorny and colleagues (1992) proposed two ways to include corrections for egg loss when analysing fecundity of caridean shrimp. The first suggested method was to plot egg mass against female mass (both logarithmically transformed) and to discard any data below the 95% confidence interval. Plotting logarithmically transformed data minimizes the variability caused by extreme data points (Somers, 1991; Sokal and Rohlf, 1995). The alternative proposed by Gorny et al. (1992) was to quantify fecundity in the pre-spawned phase, before embryo loss. The authors plotted mature ovary mass (when the weight of the ovary approximates that of newly spawned eggs) against female mass (both logarithmically transformed) and discarded all data below the 95% confidence interval. This was used to produce a corrected fecundity/size relationship. With this method, some of the data are not included, but the error is smaller than if all females with egg loss are included (Gorny et al., 1992). Egg loss during collection and storage causes problems in species that produce a few large eggs. For example, if an average brood of the mesopelagic shrimp Parapasiphae sulcatifrons comprises 20 eggs, the loss of 4 eggs represents a loss of 20% of the brood (Ramirez Llodra, 2000). In some of these decapod species, the developing oocytes are large and clearly visible in the gonads. Fecundity can be quantified by counting the number of oocytes directly from the ovaries before egg loss.

In most copepods, broods are produced at frequent intervals, from maturity to death of the adult. Many copepods produce egg sacs periodically, in which embryos develop to a hatching naupliar larval stage. In these cases, the fecundity of females and reproductive success of the population depend on a series of reproductive traits and environmental influences. These traits are (1) the number of days of egg-laying; (2) the variable period of latency between egg-laying phases; (3) the number of eggs produced per female; and (4) the effects of environmental factors, such as temperature and food quantity and quality, on the above processes (Razouls *et al.*, 1991; Laabir *et al.*, 1995; Pond *et al.*, 1996) (see section 6).

Fecundity in copepods can be quantified by direct counts of the number of eggs in the egg sacs. However, because of the quasi-continuous egg production in many copepods, the variables that give useful information on the life history of the species are: (1) daily egg production rate and its fluctuations over time; (2) lifetime fecundity; and (3) actual fecundity *sensu* Anger and Moreira (1998) (see section 2). Daily egg production rate is quantified as number of eggs copepod^{-1} day^{-1}. This gives information on changes in reproduction as a response to environmental factors such as temperature, food availability and food composition (Tester and Turner, 1990; Razouls *et al.*, 1991; Kosobokova, 1992; Jónasdóttir, 1994; Laabir *et al.*, 1995; Pond *et al.*, 1996; Kleppel and Hazzard, 2000). Lifetime fecundity is quantified as the sum of the number of eggs in each sac, from the first egg sac production to death of the adult. This measure of fecundity integrates information on the life-history characteristics, physiological constraints on reproduction and reproductive responses to environmental factors during the lifespan of the individual (Kosobokova, 1992; Williams and Jones, 1999). Actual fecundity in copepods has been quantified as the number of eggs surviving to the first naupliar larval stage. This incorporates hatching success and offspring survival, with important implications for the dynamics and demography of the population (Razouls *et al.*, 1991; Jónasdóttir, 1994; Laabir *et al.*, 1995; Pond *et al.*, 1996).

Many molluscs produce some sort of container for the eggs, such as gelatinous masses, egg jellies, egg cases or egg strings (Rupert and Barnes, 1994), and these can be attached to the substratum or brooded within the female mantle cavity (Arnold and Williams-Arnold, 1977; Beerman, 1977; Berry, 1977; Webber, 1977; Wells and Wells, 1977; Calvo *et al.*, 1998). Fecundity in molluscs has been quantified in different ways: (1) number of egg masses per female; (2) number of eggs per egg mass; and (3) number of eggs per female (product of the first and second methods). In all these cases, the quantification of fecundity is obtained by direct counts of egg masses and individual eggs within egg masses or egg cases under a microscope (Spight and Emlen, 1976; Vianey-Liaud and Lancaster, 1994; Chester, 1996; Cheung and Lam, 1999).

Most families of polychaetes exhibit a high degree of diversity of reproductive patterns. Although free spawning of small eggs that develop into planktotrophic larvae is the most common reproductive mode, there are also many examples of brooding species (Wilson, 1991). Brooding polychaetes may brood the embryos inside the body, inside the tube, or in external structures such as gelatinous masses attached to the outside of the tube. From any of the above, brooded embryos can hatch as planktotrophic larvae or as lecithotrophic larvae and also as juveniles that develop directly (Wilson, 1991; McHugh, 1993). As with molluscs, fecundity estimations in brooding polychaetes are obtained by simply counting the developing embryos in the brooding structures (McHugh, 1993).

Other taxonomic groups also include brooding species, but much less information on fecundity is available. In echinoderms, many species of the asteroid genera *Leptasterias* and *Henricia*, as well as *Asterina phylactica* spawn their eggs into a brood chamber formed by arching of their arms and disc (Menge, 1975; Emson and Crump, 1979; Strathmann et al., 1984; Chia and Walker, 1991; Chia et al., 1993). *Leptasterias ochotensis* deposit their eggs on the substratum and the adult covers them until the metamorphosed juveniles grow to a considerable size (Chia and Walker, 1991; Chia et al., 1993). *Asterina gibbosa* and *Asterina minor* deposit their eggs on the underside of rocks but leave them to develop unattended (Mortensen, 1927; Crump and Emson, 1978; Komatsu et al., 1979; Strathmann et al., 1984). Some echinoids (Pearse and Cameron, 1991; Poulin and Féral, 1995; Schatt and Féral, 1996), holothurians (Hansen, 1968; Gutt, 1991; Smiley et al., 1991; Sewell, 1994; Sewell and Chia, 1994) and ophiuroids (Byrne, 1991; Hendler, 1991; Medeiros-Bergen and Ebert, 1995) also brood their young. In these cases, the eggs are large and fecundity can be quantified by directly counting the embryos in the brood chambers or egg masses.

3.2. Broadcast spawning species

In free-spawning species, the mature eggs and sperm are released into the water column, where fertilization occurs. For most species, the timing of the spawning event cannot be predicted with exact precision and observations in the natural environment are a matter of chance. The best way to obtain fresh eggs and sperm is from conditioned specimens kept in the laboratory. When the adults have mature gonads, there are several techniques that can be used to induce spawning in order to obtain viable gametes for experimentation. Knowing when the specimens are mature is a problem in itself. If the life cycle of the species is known, the timing of gamete release can be estimated and the experiment can be designed accordingly. In some species, e.g. in gastropods such as the limpets or abalone, the gonads can be visually analysed in live specimens allowing for a rough determination of maturity (Hayashi, 1980; Baker and Tyler, 2001). In other cases, the maturity stage cannot be determined, and therefore a spawning induction might or might not result in the release of viable gametes.

The first two groups of methods described below (environmental shock and chemical induction) are used to obtain fresh gametes from live specimens. The release of eggs and sperm can be induced by different factors such as changes in the environmental conditions in the laboratory or injection of a spawning-inducing chemical (Giese and Pearse, 1974). Because many broadcast spawners produce a large number of eggs, it is necessary to spawn the specimen in a known volume of water. When all the eggs have been released, a subsample of known volume (e.g. 1 ml) is taken from a well-mixed egg suspension and the eggs are counted under a microscope. Fecundity can be estimated from the number of eggs in the subsample and the total volume of egg suspension. When fecundity is estimated from gametes obtained by induction to spawning, only ripe gametes are counted, and therefore only potential fecundity can be quantified.

An additional number of methods will be described that are used to obtain gametes and quantify fecundity in: (1) species that cannot be induced to spawn; (2) when specimens are not to be damaged by experimentation; or (3) when live specimens are not available (e.g. most deep-sea fauna).

3.2.1. Environmental shock

An environmental shock treatment is a controlled sharp change in the environment where the specimens are kept, which may simulate natural changes that induce spawning in the natural habitat (Giese and Pearse, 1974).

A thermal shock is a rapid increase or decrease in the temperature such as might occur during tidal exchanges, and which might be used as a stimulus to spawning in the natural environment. Many mollusc, including the abalone *Haliotis tuberculata*, the bivalves *Mytilus edulis*, *Placopecten magellanicus*, *Argopecten ventricosus* and *Macoma balthica* can be induced to spawn in captivity by thermal shock (Partridge, 1977 and references therein; Thompson, 1979; Langton *et al.*, 1987; Honkoop and Van der Meer, 1997; Monsalvo-Spencer *et al.*, 1997; Baker and Tyler, 2001). In the South Pacific, the holothurian *Holothuria scabra* is a commercially important aquaculture species. This species also spawns in the laboratory when exposed to a temperature shock (Morgan, 2000; Hamel *et al.*, 2001).

Light changes can also stimulate spawning. Photoperiodic cycles combined with temperature changes in the sea are known to control the onset of oocyte growth leading to a spawning event in many polychaetes (Garwood, 1980; Olive, 1980; Garwood and Olive, 1982; Clark, 1988; Fong and Pearse, 1992a, b; Olive *et al.*, 1997, 1998; Last and Olive, 1999; Watson *et al.*, 2000a, b) and this can be used under laboratory conditions to control gametogenesis and obtain eggs for experimentation and fecundity studies (Figure 2). Many species synchronize their spawning to lunar phases, notably reef corals (Rinkevich and Loya, 1979b; Wallace, 1985; Richmond, 1987). Photoperiod, tidal phase or differential tidal pressure can be the stimulant. Many ascidians spawn in response to light following darkness, and this reaction has been used in the laboratory by exposing individuals to controlled light regimes (Berrill, 1975). Also, the brooding bryozoan *Bugula neritina* has been induced to spawn under laboratory conditions by exposure to fluorescent light (Wendt, 1998). Spawning is induced by changes in light in the jellyfish *Spirocodon saltatrix*; when placed in the dark the ovaries produce a peptide that induces spawning (Ikegami *et al.*, 1978).

Figure 2 Female hesionid polychaete releasing eggs after stimulation by light and temperature shock. Image captured from video record by Dr Craig M. Young (Harbor Branch Oceanographic Institution, Ft Pierce, FL, USA).

Strong wave action is also known to stimulate spawning in the natural environment. The reef-building polychaete *Phragmatopoma lapidosa* is induced to spawn by physical disturbance. When fragments of the reef are broken by the waves and the worms are exposed, mature specimens release their gametes. This effect of wave action can be simulated in captivity by extracting the worm from its tube. The naked worms start spawning within a few minutes of having been extracted from their tubes (Mauro, 1975; E. Ramirez Llodra, personal observations).

3.2.2. *Chemically induced spawning*

Various chemicals have been shown to induce spawning, as shown by physiological studies of biochemical and hormonal pathways. Many of the chemicals act by stimulating contraction of the gonad muscles, and in some cases also by inducing the final maturation of the gametes.

Historically, echinoderms, and specifically echinoids, have been one of the most useful groups for experimental reproductive studies. Viable gametes are easily obtained and manipulated. They provide good material for fertilization and embryological studies, as well as for analysis of gamete characteristics, such as morphology, number and size. In echinoids, an injection of 2 ml of 0.55 M KCl in the coelom induces the muscles of the gonad wall to contract and ripe males and females will spawn within 5–10 minutes (Young *et al.*, 1997, 1998).

In asteroids, there is a complex hormonal system related to the maturation and release of gametes. In both males and females, the meiosis-inducing hormone 1-methyladenine (1-MA) is involved in the processes of maturation of primary oocytes and spermatocytes (Walker, 1980; Schoenmaker *et al.*, 1981; Chia and Walker, 1991; Chia *et al.*, 1993; Yamashita *et al.*, 2000). The hormone 1-MA also stimulates the contraction of the gonad wall (ovaries and testes) causing the release of mature gametes (Kanatani, 1975; Walker, 1980). The natural succession of hormonal processes involved in the maturation of reproductive cells in asteroids can be reproduced in the laboratory in order to obtain viable gametes. An injection of 0.5 ml of 10^{-4} M 1-MA into the coelom induces the gametes to finish maturation and ripe specimens spawn between 30 minutes and 18 hours after the injection (McEdward and Colter, 1987; Bosch, 1989; Shilling and Manahan, 1994).

In bivalves, the developing oocytes are arrested in prophase I of meiosis. Juneja and Koide (1996) have shown that serotonin (5-hydroxytryptamine, 5-HT) affects Ca^{2+} cellular uptake in oocytes. This increased intracellular concentration of Ca^{2+} is a signal to resume meiosis, causing germinal vesicle breakdown and spawning in bivalves such as the surf clams

Spisula solidissima and *Spisula sachaliensis*. Fong *et al.* (1996) reported data showing that in the bivalve *Mactra chinensis*, 5-HT induced spawning but not germinal vesicle breakdown. These authors proposed that there was no effect of serotonin on germinal vesicle breakdown of oocytes, because this process happens normally upon fertilization in this species (Fong *et al.*, 1996). An external application or injection of 0.4 ml of 2 mM 5-HT into the gonads or foot has been used in the laboratory to induce spawning in bivalves (Partridge, 1977; Hirai *et al.*, 1988; Ram *et al.*, 1993; Fong *et al.*, 1996; Juneja and Koide, 1996).

In many polychaetes, there is a link between environmental cues (photoperiod, temperature, salinity) and the endocrine system controlling gametogenesis and spawning. The nature of most of these hormones has yet to be understood with detail, but the existence of such a hormonal control has been described for many polychaetes (Bentley and Pacey, 1992; Hardege *et al.*, 1996; Zeeck *et al.*, 1996; Hardege and Bentley, 1997; Watson and Bentley, 1997, 1998; Watson *et al.*, 2000a, b). In some cases, an extract of the coelomic fluid containing the hormone or a homogenate of the organ that produces the hormone can be used in the laboratory to induce oocyte maturation and spawning (Olive and Bentley, 1980; Bentley *et al.*, 1984; Hardege *et al.*, 1996).

In the opisthobranch mollusc *Aplysia* sp. the egg cordons are a source of the pheromone "attractin", which induces the aggregations of individuals and mating. An injection of 0.1 ml of atrial gland extract through the foot into the haemocoel induces egg laying in *Aplysia* sp. (Painter *et al.*, 1991, 1998).

In the abalone and other molluscs, an external application of hydrogen peroxide induces spawning (Morse, 1977; Partridge, 1977; Beckevar, 1981; Moss *et al.*, 1995; Baker, 2001). Hydrogen peroxide acts as a co-substrate for prostaglandin peroxidase synthetase, which is involved in the synthesis of prostaglandin related to spawning processes (Moss *et al.*, 1995).

3.2.3. *Non-destructive weight method*

An indirect, non-destructive method to estimate fecundity has been used in some reproductive studies of free-spawning invertebrates. It consists of taking the wet weight of the same female before and after natural spawning. The difference between the pre- and post-spawned weights is used as a relative measure of gamete production. This estimation of reproductive output can be compared within species in different years, different populations or at different ages (Langton *et al.*, 1987). If the weight of an ovum is a known parameter for the species, fecundity can then be estimated from the weight of gametes produced and the average weight of ova (Barber *et al.*, 1988). This technique is used when the specimens need to be kept alive and healthy but spawning induction would cause damage or death. For example, the non-destructive method can be used when specimens from a population are being monitored over time in their natural environment and the timing of the spawning event is broadly known. A number of specimens can be collected, weighed and released again to their habitat before and after spawning, producing a minimal effect on other aspects of the individual and the population. This technique gives only an estimation of fecundity and is subject to a certain margin of error. The error is caused by weight changes in the individual other than those caused by reproductive material.

3.2.4. *Dissection of ovaries*

When specimens are preserved and freshly spawned gametes cannot be obtained, quantifying fecundity becomes an arduous task. Only an indirect estimation can be made in these cases. There are several techniques that may be employed depending on the complexity of the morphology of the ovaries and the size and location of mature oocytes. The morphology of ovaries ranges from simple aggregations of germinal cells (e.g. Cnidaria, Eckelbarger and Larson, 1988; Shick, 1991) to discrete, complex organs composed of well-structured associations of germinal and somatic cells surrounded by muscular walls and irrigated by blood vessels (e.g. Echinodermata, Chia and Walker, 1991).

In species that produce a few large eggs, or when oocytes are not strongly attached to the gonad tissues, the dissection of the ovary (strip spawning) may be sufficient to obtain the mature oocytes. This method has been used in the bivalve *Yoldia notabilis* (Nakaoka, 1994), the corals *Antipathes fiordensis* (Parker *et al.*, 1997), *Goniastrea aspera* (Sakai, 1998), *Acropora intermediata, A. millepora* and *A. hyacinthus* (Smith and Hughes, 1999), the holothurian *Actinopyga mauritana* (Hopper *et al.*, 1998) and the asteroid *Leptasterias epichlora* (George, 1994a, b). The ovaries are dissected out and opened, and the oocytes are counted directly under a microscope.

In many polychaetes, the oocytes mature free in the coelom. The oocytes can be obtained by extracting all the coelomic fluid with a pipette through a small incision in the body wall (McHugh, 1993; Zal *et al.*, 1995; Olive *et al.*, 1997, 1998). Similarly, the mature oocytes of the vestimentiferan *Riftia pachyptila* accumulate at an enlarged extremity of the oviduct and can be extracted with a pipette, allowing for the estimation of actual fecundity (C. M. Young, personal communication).

Although there is a source of error depending on the accuracy of the manipulator, who needs to ensure that no fluid is lost during the extraction, this method gives a good estimation of actual fecundity. Also, some species (e.g. the polychaete *Nephtys hombergi*) exhibit spawning failure (the gravid female spawns incompletely or not at all) in their natural environment (Bentley *et al.*, 1984; Olive *et al.*, 1997), and this leads to erratic results when such species are manipulated in the laboratory for fecundity studies.

In other organisms, fecundity may be estimated by counts of mature oocytes directly through the ovary walls. The best results are obtained in species with large oocytes and low fecundity, where the gametes can be distinctively identified. In the deep-sea deimatid holothurians such as *Oneirophanta* sp. or *Deima* sp., the oocytes grow in thin-walled elongated tubules. The walls of these ovaries are translucent and the mature oocytes grow to a maximum size of ~900 μm, allowing direct counts of oocytes (Tyler and Billett, 1987; Ramirez Llodra, personal observations). The deep-sea psychropotid holothurians such as *Psychroptes* sp. produce a few large eggs in well-structured nodular ovaries. The size of mature oocytes can reach up to 4 mm (Tyler and Billett, 1987). Other deep-sea holothurians such as the dendrochirotid *Ypsilothuria talismani* or the molpadiid *Cherbonniera utriculus* have very small body sizes (maximum lengths of ~15 mm and ~7 mm, respectively). These species have very small gonads and produce a few relatively large eggs (350 μm and 200 μm, respectively) (Tyler and Gage, 1983; Tyler *et al.*, 1987). In both cases (ovaries with very large oocytes, or very small ovaries with relatively large oocytes), the ovaries can be dehydrated with graded alcohols and the ovary tissues can be cleared with Histoclear. The ovary walls become translucent and fecundity can be quantified by directly counting the large oocytes under a stereomicroscope with transmitted light (Tyler and Gage, 1983; Tyler *et al.*, 1985; Tyler *et al.*, 1987).

In some species, fecundity can be estimated by analysis of smear slides of ovaries. For example, Sumida *et al.* (2000) estimated the fecundity of deep-sea ophiuroids by preparing smear slides of three gonads per female. The oocytes were counted on each slide under a compound microscope and the average number of oocytes per gonad was calculated. Fecundity was estimated as the mean number of oocytes per gonad multiplied by the number of gonads in each individual.

3.2.5. *Histological methods*

The most complex species to work with in terms of fecundity are those that have oocytes developing in morphologically complex ovaries, where inducing spawning is not possible and where the mature oocytes are not visible through the gonad wall. Some deep-sea asteroids and ophiuroids are good examples. The specimens are generally dead on reaching the surface when collected by deep-sea trawling gear and the oocytes develop in well-structured gonads with thick walls. A histological method has been used to estimate fecundity in such specimens, employing routine stains such as haematoxylin and eosin or Masson's trichrome (Pearse, 1961; Culling, 1974).

For example, Tyler and Gage (1980) in the brittle star *Ophiura ljungmani*, cut serial sections of three ovaries. Fecundity was quantified as the number of oocytes counted in the serial sections, averaged for the number of ovaries.

Ramirez Llodra *et al.* (2002) analysed the reproductive patterns of three species of abyssal asteroids belonging to the family Porcellan-asteridae. In these asteroids, the ovaries are large and there is a quasi-continuous production of eggs throughout the year. The large size of the ovaries, which can contain up to a few thousand eggs, makes it difficult to follow consecutive oocytes through serial sections. A different histological method was used, involving the relationship between gonad volume and mean oocyte volume (Ramirez Llodra *et al.*, 2002).

In the porcellanasterids analysed, reproduction was asynchronous and all phases of oocyte development (from oogonia to fully grown oocyte) were present in the ovary at any one time. Fecundity was quantified as total number of oocytes per female (potential fecundity) and as number of vitellogenic oocytes per female (actual fecundity). Actual and potential fecundity were estimated from the mean volume of previtellogenic and vitellogenic oocytes and the gonad volume in each female. Oocyte diameter was measured from light microscope sections (Figure 3). All oocytes that had been sectioned through the nucleus were measured with an image analysis package (Sigma Scan Pro 4). Oocyte volume (OV) was calculated assuming a spherical shape and averaged for each female.

$$OV = \frac{4 \times \pi \times R^3}{3}$$

Fecundity was estimated as follows:

V_g = volume of gonad (obtained by using a variation of the hydrostatic balance of Mohr–Westphal (Scherle, 1970)).

V_{pvo}, V_{vo} = mean volume of a previtellogenic and vitellogenic oocyte respectively.

N_{pvo}, N_{vo} = number of previtellogenic oocytes and vitellogenic oocytes respectively counted in a subsample of 100 oocytes per gonad.

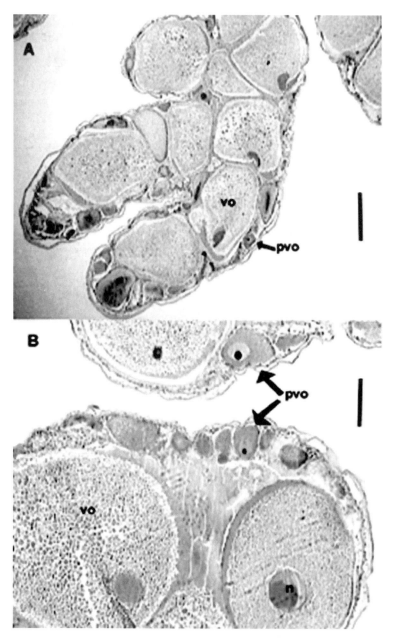

Figure 3 Light microscope sections of the ovary of a porcellanasterid asteroid stained with haematoxylin and eosin. A, section of a lobule of the ovary (4×) (scale bar 250 μm); B, detail of the section (10×). pvo, previtellogenic oocyte; vo, vitellogenic oocyte (scale bar 100 μm).

P = ratio between previtellogenic and vitellogenic oocytes:

$$P = \frac{N_{pvo}}{N_{vo}}$$

Assuming that P, the relation between the number of previtellogenic oocytes and vitellogenic oocytes counted in a subsample of the gonad is the same as the relation between these oocyte stages in the whole ovary, then:

$$P = \frac{N_{pvo}}{N_{vo}} = \frac{F_{pvo}}{F_{vo}}; \quad F_{pvo} = P \times F_{vo} \qquad \text{[Eqn 1]}$$

where F_{pvo} and F_{vo} are the total number of previtellogenic oocytes and total number of vitellogenic oocytes, respectively.

The volume of a gonad is equivalent to the number of oocytes multiplied by their volume:

$$V_g = (V_{vo} \times F_{vo}) + (V_{pvo} \times F_{pvo}) \qquad \text{[Eqn 2]}$$

Replacing F_{pvo} in Eqn 2 from Eqn 1:

$$V_g = (V_{vo} \times F_{vo}) + (V_{pvo} \times (P \times F_{vo}))$$
$$V_g = F_{vo} \times (V_{vo} + (V_{pvo} \times P))$$
$$F_{vo} = \frac{V_g}{V_{vo} + (V_{pvo} \times P)} \qquad \text{[Eqn 3: \quad Actual fecundity]}$$

From Eqns 1 and 3 we obtain the estimation for the potential fecundity (F):

$$F = F_{pvo} + F_{vo} \qquad \text{[Eqn 4: \quad Potential fecundity]}$$

This quantification of actual fecundity gives an estimation of the number of eggs that would be released in the following spawning event, but not the precise number of eggs produced. Many physiological processes are involved during gametogenesis that can change the ratio between previtellogenic and vitellogenic oocytes, such as reabsorption or phagocytosis of non-viable eggs. The real number of eggs available for fertilization might be different from the number of mature eggs in the ovaries. Nevertheless, in the case of free-spawning invertebrates from which mature eggs cannot be obtained by other laboratory methods, this is a robust and useful means of estimating fecundity.

3.3. Colonial species

Colonial species are widespread among cnidarians, bryozoans and tunicates (Rupert and Barnes, 1994). A colony is composed of interconnected individual zooids that can share resources. The zooid represents the basic unit for physiological processes, but these processes are also affected by characteristics of the colony. In some colonial organisms, there is morphological polymorphism, with different zooids performing different tasks in the colony, such as feeding, defence and reproduction.

Reproduction in modular organisms is very complex. There is an important combination of sexual reproduction through mature individuals in the colony and asexual reproduction through fragmentation of parts of the colony. In sexual reproduction, initiation of egg production, fecundity and reproductive effort are related to zooid size and position in the colony, size of the colony and growth form (Dyrynda and Ryland, 1982; Van Veghel and Bak, 1994; Van Veghel and Kahmann, 1994; Parker *et al.*, 1997; Sakai, 1998; Smith and Hughes, 1999). In asexual reproduction, the re-attachment, survival and egg production of colony fragments is related to the number of live zooids in the fragment, and therefore to the initial size of the fragment. In corals, the colony must attain a minimum size before it becomes sexually reproductive and, after reaching this critical size, fecundity increases linearly with colony size as more zooids become mature (Sakai, 1998; Smith and Hughes, 1999).

Therefore, determination of fecundity in modular organisms needs to take into account biological attributes such as: (1) the relative abundance of fertile polyps in the colony; (2) the relationship between polyp fecundity and polyp size and age; (3) the variability of fecundity with polyp position within the colony; (4) the relationship between colony fecundity and colony size and age; and (5) the indeterminate growth of the organism.

Fecundity in corals is usually quantified as the number of eggs per polyp produced in a reproductive season or event (Kojis and Quinn, 1981; Wallace, 1985; Van Veghel and Kahmann, 1994; Sakai, 1998; Smith and Hughes, 1999). Fragments of the colony are fixed in Bouin's solution or 10% formalin in sea water and decalcified with formic acid and sodium citrate (Rinkevich and Loya, 1979a, b; Van Veghel and Kahmann, 1994) or with formalin and acetic acid (Sakai, 1998). After decalcification, individual polyps can be cut from the fragment and dissected under a stereomicroscope. The number of gonads per polyp and number of eggs per gonad are counted. Fecundity is then quantified as the number of eggs per polyp (number of gonads per polyp × number of eggs per gonad) (Van Veghel and Kahmann, 1994; Sakai, 1998; Smith and Hughes, 1999). In the case of brooding corals, the number of developing embryos or planulae can be quantified instead (Rinkevich and Loya, 1979a, b; Chornesky and Peters, 1987; Richmond, 1987).

Fecundity in corals has also been quantified as total egg production of the colony. For this, three parameters need to be estimated: (1) total number of polyps in the colony; (2) percentage of gravid polyps in the colony; and (3) number of eggs per polyp. Colony fecundity is then obtained by multiplying the number of eggs per polyp by the percentage of mature polyps (Parker *et al.*, 1997).

In bryozoans and ascidians, there are broadcast spawning species and brooding species. Some marine bryozoan species produce small eggs that are shed into the sea water, but most species brood their eggs. Brooding bryozoans usually produce a few large eggs and release short-lived lecithotrophic larvae (Reed, 1991 and references therein). Most colonial ascidians are ovoviviparous, producing a few large eggs, which are typically brooded in the oviduct, atrium or colony matrix (Berrill, 1975 and references therein). To quantify fecundity in corals, biological patterns related to their modular structure need to be taken into account. The appropriate methodology for counting the oocytes, embryos or larvae will depend on the organism's reproductive patterns.

4. FECUNDITY IN LIFE-HISTORY STRATEGIES

4.1 Life-history theory

The life-history strategy of an organism represents the acquisition, over evolutionary time, of a series of co-adapted traits. The characteristics of these traits and of the relationships among traits are designed by natural selection to solve a particular set of environmental variables (Wilbur *et al.*, 1974; Stearns, 1977, 1980; Southwood, 1988; Eckelbarger, 1994). Natural selection is defined as the differential reproduction and survival correlated with heritable traits, or in other words, the variation of fitness among individuals (Stearns, 1980, 1992). Fitness in life-history theory is the expected contribution of an allele, genotype or phenotype to future generations. The fitness of genes or organisms is related to other genes or organisms in the population and it is a function of the environment in which it is measured (Stearns, 1992). Fitness in life-history models has been represented as the number of offspring produced by the end of a lifetime, or as the rate of increase reached through reproduction during a lifetime of variable duration (Bell, 1980).

Life-history theory analyses the causes of observed differences in fitness among life-history traits and predicts the phenotype at equilibrium

(Stearns, 1992). Two conditions are necessary for natural selection to occur. Firstly, the genotypic condition gives heritable variability for the trait in question determining whether there will be a response to selective pressures. Secondly, individuals must vary in fitness, and this is the phenotypic condition. Population genetics studies the genotypic condition, while life-history evolution analyses the evolution of fitness components (Stearns, 1992).

Studies on life-history evolution aim to predict the life-history pattern that an organism will display in any given environment, and to understand the selective forces that shape the evolution of the different strategies observed in nature (Eckelbarger, 1994). The variables composing a life-history strategy are: somatic growth rate, age and size at maturity, age- and size-specific fecundity, egg size, age- and size-specific reproductive effort, age- and size-specific mortality, larval developmental type and time, and adult lifespan (Wilbur et al., 1974; Rose, 1983; King and Butler, 1985; Olive, 1985; Stearns, 1992; Winemiller, 1992; Winemiller and Rose, 1992; Eckelbarger, 1994; Jaeckle, 1995; Hadfield and Strathmann, 1996; McCann and Shuter, 1997). The different variables of a life-history strategy are not independent. There are important trade-offs (see sections below) and natural selection acts on groups of traits that co-evolve (Rose, 1983; Stearns, 1992; Giangrande et al., 1994; Eckelbarger, 1994).

Four main elements play a central role in the analysis of life-history evolution: (1) demographic parameters of the population; (2) quantitative genetics; (3) the trade-offs between life-history traits; and (4) species-specific design constraints (Stearns, 1992). Demography relates age- and size-specific fecundity and mortality to variation in fitness. Quantitative genetics describes the genetic variation available in the individuals upon which natural selection will act. The physiological trade-offs among reproductive variables determine the co-variation and co-evolution of life-history traits under selective pressure. And finally, the species-specific design constraints delimit the range of variation within traits and create barriers to reaching optimal stages. The complexity and variability of life-history strategies in marine invertebrates arises because the organisms have evolved many different combinations of life-history traits to affect fitness. These strategies are moulded through the selective pressures of demographic characteristics and environmental conditions, biotic and abiotic factors and the limitations imposed by morphological and functional constraints (Caswell, 1980a; Stearns, 1980, 1992; Rose, 1983; Tuomi et al., 1983; Olive, 1985; Giangrande et al., 1994; Ebert, 1996; McCann and Shuter, 1997). Figure 4 shows a proposed scheme for the evolution of life-history traits.

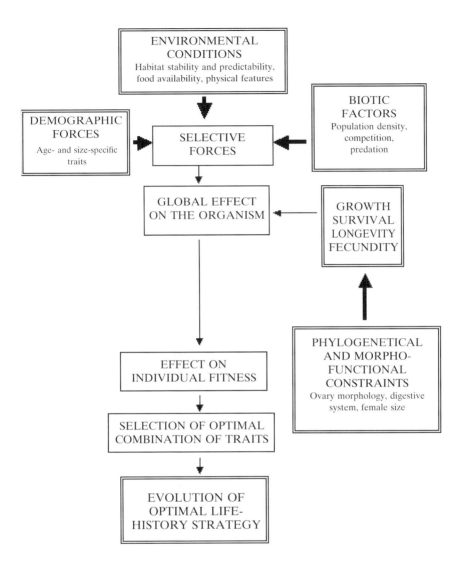

Figure 4 Diagrammatic representation of possible evolution of life-history strategies in marine invertebrates. Modified with permission of the editors from Figure 7 in Giangrande, A., Geraci, S. and Belmonte, G. (1994). Life-cycle and life-history diversity in marine invertebrates and the implications in community dynamics. *Oceanography and Marine Biology: An Annual Review* **32**, 305–333, with permission.

A sound knowledge of the life-history traits of a species is imperative to be able to understand the effects of natural selection (Stearns, 1992; Giangrande *et al.*, 1994; McCann and Shuter, 1997). The evolution of life-history variables and the plasticity of reproductive traits determine population dynamics and affect interactions between species in communities (Stearns, 1992; Hadfield and Strathmann, 1996). Accordingly, many studies have focused on the analysis of life-history evolution and its consequences in the adaptation of a species to its habitat. In the natural environment, resources of time and energy available for an organism are limited. There is, therefore, competition for these resources among maintenance, growth and reproduction (Gadgil and Bossert, 1970; Olive, 1985; Giangrande *et al.*, 1994). The life-history strategy of a species is composed of a series of reproductive and growth patterns that are the result of selection for optimal energy allocation between reproduction and somatic growth (Olive, 1985; Giangrande *et al.*, 1994).

A study of the evolution of life-history strategies requires: (1) the identification of the minimum number of selective pressures (e.g. population density, food resources, demography) that explain the acquisition of the patterns observed; and (2) identifying what causes differences in fitness among life-history variants (Wilbur *et al.*, 1974; Stearns, 1992). To understand a pattern of variation in a life-history trait and follow its evolution, information is needed on the selective pressures generated by demography (mortality and fecundity), phenotypic and genotypic variation, the physiological trade-offs found amongst reproductive parameters and functional and morphological constraints (Stearns, 1992). However, this amount of knowledge has yet to be obtained for any species, and studies concerning the evolution of life-history strategies vary considerably depending on the parameters being analysed (Wilbur *et al.*, 1974; Rose, 1983; Stearns, 1992; Giangrande *et al.*, 1994; Podolsky and Strathmann, 1996).

Studies on fecundity are based on the egg, or in a broader approach, on the offspring. Because producing eggs is expensive for the adult in units of energy allocation, and the allocation of energy to reproduction is directly linked to fitness, the analysis of fecundity is at the centre of many studies of life-history evolution. There is a direct relationship between fecundity and other life-history traits such as egg size, age at maturity and age-specific reproductive effort. These traits are connected through the bioenergetics of the organism and selective pressures of the environment, and evolve to maximize fitness. Fecundity is not, however, a highly conservative parameter within species, and varies with nutrition, population density and adult age and size (Eckelbarger, 1986). The data obtained from different taxonomic groups are very diverse. Fecundity can range from the production of as few as two eggs in some small infaunal brooding species (such as in the infaunal deep-sea bivalve *Microgloma* sp. (Scheltema, 1994)) to the production of thousands of small eggs that are shed into the water column by some bivalves, polychaetes or echinoids.

4.2. Life-history models and the role of fecundity

The evolution of life histories is seen as the response to conflicting demands for the allocation of limited resources to competing functions: growth and reproduction both present and future (Gadgil and Bossert, 1970; Olive, 1985; Giangrande *et al.*, 1994; Olive *et al.*, 1997, 2000). Models of life-history theory are developed in order to predict the combination of traits that will evolve in an organism living in a specific environment (Giangrande *et al.*, 1994). Different models and theories of life-history evolution (deterministic model, stochastic model, demographic theory, Winemiller–Rose model) have been developed to give alternative explanations to the adaptation of life-history traits to the environment. In all these models, fecundity plays a central role.

4.2.1. *The deterministic model: r–K selection*

In the deterministic model (MacArthur and Wilson, 1967; Pianka, 1970), the density of a population with respect to resources is the main selective pressure that explains the selection for the major life-history traits. Pianka (1970) suggested a continuum of life-history strategies between two extremes r selection and K selection. The term r refers to the maximal intrinsic rate of natural increase of the population from the Euler–Lotka demographic equation (see section 4.2.3). The parameter r is a measure of fitness in life-history theory. The term K refers to the carrying capacity of the population. The two endpoint strategies of the $r–K$ model are described by specific combinations of life-history traits (Table 2). The life-history strategy of any organism is seen as a compromise between these two extremes.

The r endpoint represents the ecological vacuum, where there are no density effects and there is no competition. In these conditions, the optimal strategy would be to allocate all energy into reproduction. This results in a high fecundity with the production of small eggs. The associated traits are short lifespan, early maturity, small adult body size, semelparity and no parental care. It is the quantitative extreme (favours the production of high number of offsprings) leading to high productivity (Pianka, 1970).

Table 2   Summary of environmental and population conditions and characteristic life-history traits of the *r* and *K* selection strategies.

Parameters	Strategy	
	r selection	*K* selection
Environment	Variable/unpredictable	Constant/predictable
Population	Density-independent: variable size, below carrying capacity, unsaturated, low competition	Density-dependent: constant size, at carrying capacity, saturated, high competition
Life-history traits		
Growth	Fast	Slow
Death rate	High	Low
Adult size	Small	Large
Lifespan	Short	Long
Age at maturity	Early	Delayed
Spawning frequency	Semelparity	Iteroparity
Fecundity	High	Low
Size of offspring	Small	Large
Juvenile survivorship	Low	High

The opposite extreme is the *K* endpoint, where density effects on the
population (relative to factors such as food supply, territory or mating
partners) are maximal and competition is high. Here, the strategy that
would maximize individual fitness would be to allocate most energy to
growth and maintenance. This results in a reduced fecundity, but a high
allocation of energy per egg, producing highly competitive offspring. The
main associated traits are a longer lifespan, delayed maturity, larger adult
size and iteroparity with common parental care. This is the qualitative
extreme (favours high quality of offspring) leading to efficiency in a saturated and highly competitive environment (Pianka, 1970).

Pianka (1970) also suggested that *r* strategies will evolve in variable or
unpredictable environments while *K* strategies will be representative of
relatively stable or predictable environments.

An important prediction of this approach is that species with *K*
strategies have a lower individual annual reproductive effort than
organisms exhibiting *r* strategies (Clarke, 1979; Giangrande *et al.*, 1994).
Numerous examples of species showing *K*-strategy traits such as slow
growth, low fecundity with the production of large and rich eggs are
found in the Antarctic (Clarke, 1979; Clarke and Gore, 1992; Gorny *et
al.*, 1992) and deep-sea benthos (Pain *et al.*, 1982a, b; Tyler and Billett,
1987; Gage and Tyler, 1991; Eckelbarger and Watling, 1995; Ramirez

Llodra, 2000). In these environments, food availability is uniform or has a predictable seasonal pattern and the life-history strategies have evolved under a saturated habitat. However, the observation that reproductive traits in these species correlate with traits predicted by the K strategy is only a qualitative description of the findings, and more information on the competition and demography of the populations is needed (King and Butler, 1985; Stearns, 1992; see below).

The $r–K$ theory was very popular during the 1970s and 1980s, and has been used successfully to describe life-history patterns. However, this model is too simple to describe the variability found in life histories and many populations do not match the expectations (Wilbur et al., 1974; Stearns, 1977; Whittaker and Goodman, 1979; Southwood, 1988; Stearns, 1992; Winemiller and Rose, 1992; Giangrande et al., 1994; Benton and Grant, 1996; McCann and Shuter, 1997). There are two major groups of criticisms to the $r–K$ model that explain why the life-history strategies of many organisms do not satisfy the predictions.

The first criticism is that the deterministic model does not include phylo-genetic and morpho-functional constraints (Stearns, 1977; Giangrande et al., 1994; McCann and Shuter, 1997). Design constraints act as barriers preventing the evolution of life-history traits to a theoretical optimum (see section 5.1.3). Secondly, the $r–K$ model is based at the level of population regulation and ignores hypotheses based on demographic aspects of the population (age-specific models). Studies in which results agree with the model predictions describe patterns of reproduction consistent with r and K selection, but fail to identify the mode of popu-lation regulation (Stearns, 1977, 1992; Whittaker and Goodman, 1979; Giangrande et al., 1994). Tuomi et al. (1983) suggested that the patterns observed in the $r–K$ continuum are the result of intercoupling of life-history traits and are not forced by ecological pressures. Demographic parameters (age-specific traits) and their trade-offs play a central role in the evolution of life-history strategies (see section 4.2.3).

The deterministic model has, however, provided a useful framework in which to interpret observations of life-history strategies in many studies. The error that must be avoided is the association of a described pattern with the effects of population density when this relationship has not been tested (Schaffer, 1974; Stearns, 1992).

4.2.2. The stochastic models

The stochastic models (Cohen, 1968; Murphy, 1968; Schaffer, 1974) predicts the evolution of similar combinations of life-history traits as for the $r–K$ model, but for different reasons. The selection of life-history strategies is explained in the stochastic model on the basis of uncertainty of survival from zygote to maturity and survival of adults to subsequent reproduction events (Murphy, 1968). The evolutionary pressure is here generated by the effects of environmental variations on the differential mortality of pre-reproductive ($\sim$ juveniles) and reproductive ($\sim$ adults) individuals. If environmental fluctuations cause highly variable juvenile mortality, then delayed maturity, low reproductive effort and the production of small broods in several episodes over a long lifespan will evolve. Iteroparity with breeding over a long period may be a tactic that compensates for bad years with good years in a variable environment. The opposite pattern, the selection of semelparity with early maturity and high fecundity, evolves when adult mortality is high or variable (Cohen, 1968; Murphy, 1968; Schaffer, 1974; Bell, 1980).

4.2.3. Demographic theory

In the 1990s, acknowledging the weaknesses of the deterministic model (see section 4.2.1), many studies shifted from r–K theory to models using demographic parameters as the basis of life-history evolution (Stearns, 1992; Winemiller, 1992; Winemiller and Rose, 1992; Giangrande et al., 1994; Ebert, 1996; McCann and Shuter, 1997). Demography is the analysis of the effects of age structure on population dynamics and on natural selection (Stearns, 1992). The selection of certain life-history traits is directly related to the demographic patterns of a population (Wilbur et al., 1974; Stearns, 1977, 1992; Whittaker and Goodman, 1979; McCann and Shuter, 1997; Olive et al., 2000). Two conditions are required for natural selection and evolution to occur (Stearns, 1992). First, there must be heritable variability (the genotypic condition). Second, there must be variation in fitness among individuals (the phenotypic condition).

The demographic theory for the evolution of optimal reproductive strategies was first suggested by Williams (1966). This theory is based on the assumption that reproductive effort and related life-history traits are optimized by maximizing fitness under purely demographic selective pressure. The demography model uses age- and size-specific fecundity and mortality as the basis of variation in fitness and natural selection. Life-history evolution under demographic theory assumes that the phenotype possesses demographic traits that are connected and constrained by trade-offs. These traits are birth, age and size at maturity, number and size of offspring, growth and reproductive investment, lifespan and death (Stearns, 1992; Winemiller, 1992; Winemiller and Rose, 1992). The Euler–Lotka equation is the basic equation of demography and life-history evolution:

$$1 = \int_{\alpha}^{\omega} e^{-rx} l_x m_x d_x$$

This equation identifies the relationships of age at maturity (α), age at last reproduction (ω), probability of survival to a given age class (l_x) and number of offspring expected in a given age class (m_x) to the intrinsic rate of growth (r). This model assumes fixed birth and death rates, which determine the intrinsic rate of increase of the population (r). In life-history theory, the solution to the Euler–Lotka equation, r, measures the fitness of genes and is important because it weights the contribution of each age class to fitness (Stearns, 1992; Olive et al., 2000). In populations at or near the evolutionary stable state (ESS), the exponent term (rx) of the Euler–Lotka equation approaches unity and fitness can be estimated as the lifetime product of survival and age-specific fecundity (or net reproductive rate R_0).

When analysing the evolution of the life-history strategy of an organism, we need to know what traits have heritable variability, and how the variable traits trade-off. Then we can apply the demography theory to understand the selection pressures affecting the phenotype (Stearns, 1992). The demography theory explains the selection of optimal reproductive tactics by the balance between current reproductive output and residual reproductive value ($\sim$ lifetime product of survivorship and fecundity) (Williams, 1966; Wilbur et al., 1974; Stearns, 1980, 1992; Olive, 1985; Giangrande et al., 1994; McCann and Shuter, 1997; Olive et al., 2000) (see sections 5.3 and 5.4).

However, there are some criticisms of the demography theory. The model assumes a stable age distribution, and therefore has little predictive power for organisms living in unpredictable environments. The demographic model also assumes the co-adaptation of life-history traits, but some traits have strong relationships with physiological and developmental processes. The demography theory ignores the effects of physiological, structural, morphological and developmental constraints to evolution. This suggests the need to develop models under constrained optimization, where there are explicit barriers (constraints to variation of a particular trait) that determine the boundaries of opportunities for the different traits (Stearns, 1977, 1980; Caswell, 1980a; Tuomi et al., 1983; Olive, 1985; Giangrande et al., 1994) (see section 5.1.3). Also, Tuomi et al. (1983) argued that the basic trade-off suggested by the demographic theory between reproductive effort and somatic investment is not always the case. If resources are not limiting, the allocation of energy to reproduction might not have a cost effect on somatic investment, and therefore might not affect survival and residual reproductive value (Tuomi et al., 1983). Caswell (1983) emphasized the importance of phenotypic plasticity in the evolution of demographic life-history models. This author suggested that plasticity of a character is a life-history trait in itself. By increasing the probability of survival under differing environments, plasticity affects fitness, and is therefore an important trait to be included in demographic models (Caswell, 1983).

To understand the variation and evolution of complex life-history strategies, information is needed on genetic variation (responses of groups of genes to selective pressure), phenotypic variation (differential fitness), trade-offs among important life-history traits, physiological and morpho-logical constraints, and the selective pressures generated by demography (Stearns, 1992). Such a volume of information is not available for any organism, and because of the diversity of organisms and environments no single life-history theory explains the life-history evolution of all species (Rose, 1983; Tuomi *et al.*, 1983; Stearns, 1992). Nevertheless, the demo-graphic theory gives basic parameters from which to calculate the strength of selection on life-history traits, and is a good approach to the problem of life-history evolution. This needs to be complemented by studies of genetic variation, physiological trade-offs and design constraints (Tuomi *et al.*, 1983; Stearns, 1980, 1992).

4.2.4. *The Winemiller–Rose demographic model*

Winemiller (1992) and Winemiller and Rose (1992) proposed a simple demographic model to explain life-history evolution. Winemiller (1992) described the three main components of fitness as being: (1) fecundity; (2) survivorship of juveniles; and (3) age at maturity. These three life-history attributes are contained in the two possible demographic estimates of fitness in life-history theory, which are V_x (the reproductive value of an organism or age class) and r (the intrinsic rate of natural increase of a genotype or population) (Winemiller, 1992). The Winemiller–Rose model is based on an amplification of the two-dimensional r–K model, by incorporating the three major demographic components of fitness. The model implies that trade-offs among life-history traits are based on physiological and ecological constraints that ultimately select for genetic correlations.

The Winemiller–Rose model proposes a continuum in a three-dimensional triangle where each vertex reflects a boundary life-history strategy. Each one of these three endpoint strategies is defined by three demographic parameters: age at maturity, fecundity and juvenile survivorship (Winemiller, 1992; Winemiller and Rose, 1992; McCann and Shuter, 1997). The resulting endpoint strategies are: (1) the periodic strategy, with late maturation, high fecundity and low juvenile survivorship; (2) the opportunistic strategy, with early maturation, low fecundity but repeated breeding events and low juvenile survivorship; and (3) the equilibrium strategy, with moderate maturation age, low fecundity and high juvenile survivorship (Table 3, Figure 5).

Table 3 Summary of reproductive traits in the opportunistic, equilibrium and periodic life-history strategies of the Winemiller–Rose model.

Life-history traits	Strategy		
	Opportunistic	Equilibrium	Periodic
Adult size	Small	Large	Large
Lifespan	Short	Long	Long
Age at maturity	Early	Moderate	Late
Spawning frequency	Multiple	Single	Single
Fecundity per spawning event	Low	Low	High
Size of offspring	Small	Large	Small
Juvenile survivorship	Low	High	Low

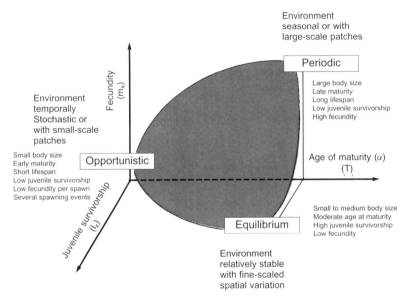

Figure 5 The Winemiller–Rose model. The three endpoint strategies (opportunistic, periodic and equilibrium) are shown with their characteristics and the related environmental conditions. lx, probability of survival to a given age class; mx, number of offspring expected in a given age class; T, lifetime. Redrawn from Figure 6 in Winemiller, K. and Rose, K. (1992). Patterns of life history diversification in North American fishes: implications for population regulation. *Canadian Journal of Fisheries and Aquatic Science* **49**, 2196–2218, with permission.

In the Winemiller–Rose model, the *r* strategy is split into periodic and opportunistic strategists. The periodic strategists share with the classical *r* strategists high fecundity and low juvenile survivorship. They differ, however, in that they are large, long-lived and have late maturation. The periodic strategy maximizes age-specific fecundity at the expense of turnover time (delayed maturity in order to attain a size sufficient for production of a large clutch) and juvenile survivorship (small eggs). They inhabit predictable and seasonal environments (Winemiller and Rose, 1992; McCann and Shuter, 1997).

The opportunistic strategists have most of the classical *r* strategist's traits, with small body size, early maturation, low juvenile survivorship and short lifespan. But they differ in that fecundity per spawning event is low (although there are multiple spawning events spread over a prolonged annual spawning season). The opportunistic life-history strategy maximizes population growth through a reduction in the mean generation time. Early maturation allows for short lifespan and high population turnover, but diminishes the capability to produce large clutches and large eggs. However, because of their multiple spawning events, annual fecundity is high and allows for recolonization of new habitats. These organisms inhabit highly disturbed and unpredictable environments (Winemiller and Rose, 1992; McCann and Shuter, 1997).

K strategy is redefined in the Winemiller–Rose model as the equilibrium strategy and its boundaries are narrowed. The equilibrium strategists have moderate age at maturity, low fecundity and high juvenile survivorship. They differ from the classical *K* strategists in that they have small to medium body sizes. The species that evolve in the equilibrium strategy maximize juvenile survivorship at the expense of fecundity. They inhabit constant environments (Winemiller and Rose, 1992; McCann and Shuter, 1997).

McCann and Shuter (1997) tested the hypothesis that allometric relationships with body size for fecundity and age at maturity in fish differ across the endpoint strategies of the Winemiller–Rose model. They showed that for a given adult weight, the range of ovarian production reflects the range in reproductive strategies available for a population with that average adult weight. These authors proposed a series of constraints that could be limiting the observed life-history strategies. The lower constraints of energy allocation to reproduction were probably determined by demographic traits of population persistence. There is a minimum investment in eggs necessary to replace exactly the parental pair, and a minimum investment in egg biomass required to ensure recruitment above a critical value. The upper constraints were determined by energetic and physical factors related to adult survivorship. There is a limited amount of energy that can be allocated to the ovary, and there are morphological design barriers such as body cavity space for ovarian growth or egg mass attachment (McCann and Shuter, 1997).

In summary, the most recent models of life-history theory explain the evolution of patterns of covariation in reproductive traits based on demographic parameters and in relation to variable environmental conditions. The selection of life-history strategies will be founded on the optimization of different combinations of traits to maximize fitness. For example, the benefits of early maturation are short generation times and higher survival rate to maturity, while the benefits of delayed maturity are higher lifetime fecundity, higher initial fecundity, and better quality of offspring accounting for a higher larval survival. Selection for early or delayed maturity would depend on an optimal balance between the benefits and costs of the life-history traits related to age at first maturity (Olive, 1985; Stearns, 1992; Giangrande *et al.*, 1994; McCann and Shuter, 1997).

5. RELATIONSHIP OF FECUNDITY TO OTHER LIFE-HISTORY TRAITS

In the life-history strategy of a species, the different variables are interconnected and there is co-evolution of reproductive traits (see section 4.1). These relationships are characterized by covariations and trade-offs between pairs of characters (Gadgil and Bossert, 1970; Wilbur *et al.*, 1974; Caswell, 1980a; Southwood, 1988; Stearns, 1992; Eckelbarger, 1994). The ultimate cause of these relationships is the allocation of energy to reproductive processes. The result is the maximization of fitness in a specific environment and within certain limits (Gadgil and Bossert, 1970; Olive, 1985; Stearns, 1992; Ebert, 1996) (see section 4).

In the following sections, the main relationships of fecundity with life-history traits such as egg size, female size and age, age at first reproduction and age-specific reproductive effort will be described and illustrated with examples.

5.1. Fecundity and egg size

Of egg characteristics, size is probably the parameter that has received most attention in reproductive studies (McEdward and Carson, 1987). Eggs are amongst the largest cells in an organism and represent the energetic unit invested in the next generation (Eckelbarger, 1986). Eggs show a wide size range within and between species of marine invertebrates, varying from <100 μm in some planktotrophic species to 4 mm in some lecithotrophic species (Herring, 1974b; Clarke, 1979; McEdward and Carson, 1987; Mauchline, 1988; McEdward and Chia, 1991; Clarke, 1993a; McHugh, 1993; Jaeckle, 1995; McEdward, 1997; Krug, 1998).

Fecundity and egg size are intimately related in life-history strategies, because the combination of these two parameters represents the amount of energy that has been allocated to production of oogenic material. The evolution of life-history strategies in different species has resulted in different patterns of packing this oogenic material in order to maximize fitness (see section 5.1.1 below). Accordingly, the relationships between egg size and fecundity, fertilization success, energy content, parental investment, larval development and larval mortality have interested life-history biologists for decades (see section 5.1.2 below).

5.1.1. *Trade-offs between fecundity and egg size*

The trade-off between fecundity and egg size is one of the most important in life history theory because of its direct relationship with maternal investment in reproduction (McGinley *et al.*, 1987). At any given time, a set amount of energy is available to an individual to invest in reproduction. Because the energy intake by an organism is limited, there is a trade-off between the number of eggs produced and their size. If there is an increase in the energy assigned to an individual offspring, the number of offspring that can be produced decreases (Smith and Fretwell, 1974; McGinley *et al.*, 1987). There are, therefore, different ways of packaging the oogenic material that can be produced by the adult. Animals may produce a large number of small eggs or a small number of large eggs. The former is typical of species with planktotrophic development and the latter is typical of species with lecithotrophic or direct development (Thorson, 1950; Menge, 1975; Tyler and Pain, 1982; Olive, 1985; Gorny *et al.*, 1992; Stearns, 1992; Clarke, 1993a; Jaeckle, 1995; Levitan, 1993, 1996; Podolsky and Strathmann, 1996; McEdward, 1997; Ohtomi, 1997; Krug, 1998). This trade-off between fecundity and egg size is a basic component of all life-history models (see section 4.2). It represents different ways of partitioning a limited energy resource into offspring production. If fitness is determined as the number of surviving offspring (fecundity × survival probability (Caswell, 1980b)), the fecundity-egg size trade-off is affected by selective pressures through larval mortality (Vance, 1973a, b; Christiansen and Fenchel, 1979), fertilization success (Levitan, 1993, 1996; Styan, 1998) larval developmental time and survival (Podolsky and Strathmann, 1996). There are also physiological and morphological constraints, such as ovary structure, oogenic patterns and adult size, that determine the range limits of egg number and size (Stearns, 1977, 1992; Caswell, 1980a; Eckelbarger, 1994; Giangrande *et al.*, 1994) (see section 5.1.3).

There are numerous observations of the trade-off between number and size of eggs in marine invertebrates. Especially illustrative examples are found in studies of reproductive output in poecilogonous species with sympatric individuals that have mixed larval strategies (e.g. having plankto-trophic and lecithotrophic development). Only a few species of polychaetes and opisthobranch molluscs are considered to be truly poecilogonous (Chia *et al.*, 1996). Species of poecilogonous polychaetes such as *Streblospio benedicti* and *Capitella* sp. with different egg sizes and number, can have the same overall reproductive investment with a different packaging of the egg material. The polychaete *Streblospio benedicti* has siblings with lecitho-trophic development sympatric with other individuals of the same species that produce planktotrophic larvae (Levin, 1984). While lecithotrophic forms produce eggs six times larger than those produced by the plankto-trophic ones, the former produce six times fewer eggs. As a result, the reproductive investment is similar in both strategies (Levin and Creed, 1986; Levin *et al.*, 1987). Similarly, studies on the fecundity and egg size of *Capitella* sp. showed that individuals from different environments made similar reproductive investment but differed in their fecundity-egg size trade-off (Qian and Chia, 1991). Also, in a poecilogonous population of *Capitella* sp. reared from a single female with lecithotrophic development, some individuals produced a large number of small eggs that underwent planktotrophic development, while other adults produced a small number of large eggs that underwent lecithotrophic development (Qian and Chia, 1994). In molluscs, the opisthobranch *Alderia modesta* is a rare example exhibiting poecilogony in natural populations. In a natural population studied in San Diego, half of the adults spawned masses containing ~300 small eggs (68 μm in diameter) while the other half of the adults spawned ~30 eggs of a larger size (105 μm in diameter) (Krug, 1998).

Offspring size is related to offspring survival through adult investment in the individual eggs. Larvae hatching from larger eggs are fitter and hatch at a more advanced stage of development, which diminishes the period spent in the water column and increases survival probability (Vance, 1973a, b; Smith and Fretwell, 1974; Stearns, 1992). There is, however, a cost in producing large, "better" offspring, and that is the decreased fecundity. The factors affecting the selection of an optimal egg size (the egg size and fecundity that will maximize parental fitness) have been studied for decades. Several theoretical models have been proposed. These are described briefly in the next section.

5.1.2. *Optimal egg size models*

In 1973, Vance developed the fecundity-time hypothesis, a theoretical model to predict optimal egg size (Vance, 1973a, b). This model relates the reproductive energetic efficiency to egg size in marine benthic invertebrates. Selection will favour the reproductive strategy that results in the greatest number of offspring per caloric investment in reproduction. In Vance's model, fitness was measured as number of settled offspring (metamorphs). The model predicts a bimodal distribution of egg sizes corresponding to the observed larval developmental types in nature, planktotrophy and lecithotrophy. The energy content of the egg was assigned a number between zero for eggs that feed on plankton (plankto-trophy) and one for eggs that support larval development without feeding (lecithotrophy).

The model makes four assumptions: (1) the pre-feeding period increases with increasing egg size; (2) the feeding period decreases with increasing egg size; (3) there is a constant mortality; and (4) there is 100% fertilization success. If energy invested in reproduction is constant, and we assume a constant mortality and linear relationship between egg size and developmental time, Vance's model predicts extreme egg sizes. Offspring from smaller eggs have longer developmental times. This implies longer exposure in the water column and therefore higher mortality. On the contrary, lecithotrophic larvae hatch at a more advanced stage and have higher survival probabilities. If losses in long planktonic development (production of a high number of small eggs that develop into planktotrophic larvae) equals the losses of fecundity in the production of few large eggs that develop into lecithotrophic larvae, then both strategies (planktotrophy and lecithotrophy) result in an equal number of settled offspring and are evolutionarily stable. The optimization of energy allocation into offspring selects for extreme egg sizes (Vance, 1973a, b).

Intermediate egg sizes are the rule rather than the exception in many planktotrophic species (Hadfield and Strathmann, 1996; Levitan, 1996; McEdward, 1997). Sewell and Young (1997) conducted a study to test the hypothesis of a bimodal distribution of egg sizes in echinoderms. These authors showed that, although the species with planktotrophic development formed a distinctive group with smaller egg sizes, there was some overlap in egg size between planktotrophic and lecithotrophic/brooding species in the holothurians, ophiuroids, echinoids and asteroids. Sewell and Young (1997) concluded that echinoderm egg sizes need to be seen as a continuum and not as a bimodal distribution.

In 1974, Smith and Fretwell proposed a graphical model to explain the relationship between energy allocated to individual offspring (or egg size) and parental fitness, in order to estimate optimal offspring size. The model is based on two assumptions related to offspring size and offspring fitness. First, a minimum energy allocation in the egg is necessary for an offspring to survive. Second, there is a positive and convex relationship between offspring size and offspring fitness (Smith and Fretwell, 1974; McGinley et al., 1987). The Smith–Fretwell model proposes selection for a single optimal level of investment in offspring, and therefore a determined number of eggs produced, that maximizes parental fitness. In the model, the way in which energy is expended or the advantages gained from it are not specified, allowing effects of competition, predation and the physical environment to be represented simultaneously. This model has been criticized because it does not consider environmental factors affecting offspring fitness, and because of the variability of egg sizes observed in natural populations. McGinley et al. (1987) concluded from theoretical mathematical models that environmental heterogeneity rarely favours the production of variable offspring, supporting the Smith–Fretwell model. These authors suggested that the observed variability is caused by temporal and environmental effects during offspring development, which constrain the ability of the adult to allocate equal amounts of energy to each egg (McGinley et al., 1987). The Smith–Fretwell model was also supported by a recent study on the fecundity-egg size trade-off (Einum and Fleming, 2000). These authors suggested that there is a strong directional selection for single egg size. They suggested that the observed variability in egg sizes in nature is caused by morpho-functional constraints and the adaptive phenotypic plasticity caused by the relationships between egg size and related maternal traits with offspring fitness (Einum and Fleming, 2000).

In studies following Vance's model, the evolution of egg size in invertebrate species has been related to different pre- and post-zygotic factors, where optimal egg size is determined by the relationship between offspring size and offspring fitness. Offspring fitness has traditionally been viewed as offspring survival. However, fertilization success related to gamete characteristics can also be a major factor determining fitness (Levitan, 1993), causing strong selective pressure on the evolution of egg size and related reproductive traits such as fecundity. Levitan (1996) suggested that selective pressures could affect egg size evolution at three different phases of a species' life history: (1) before the embryos or larvae enter the plankton; (2) during the planktonic phase; and (3) just after metamorphosis and settlement. Working with three sympatric species of the echinoid *Strongylocentrotus*, Levitan (1993, 1996) rejected the idea of selection during the planktonic or the newly metamorphosed phases. This author proposed a hypothesis to explain the evolution of egg size in marine invertebrates in terms of fertilization success. According to this hypothesis, larger eggs would be fertilized more effectively, and variation in factors such as adult body size, population density, microhabitat distribution and fertilization kinetics, which all influence fertilization success, would allow for different sperm limitation and fertilization patterns. The characteristics of the gametes, both eggs and sperm, influence fertilization success, zygote production and ultimately fitness. Levitan (1993, 1996) concluded that fertilization success plays a major role in evolution of marine invertebrate egg sizes and related life-history traits.

In answer to Levitan's hypothesis of a prezygotic selection for optimal egg sizes, Podolsky and Strathmann (1996) proposed postzygotic selection forces. These authors argued that with a given allocation of energy for reproduction, the increase in fertilization obtained from having larger eggs cannot alone compensate for the decrease in fecundity resulting from the production of larger eggs. Although larger eggs would be fertilized at a greater rate, total zygote production would decline because of the loss of fecundity. The fecundity-fertilization trade-off would, therefore, favour the division of resources into small eggs. Moreover, factors other than gamete encounter related to egg size – such as egg organic content, sperm life, parental reproductive behaviour or egg chemical attraction for sperm – can play an important role in fertilization success. Fertilization would then have a small selective pressure on egg size evolution. Podolsky and Strathmann (1996) suggested that the advantages of maintaining large eggs are postzygotic, such as shorter larval developmental time and higher survival probability of the offspring. In a subsequent study, Styan (1998) proposed that polyspermy can also affect the evolution of egg size. Most eggs are spherical, minimizing the surface area related to volume available for sperm attachment. This author suggested that larger eggs can result in higher rates of polyspermy, which could have a selective pressure on the evolution of egg sizes as strong as increasing the chance of sperm-egg collisions (Styan, 1998).

The fecundity-egg size relationship plays a major role in life-history theory, because it is strongly related to the energetic characteristics of the organism and influenced by the environmental conditions. Factors such as parental investment (Jaeckle, 1995), food availability and quality (Nichols et al., 1985; Tyler et al., 1993; Campos-Creasey et al., 1994; Brey et al., 1995; Jaeckle, 1995), fertilization success (Levitan, 1993, 1996), egg energy content (McEdward and Carson, 1987; McEdward and Chia, 1991; Clarke, 1993a, b; Jaeckle, 1995) or larval developmental time and mortality (McEdward and Carson, 1987; Stearns, 1992; Young and Tyler, 1993; Jaeckle, 1995; Podolsky and Strathmann, 1996; Hoegh-Guldberg and Emlet, 1997) are all related in one way or another to egg size and therefore to fecundity. For life-history models to have a greater predictive power, it would be desirable to incorporate all of the above factors, as well as environmental heterogeneity and variability and morpho-functional constraints that limit evolution of certain traits (Wilbur *et al.*, 1974; Hadfield and Strathmann, 1996; Podolsky and Strathmann, 1996).

5.1.3. *Design barriers to evolution of egg size and fecundity*

Design barriers are not commonly included in life-history models. These constraints define the upper and lower limits of the range of variation of certain life-history variables, and keep life-history traits from reaching predicted optimal stages (Stearns, 1977, 1980). Such design barriers may be physiological and/or morphological.

Examples of physiological constraints can be found in a variety of processes. For example, different responses to availability patterns of organic matter in the environment can reflect phylogenetic constraints based on feeding biology, digestive efficiency, nutrient storage and mobilization and vitellogenic mechanisms. The efficiency with which an organism utilizes ingested nutrients and makes them available for growth and reproduction is determined by physiological processes in the digestive system, limiting the amount of energy that can be allocated to produce eggs (Stearns, 1992; Eckelbarger, 1994; Eckelbarger and Watling, 1995).

Vitellogenic mechanisms in invertebrates are often constrained phylogenetically (see section 1), and determine the quantity, quality and rate of energy incorporation in the egg (Eckelbarger, 1986, 1994; Eckelbarger and Watling, 1995). For example, the ability to respond to unpredictable resources by producing eggs rapidly is characteristic of species with heterosynthetic vitellogenesis, which allows for the fast incorporation of nutrients into developing oocytes. This rapid response to resources and incorporation of energy in growing oocytes in these species allows an increasing fecundity when food quality and/or quantity increase (Eckelbarger, 1986) (see section 6.1). On the contrary, species with nutrient storage, transfer organs and autosynthetic or mixed vitellogenic pathways have a slow and long production of eggs, which results in a relatively predictable timing of spawning. Therefore, semelparity and iteroparity, together with associated reproductive traits such as number and size of eggs (see section 4), are manifestations of different vitellogenic mechanisms (Eckelbarger, 1983, 1986, 1994). Also, complex hormonal controls of physiological processes can limit the variability of life-history traits such as the gametogenetic and spawning cycles controlled by endocrine systems (Olive, 1985; Bentley and Pacey, 1992; Yamashita *et al.*, 2000).

Examples of morphological constraints are also numerous and diverse. Ovarian morphology is directly related to oogenic processes and therefore ultimately limits life-history traits such as rate of egg production, frequency of breeding, size and energy content of eggs, and derived larval forms (Eckelbarger, 1994). These morphological barriers are phylogenetic constraints determined by the ancestral history of the species. The variety of metazoan ovaries and vitellogenic mechanisms provides a wide range of possibilities for the evolution of different life-history strategies (Eckelbarger, 1983, 1994; Eckelbarger and Watling, 1995).

A major design barrier to reproductive traits is defined by adult body size. The hard exoskeleton of some brooding invertebrates, such as crustaceans, limits the physical space where the embryos are carried (Barnes and Barnes, 1968; Hines, 1982; King and Butler, 1985; Corey, 1987; Corey and Reid, 1991; Reid and Corey, 1991; Somers, 1991; Clarke, 1993b). These species display a linear relationship between body size and fecundity (see sections 3.1 and 5.2). Also, large broods may experience significant egg loss when embryo size increases during development and the egg mass becomes too large for the space available (see section 3.1).

A small body size can limit the number of eggs that can be produced, resulting in the observed patterns of small organisms producing a few large eggs that are brooded. The small size of some invertebrates imposes a design barrier to the production of a large number of eggs. The number of eggs produced in a single spawning event would not be large enough to maintain the population, and therefore semelparity is not possible. These species produce a few large eggs with high parental care (Hines, 1986; Cassai and Prevedelli, 1999). Cassai and Prevedelli (1999) working with polychaetes, suggested that the very small body size of *Ophryotrocha labronica* imposes a design constraint, causing selection for the production of a few large eggs at a semi-continuous rate. In congeneric scleractinian corals of the genus *Porites* sp., different species produce large and small colonies. The small colonies reproduce by brooding a small number of large eggs in quasi-continuous patterns throughout most of the year. On the contrary, the large colonies produce a high number of smaller eggs, which are released into the water column during a specific annual spawning event (Richmond and Hunter, 1990).

A different hypothesis to explain the relationship between small body size and the production of few large brooded eggs was proposed by Strathmann and Strathmann (1982) and Strathmann et al. (1984). The allometry hypothesis suggested by these authors is that as adult size increases, fecundity increases with body volume while brooding space increases with body area. In this case, the capacity to produce eggs will be greater than the capacity to carry them. The hypothesis was tested with the small brooding asteroid *Asterina phylactica*. In the description of this species, Emson and Crump (1979) suggested that the small size of the individuals was the cause for their brooding patterns. The data from Strathmann *et al.* (1984) showed that larger asteroids suffered a higher rate of egg loss from their broods and supported the allometry hypothesis for the evolution of brooding. This relationship of small body size and low fecundity with brooding has been demonstrated in a number of groups, including polychaetes (Knight-Jones and Bowden, 1984; Cassai and Prevedelli, 1999), molluscs (Lalli and Wells, 1973; Pearse, 1979; Sastry, 1979) and echinoderms (Menge, 1975; McClary and Mladenov, 1989).

5.2. Fecundity and female size and age

The relationship between female size and fecundity is a major characteristic of reproduction in many taxa, and is related to morphological and physiological constraints in energy allocation and gonad production.

The relationship between body size and fecundity plays a central role in the reproductive patterns of crustacean decapods. There is a positive correlation between the number of eggs carried on the pleopods and the female's carapace length, with larger females carrying more eggs (Hines, 1982, 1988; King and Butler, 1985; Ivanova and Vassilenko, 1987; Corey and Reid, 1991; Reid and Corey, 1991; Somers, 1991; Van Dover and Williams, 1991; Gorny *et al.*, 1992; Hartnoll *et al.*, 1992; Clarke, 1993b; Hannah *et al.*, 1995; Company and Sardà, 1997; Lardies and Wehrtmann, 1997; Ohtomi, 1997; Thessalou-Legaki and Kiortsis, 1997; Anger and Moreira, 1998; Fonseca-Larios and Briones-Fourzán, 1998; Ramirez Llodra, 2000; Ramirez Llodra *et al.*, 2000). The hard exoskeleton of decapods limits the physical space available between the pleopods for the attachment of eggs, constraining the number of embryos that can be brooded (Hines, 1982; Somers, 1991; Clarke, 1993b). If the number of eggs produced is quantified as egg mass weight, this egg mass weight correlates with carapace volume and female weight (Corey and Reid, 1991). Therefore, apparent variability in fecundity of females of different populations or species might simply be a consequence of variability in female size and this factor needs to be taken into account in comparative studies (see section 3.1).

A positive correlation between brood size and female size is also found in some opportunistic polychaetes such as *Streblospio benedicti*, *Ophryotrocha puerilis puerilis* and *Capitella* sp. (Levin and Creed, 1986; Qian and Chia, 1991; Bridges *et al.*, 1994; Bridges, 1996). Opportunistic species have the ability to respond at a population level to environmental disturbance, organic enrichment or unexploited habitats, by rapidly converting nutrients into egg production (Eckelbarger, 1986). An increase in food availability can fuel an increase in adult body size. Larger females produce more eggs through a higher number of fertile segments, or through a size-dependent increase in resource acquisition and allocation to reproductive processes (Bridges *et al.*, 1994). *Capitella* sp. and *Streblospio benedicti* reach maximum body size before first spawning. The subsequent decrease in body size throughout the adult life may account for the decrease in fecundity of *Capitella* sp. and *Steblospio benedicti* (Qian and Chia, 1992a). On the contrary, the number of eggs produced per spawning event in the polychaete *Ophryotrocha labronica* remains more or less constant throughout its lifespan, decreasing only 12% from first maturity to death. This constant fecundity with age is related to the constant body size of adults from first maturity to death (Cassai and Prevedelli, 1999).

In opportunistic polychaetes, there is a greater egg production at early ages, a decrease in age at first maturity and an increase in adult body size, causing exponential population growth when resources are being exploited (Bridges *et al.*, 1994; Bridges, 1996; Levin *et al.*, 1996; Linton and Taghan, 2000; Mendez *et al.*, 2000). However, Bridges and co-workers (1994) found different reproductive responses to enriched sediment in *Streblospio benedicti* and *Capitella* sp. They suggested that even though these opportunistic species share similar growth and repro-ductive rates, they display differing life-history traits. The selection for certain life-history patterns may depend on species-specific sensitivity to environmental factors such as oxygen level, tolerance to toxic elements, feeding behaviour and digestive physiology (Bridges *et al.*, 1994; Bridges, 1996).

Bivalve molluscs have been studied in detail because of the commercial importance of many species. In several scallops, mussels and oysters, rapid growth in the early years is followed by a decline in growth but production of gametes increases throughout their lifetime. With increasing age, there is a gradual transition in the allocation of energy from growth to reproduction (Rodhouse, 1978; Griffiths and King, 1979; Thompson, 1979; Kautsky, 1982; MacDonald and Thompson, 1985a, b; Langton *et al.*, 1987; Honkoop and Van der Meer, 1997; Honkoop *et al.*, 1998). The inner volume of the shell represents the upper limit for reproductive production so fecundity is positively correlated to shell size (Rodhouse, 1978; Nakaoka, 1994; Honkoop and Van der Meer, 1997). Because of the significant relationship between body size and fecundity, Honkoop and Van der Meer (1997) proposed the use of the easily-quantified body mass index (BMI ($mg\,cm^{-3}$) = ash-free dry weight/shell volume) to estimate the reproductive output of an individual.

In the protobranch bivalve *Yoldia notabilis*, the increase in fecundity is more strongly related to female size than to age. The individuals of *Y. mutabilis* from populations living at different depths have a different age at first maturity. The specimens of *Y. mutabilis* from the deeper population (14 m) mature 1 year later than the specimens from the shallower population (10 m). However, size at first maturity is similar in both populations and coincides with the size at which these bivalves escape predation pressure by the crab *Paradorippe granulata* (Nakaoka, 1994). The production of eggs is then optimized when adult mortality is reduced, increasing female fitness. The fecundity of the mussel *Mytilus edulis* also depends more on size than on age, growth or production, resulting in maximum gonad production when excess food is available (Kautsky, 1982).

In colonial species such as corals, analysing reproductive traits can be more arduous because of the modular nature of the organisms (see section 3.3). The morphology of the colonies and the size of polyps can affect life-history traits such as fecundity, egg size and parental care. There is a tendency among massive growth forms or corals with large polyps to develop gonads in their mesenteries and produce a large number of eggs, which are spawned into the water column. On the contrary, branching forms or corals with small polyps usually develop gonads in their body cavities and reduce the number of eggs during oogenesis. These species brood a small number of embryos to the planula stage (Rinkevich and Loya, 1979a).

Because of the modular characteristics of colonial organisms, several traits need to be taken into account when analysing the relationship between fecundity and adult size. The main traits to consider are: (1) indeterminate growth of the organism, (2) polyp heterogeneity within the colony, and (3) different relationships of fecundity and size in individual polyps and in the colony as a whole. In most corals, first maturity is attained when the colony reaches a critical size related to a minimum number of polyps. Fertility can be variable within a colony, and colony fecundity is related to both individual polyp fecundity and number of fertile polyps. Colony fecundity in corals usually increases with colony size, while polyp fecundity can increase with polyp size, age and/or position (Rinkevich and Loya, 1979b; Kojis and Quinn, 1981; Wallace, 1985; Chornesky and Peters, 1987; Parker *et al.*, 1997; Sakai, 1998; Smith and Hughes, 1999).

In the reef-building corals *Monastrea annularis* and *Antipathes fiordensis* there is no homogenous fecundity within a colony, but at the individual level fecundity increases with polyp size (Van Veghel and Kahmann, 1994). Also, larger colonies produce a higher overall number of eggs through a higher number of fertile polyps (Parker *et al.*, 1997). In the massive coral *Goniastrea aspera*, there is a difference in the fertility of polyps depending on their position within the colony. In the scleractinian coral *Porites astreoides*, there is also a low density of reproductive polyps around the edges of the colony (Chornesky and Peters, 1987). These authors suggested three possible explanations for the variation in fertility. Firstly, there could be an energetic constraint, where energy allocation to growth around the edges would result in less energy available for reproduction and the subsequent decrease in gonad development. Secondly, the edges being more likely to experience injury or predation, it would be advantageous to limit the production of expensive reproductive material to the more protected central polyps. Finally, in *Porites astreoides* where polyp fecundity is related to polyp age, the edge polyps might simply be too young to have mature gonads (Chornesky and Peters, 1987). Kojis and Quinn (1981) found that there was no heterogeneity in fertility of the massive colonies of the hermatypic coral *Goniastrea australensis*. These authors suggested that in massive corals, growth is homogenous through-out the colony and all polyps can allocate similar amounts of energy to growth and reproduction.

In corals, an important aspect of life history is the dispersal and survival of colony fragments. Because of the relationship among colony size, food intake and fecundity, small fragments have a lower probability of survival and of reproduction. Corals must reach a minimum colony size for the onset of maturation, and this minimum size is species-specific (Rinkevich and Loya, 1979b; Kojis and Quinn, 1981; Wallace, 1985; Chornesky and Peters, 1987; Parker *et al.*, 1997; Sakai, 1998; Smith and Hughes, 1999). Smith and Hughes (1999) suggested that the loss of fecundity in small fragments can be a temporal adaptation for survival, with a re-allocation of energy towards growth and repair before the new colony can be reproductively active again.

5.3. Fecundity and age at first maturity

Age at first maturity is defined in life-history theory as the age at which the female produces the first batch of eggs. This age plays a pivotal role in the life of an individual, dividing the lifespan of the organism in two phases, preparation and fulfilment (Stearns, 1992). Fitness is extremely sensitive to age at maturity, because selective pressures and trade-offs change dramatically at this time (Stearns, 1992; Olive *et al.*, 2000). Selection for age at first maturity results from the balance of benefits and costs of starting reproduction at that age. This balance is affected by the interaction between age and size, which play an important role in the determination of the onset of gametogenesis.

The relationship between fecundity and age depends on the life-history strategy of a species. Early maturing species have a shorter generation time and a higher probability of reaching maturity simply because of the shorter time needed to reach first reproduction. The cost is a reduction in future fecundity. In contrast, species with delayed maturity live longer, allowing them to grow larger and therefore have a higher initial fecundity. The longer lifespan also gives the possibility of undergoing a higher number of reproductive events and therefore a higher lifetime fecundity. And finally, species with delayed maturity and its associated characteristics (larger size, longer lifespan) are able to produce offspring of higher quality and in some cases to provide parental care (Stearns, 1980, 1992). Accordingly, maturity can be delayed until fitness gained by producing more offspring with a higher survivorship is balanced by the loss of fitness caused by a longer generation time and lower survival to maturity (Bell, 1980; Stearns, 1980, 1992). The demography theory predicts an age at maturity that optimizes lifetime reproductive success (Stearns, 1992; Winemiller, 1992; Winemiller and Rose, 1992). The optimal age at maturity will evolve under selective pressures affecting juvenile and adult survivorship and their relation with age-specific fecundity and mortality. If adult mortality is low, a long lifespan allows for a longer growth period. In this case, delayed maturity with the production of more and better quality eggs can evolve. If adult mortality is high, early maturity with a high reproductive effort soon after maturity will evolve (Murphy, 1968; Wilbur et al., 1974).

In a study of the reproductive traits of Antarctic caridean decapods, Clarke (1979) suggested that species living in cold water show typical K strategies, with delayed maturity, leading to a larger size that allows for the production of larger eggs. These species live in a relatively homogenous and predictable environment, where adult mortality is low and the production of larger eggs hatching into better quality larvae confers a higher survival probability (Clarke, 1979). Similar results were found when comparing caridean shrimp from the Weddell Sea to related subAntarctic species. The species from the high Antarctic had a delayed age at maturity and reached larger sizes, which allowed them to produce eggs of better quality than their subAntarctic counterparts (Gorny et al., 1992).

Olive et al. (2000) studied selection for age at first maturity in an experimental study of fitness components of seasonal reproduction in the polychaete Nereis virens. This species is semelparous and the specimens have variable age at first maturity, the latter being related to growth rate. Because individuals of N. virens die after breeding, there are two elements of mortality in their life cycle: breeding mortality and non-breeding mortality. Olive *et al.* (2000) showed that when the non-breeding mortality rate is greater than a certain value ($z_x > 0.5$), the risk of mortality is too high to delay maturity and reproductive value at age 1 is maximized. When the non-breeding mortality is lower than this value, the increase in fecundity with age compensates for external mortality and reproductive value is maximized at ages greater than 1 (Olive *et al.*, 2000).

In populations of the poecilogonous polychaete *Streblospio benedicti*, worms producing planktotrophic larvae matured at a younger age than worms producing lecithotrophic larvae but also suffered from a higher mortality early in life (Levin *et al.*, 1987; Levin and Creed, 1991). These authors suggested that the ability to have similar population growth rates with different combinations of life-history traits confers an adaptive response to different environmental conditions in the natural habitat.

In a study of the reproductive strategies of asteroids, Menge (1975) suggested that *Pisaster ochraceus* delays maturity until a certain body weight is reached, when the individuals are sufficiently robust to survive long enough to produce the minimum number of offspring required to replace themselves (Menge, 1975).

5.4. Fecundity and reproductive effort

Reproductive effort is the total energy used in reproduction. This includes the energy allocated to gonad production and egg packaging structures (egg cases or egg masses), the energy spent in collecting supplementary food during the reproductive period, territorial and mating behaviour and parental care (Clarke, 1987; Thessalou-Legaki and Kiortsis, 1997). Reproductive effort is very difficult to quantify with precision and reproductive output has been used as a useful estimate. Reproductive output is quantified as the biomass of reproductive products per unit biomass of the female (Clarke, 1987; Thessalou-Legaki and Kiortsis, 1997; Anger and Moreira, 1998). When the energy available for organisms is limiting, an increase in reproductive effort results in an increase in reproductive output and a reduction in somatic investment. Moreover, because of the trade-off between egg size and number, an increase in reproductive output would be reflected by an increase in fecundity or by an increase in offspring survivorship resulting from a higher investment per individual egg. The reproductive output is therefore defined by two aspects, the quantity of eggs (or fecundity), and the quality of eggs (or egg size) related to offspring survival (Olive, 1985; Bell and Fish, 1996).

Poecilogonous species can have the same overall reproductive invest-ment and output. The polychaetes *Streblospio benedicti* and *Capitella* sp. have sympatric populations producing either planktotrophic or lecitho-trophic larvae. The reproductive output of both strategies is similar. However, species developing through lecithotrophic larvae produce a small number of large eggs, while species developing through plankto-trophic larvae pack their oogenic material in a high number of small eggs (Levin and Creed, 1986; Levin *et al.*, 1987; Qian and Chia, 1991) (see also section 5.1.1). Another example of comparable reproductive output in species with different life-history strategies is that of the terebellid polychaetes *Ramex californiensis* and *Nicolea zostericola*. These two polychaetes grow to a similar size, but while the former produces ~44 larvae per cocoon with a diameter of 410 μm, the latter produces ~665 larvae in its single spawning period. The number of larvae produced over a lifetime is similar because *R. californiensis* has a lower initial fecundity, that has a smaller cost to future fecundity, allowing for a production of up to 11 cocoons over a lifetime (McHugh, 1993).

An important aspect of life-history theory is the relationship between current reproduction and survival. The allocation of energy into reproduc-tive processes can diminish survival of the adult to the next reproductive event (Williams, 1966; Cohen, 1968; Christiansen and Fenchel, 1979; Bell, 1980; Charlesworth, 1990; Stearns, 1992; Ebert, 1996). This has been called the "survival cost" (Bell, 1980). For example, Prevedelli and Vandini (1999) studied the effects of diet on the survival and reproduction of the polychaete *Dinophilus gyrociliatus*, and suggested that the higher mortality observed in the specimens fed with the higher quality diet could be related to their higher fecundity.

The theoretical optimal life-history strategy will optimize lifetime reproductive success by maximizing the allocation of resources to growth, survival and reproduction from birth to death. For this, the reproductive effort model determines the age at maturity and the number of offspring produced during the lifetime of the organism under the trade-off between actual reproductive output and residual reproductive value (Williams, 1966; Bell, 1980; Charlesworth, 1980, 1990; Stearns, 1992; Olive *et al.*, 1997) (Figure 6). Fecundity of an individual that has already reproduced might be lower than fecundity of individuals maturing for the first time. Because the total energy that an organism can assimilate and allocate to growth and reproduction is limited, the allocation of energy to repro-duction will be associated with a decrease in somatic investment. This reduced somatic investment can affect the survival and growth of the adult, which in turn will affect subsequent reproductive events. The residual reproductive value is the lifetime product of survivorship and fecundity. The allocation of resources to growth, survival and reproduction

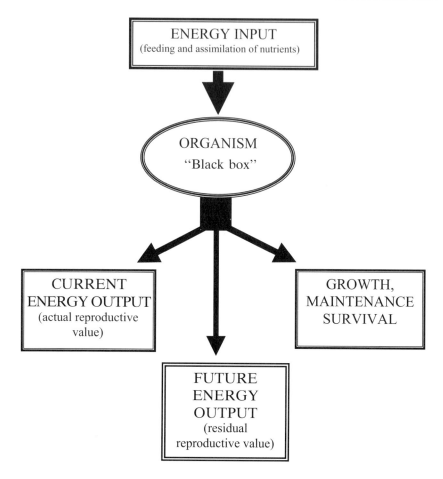

Figure 6 Allocation of energy to actual reproductive value and residual reproductive value. Drawn from data in Olive *et al.*, 1997.

has to be optimized over the lifespan of the organism, and this is affected by the trade-off between current reproduction and the subsequent adult reproductive success (Williams, 1996; Bell, 1980; Charlesworth, 1980, 1990; Stearns, 1992; Giangrande *et al.*, 1994; Olive *et al.*, 1997). For example, the opportunistic polychaete *Capitella* sp. has a low initial reproductive effort allowing for an enhanced adult survivorship, which in turn results in a longer lifespan and a higher number of reproductive events during its life (Qian and Chia, 1992a). Olive and colleagues (1997) studied the variable spawning success of the polychaete *Nephtys hombergi* and showed that in good spawning years the ratio "energy available (prey density and calorific content):energy required (for gamete production)" was high, but it was low in the bad spawning years. The authors suggested that reproductive variable spawning is a homeostatic mechanism for alternative routes to fitness, where energy can be allocated to current reproduction, or to growth and survival, and therefore to future reproductive success.

The number and quality of eggs produced by an organism during its lifetime vary with adult age. The reproductive effort models of life-history evolution optimize age and size at maturity and the number of offspring produced at each age under the assumption of a trade-off between current reproduction and subsequent adult reproductive success (Charlesworth, 1980, 1990; Stearns, 1992). The reproductive effort is maximized at a certain age, depending on the life-history strategy of the species. In species with decreasing profit at high levels of reproductive effort (convex profit curve), or when there is a higher mortality risk related to investment in reproduction (concave cost curve), there would be selection for intermediate levels of reproductive effort (Figure 7). In these cases, iteroparity

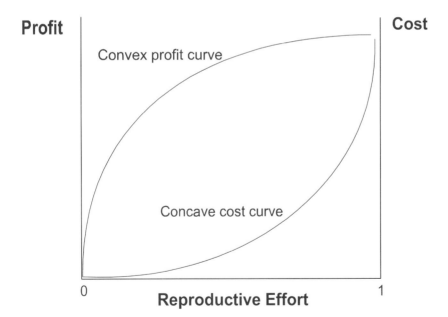

Figure 7 Profit and cost curves of the reproductive effort model for evolution of life-history strategies. Modified from Stearns, S. C. (1992). "The Evolution of Life Histories". Oxford University Press, Oxford, with permission.

is expected. Conversely, when mortality increases at a certain age class, fecundity will be optimized at the previous age range resulting in semelparity (Gadgil and Bossert, 1970; Schaffer, 1974; Stearns, 1992). The nudibranchs *Aldalaria proxima* and *Onchidoris muricata* are ecologically, morphologically and taxonomically close species but show distinct life-history strategies. Both species are semelparous and lay gelatinous coiled egg masses in several spawns over the unique spawning period. Reproduction incurs a cost that is met partially by loss of somatic mass. *A. proxima* grows to a large size allowing for a first large spawn, which maximizes reproductive output at a time of declining life expectancy (degrowth). On the contrary, *O. muricata* grows to a smaller size and produces a smaller first spawn. In this species, the current reproductive costs are met by current energy supply from ingestion, which allows for a slower degrowth and provides the energy to produce a larger number of temporally and spatially separate spawn masses (Havenhand and Todd, 1988; Todd and Havenhand, 1988). Also, when an unstable environment increases the adult mortality risk at any time, as in the gravel amphipod *Pectenogammarus planicurus*, point reproductive success will be maximized (Stearns, 1992; Bell and Fish, 1996).

Ultimately, it is not the number of surviving offspring in a single brood that is important, but the total number of offspring surviving in the lifetime of the adult, and therefore lifetime fecundity is optimized (Gadgil and Bossert, 1970; Stearns, 1992; Bell and Fish, 1996).

6. VARIABILITY OF FECUNDITY WITH ENVIRONMENTAL FACTORS

Most variables of a life-history strategy in marine invertebrates are characterized by a range of values within a phylogenetic and environ-mental constrained spectrum of possibilities. To be able to respond to changing environmental factors, these variables show phenotypic plasticity. That is, most life-history variables are highly polytypic (Caswell, 1983; Hadfield and Strathmann, 1996). There are two types of polytypy: plasticity and flexibility. Plasticity is an environmentally induced polytypy reflected in morphological differences among siblings. Flexibility is the capacity to display different functional abilities as a response to different environmental situations (Hadfield and Strathmann, 1996). The term variability should only be used for traits whose variance is genetic or stochastic, when variance is always present or when variance is not an individual response to environmental conditions (Hadfield and Strathmann, 1996). Flexibility and plasticity provide an increased probability of survival and future fecundity of individuals. Life-history variability gives the population and species a greater probability of reproductive success, dispersal, recruitment and persistence in geological times, and therefore does affect fitness (Caswell, 1983; Hadfield and Strathmann, 1996; Olive *et al.*, 2000).

Within life-history variables, fecundity is one of the most flexible. The order of magnitude of the number of eggs produced by a species is ultimately determined by natural selection in relation to: (1) evolution of its life-history strategy and related trade-offs with egg size and reproductive effort; and (2) design constraints (see sections 4 and 5). But there are environmental factors, such as food availability, temperature, salinity or presence of toxic elements, that influence the number of eggs produced by an individual within determined species-specific limits (Stearns, 1992; Eckelbarger, 1994).

6.1. Fecundity and food availability

The production of eggs is an energy-demanding process (see section 1). Accordingly, food availability plays a central role before and during oogenesis (Eckelbarger, 1986; Chia and Walker, 1991). The processes of vitellogenesis use up a large proportion of an organism's energy intake, and this requirement varies depending on the vitellogenetic pathway used by the species (see section 1). A continuum is found between species that have heterosynthetic mechanisms of yolk production, allowing for fast egg production and short periods between breeding events, and species with autosynthetic vitellogenesis causing slow egg production and long periods between breeding events (Eckelbarger, 1983, 1994; Jaeckle, 1995). Long-lived species with slow egg production have vitellogenic strategies that are consistent with a continuous or predictable food supply and a relatively stable environment such as temperate latitude habitats, the Antarctic benthos or the abyssal plains (Clarke, 1979; Gage and Tyler, 1991; Eckelbarger, 1994; Eckelbarger and Watling, 1995). In contrast, unstable environments or unpredictable food supply, such as large food falls in the deep sea or the ephemeral hydrothermal vents, would select for opportunistic strategies with fast egg production capabilities (Eckelbarger, 1994).

The oogenic processes of any species are phylogenetically constrained by the morphology of the ovary, the species-specific vitellogenetic pathway and the digestive structure related to the transfer of nutrients from the somatic organs to the ovaries (see section 1). Nevertheless, fecundity and the quality of offspring are directly related to the nutritional state of the adult and its resource allocation. An increase in food quantity or quality will not affect the basic gametogenic pathways of the individual. However, if food quantity or quality increases, more energy can be allocated to reproduction, resulting in an increase in fecundity or offspring quality (Olive, 1985; Eckelbarger, 1986; Jaeckle, 1995; Eckelbarger and Watling, 1995; Bertram and Strathmann, 1998). Shortage of food may cause a decrease or even an interruption of yolk synthesis and a lowering of fecundity or even a failure to reproduce (Barber *et al.*, 1988; Olive and Morgan, 1991; Qian and Chia, 1991; Bridges *et al.*, 1994; Eckelbarger, 1994; Levin *et al.*, 1994; Eckelbarger and Watling, 1995).

Food quality or food quantity, or both, can affect the reproductive output of a species. An experimental study of compensatory feeding of three sympatric amphipods showed interspecific differences in fitness related to physiological rather than behavioural constraints. All three species of amphipods showed compensatory feeding when exposed to low food quality. Low food quality resulted in lower survivorship, growth and fecundity in *Elasmopus levis* and *Gammarus mucronatus*, while *Amphithoe longimana* was able to circumvent the effects of low food quality by eating more food (Cruz-Rivera and Hay, 2000).

There is good evidence that food quantity and quality are linked to egg production rate, fecundity and egg quality in copepods. Many copepods have opportunistic life-history traits, with iteroparity, rapid growth and high fecundity in successive broods. In these species, no food reserves are built up and energy for reproduction is directly derived from food supply (Jónasdóttir, 1994; Williams and Jones, 1999). Laboratory and field experiments have shown that food availability plays a major role in the number of eggs produced, the interval between broods and the size of eggs. An increase in food quantity and/or food quality results in an increase in reproductive output. This is expressed as an increase in fecundity and a reduction in the time between breeding events (Tester and Turner, 1990; Razouls *et al.*, 1991; Jónasdóttir, 1994; Laabir *et al.*, 1995; Bell and Fish, 1996; Pond *et al.*, 1996; Williams and Jones, 1999; Kleppel and Hazzard, 2000). Fast egg production in species with heterosynthetic vitellogenesis allows for a rapid use of energy intake. An increase in food quantity and quality in the environment maintains a high reproductive rate in female copepods by fuelling vitellogenesis and therefore sustains the maturation of successive batches of eggs (Razouls *et al.*, 1991).

This trend towards a higher fecundity and smaller eggs at increasing levels of energy availability is found in many crustaceans (copepods, amphipods, caridean shrimp; see examples below) and reflects the adaptation of the reproductive investment to variable environmental factors. While the overall investment in reproduction is set by the energy available to the female, the investment per embryo is related to the conditions awaiting the pelagic larvae. When food quantity or quality are poor in the environment, the investment per offspring is high, allowing for higher survival probability of the larvae. Conversely, when the energy available in the water column is high, the females invest less per offspring, producing a larger number of lower quality embryos that hatch and develop in a more favourable environment. Gorny and co-workers (1992) showed that the Antarctic caridean shrimps *Chorismus antarcticus* and *Notocrangon antarcticus* in the Weddell Sea produce fewer but larger eggs than do the lower latitude populations around South Georgia. Many authors have suggested that there is a need to provide more energy for larvae hatching in environments with a high mortality risk. Greater energy allocation in the egg increases offspring fitness by increasing the larval survival probabilities (Thorson, 1950; Herring, 1974b; Omori, 1974; King and Butler, 1985; Clarke and Gore, 1992; Gorny *et al.*, 1992; Clarke, 1993a; Company and Sardà, 1997; Lardies and Wehrtmann, 1997). The reproductive processes of deep-sea caridean shrimp have been compared with mesopelagic species from the north-east Atlantic, benthic species from an upwelling area in the Indian Ocean and hydrothermal vent shrimp from the Mid-Atlantic Ridge vents (Ramirez Llodra, 2000). The data show a higher fecundity in the species from the hydrothermal vents and the upwelling area compared with the mesopelagic species, possibly related to a higher food availability in the vent and upwelling habitats (Ramirez Llodra, 2000).

Polychaetes have often been used for experiments on the effects of environmental factors such as temperature and food in reproduction. Within the family Spionidae several species exhibit poecilogony. The already cited *Streblospio benedicti* exists as sympatric forms that produce either lecithotrophic or planktotrophic larvae. Experiments by Levin and Creed (1986) showed that changes in temperature or food did not cause a switch from one mode of development to the other, implying that larval type is genetically determined and polymorphic in this species. The genetic basis of this dichotomy between planktotrophic and lecithotrophic development, and their respective life-history traits was later determined in the laboratory (Levin and Creed, 1991). Under uniform laboratory conditions, the two strains (specimens with planktotrophic and lecithotrophic development) of *S. benedicti* achieved a similar production of oogenic material and growth (Levin *et al.*, 1987). Conversely, food variability has an important and differential effect on the reproductive output of *S. benedicti*. An increase in food quantity and quality resulted in a higher production of eggs in the specimens with lecithotrophic development. These have more heterosynthetic yolk bodies than the specimens with planktotrophic development, allowing for a faster allocation of nutrients to the production of oocytes (Levin and Creed, 1986; Levin and Bridges, 1994).

Experiments on the short-term reproductive response of the opportunistic polychaete *Capitella* sp. fed on different diets also showed that
an increase in food ration or quality results in increased body size, higher
fecundity and higher reproductive output (Grémare *et al.*, 1988; Linton and
Taghan, 2000). Qian and Chia (1991) suggested that opportunistic
polychaetes have a high genetic variation in reproductive traits such as
fecundity, egg size and adult size, providing flexibility in reproduction
and growth in variable environments. These authors presented evidence
for the production of a higher number of smaller eggs in a population of
Capitella sp. feeding on high quality food (squid egg capsules) compared
with a population feeding on a lower quality food source (detritus). In a
later study, the same authors showed a differential growth and reproductive response to diet in siblings of *Capitella* sp. with either planktotrophic
or lecithotrophic development (Qian and Chia, 1992b). There was a positive relationship between growth and the quality of the diet. Planktotrophs
fed with high quality food matured earlier and spawned more often,
resulting in an overall increase of lifetime fecundity. In lecithotrophs, a
higher quality diet increased the number of eggs per spawning event but
the eggs were smaller, while a lower food quality resulted in a lower
fecundity but larger eggs. *Capitella* sp. can maintain its population at a
certain density in times of lower food availability by producing fewer but
higher quality eggs. The authors proposed that this plasticity in reproduction could be adaptive to a dynamic and fluctuating environment (Qian
and Chia, 1992b, 1994). There is, however, a critical food level below which
Capitella sp. stops producing gametes. During the ecological process of
succession in a new habitat, the disappearance of the opportunistic
Capitella sp. was related to a decreased fitness produced by a decrease
in food concentration in the sediment and the inability to reproduce
below a threshold protein concentration (Linton and Taghan, 2000).

A study on the effects of food supply and population density in growth
and reproduction of the infaunal polychaete *Polydora ligni* showed a
negative effect of low food supply in reproduction. A decrease in food
supply associated with an increase in population density resulted (in
both laboratory and field) in a longer maturation time at smaller individual
body size, longer time between broods, a decrease in the number of
gametogenic segments, a decrease in the number of egg capsules and the
associated decrease in fecundity (Zajac, 1986).

Salinity and diet affect the survival, fecundity and sex determination of
the polychaete *Dinophilus gyrociliatus*. Greater fecundity (higher number
of eggs per week) was found in the group fed with the higher quality food.
The higher fecundity was related to a greater and earlier mortality. This
higher mortality rate could be related to the greater amount of energy
allocated to reproduction in specimens fed with high quality food. Also,

individuals kept in low food conditions showed a bias in sex ratio towards males, probably because of the lower energy requirement for the production of male gametes (Prevedelli and Vandini, 1999; Prevedelli and Simonini, 2000). Similarly, in the reproductive effort of the small iteroparous polychaete, *Ophryotrocha labronica*, food quality affects fecundity. Specimens fed with higher quality food spawned more frequently and produced more eggs per spawning event than those given lower quality diets. The former had therefore a higher lifetime fecundity than the latter (Prevedelli and Vandini, 1998).

In marine bivalves, a decrease in nutritive conditions during gametogenesis results in a decrease in the egg production, either lower fecundity or production of smaller eggs (Bayne *et al.*, 1983; Barber *et al.*, 1988). In the giant scallop *Placopecten magellanicus* in the Gulf of Maine, a deep population at 76 m had a lower production of eggs compared with a shallow-water population at 6 m (Barber *et al.*, 1988). These differences were attributed to differences in food availability in the two populations. As the egg size was similar in both, the result was a reduction in reproductive output of the deep-water scallops without affecting the survival of the larvae. Similar results were obtained by MacDonald and Thompson (1985a, b) with different populations of *P. magellanicus* from Newfoundland and New Brunswick. The adult size, somatic growth and reproductive output were lower in the deeper water population where temperature, food quantity and quality are lower. These authors proposed that growth rate and reproductive output of an individual are good indicators of suitability of the environment, because these indices integrate the physiological processes. In the intertidal bivalve *Macoma balthica*, extra energy intake allows for higher somatic and reproductive production. Food quantity, quality and temperature all affect reproductive output (Honkoop and van der Meer, 1997). The mussel *Mytilus edulis* shows high variability in fecundity between and within populations in different seasons, related to geographical or seasonal changes in food availability (Bayne and Worrall, 1980; Kautsky, 1982). The bivalve *Pecten maximus* shows differences in egg quality related to variability in environmental conditions during gametogenesis (Paulet and Boucher, 1991). Eggs showing ultrastructural abnormalities and lower lipid content had lower hatching success and produced abnormal larvae.

Egg production and packing in the intertidal scavenging gastropod *Nassarius festivus* are influenced by food availability. The number and size of egg capsules, the number of eggs per capsule, the energetic contents of eggs and the proportion of energy allocated to reproduction decreased with decreasing food availability. Absence of shell growth and a decrease in body weight indicate that the bulk of energy intake of adult *N. festivus* is used for reproduction not growth (Cheung and Lam, 1999). However, mortality was lower in specimens with less food. In an iteroparous species, it would be advantageous, in times of food scarcity, to allocate more energy to maintenance in order to survive until the next reproductive event when food availability might be higher (Cheung and Lam, 1999). In the estuarine nudibranch *Tenellia aspera*, starvation produces a decrease in adult size and reproductive output. This gastropod has a plastic egg size that is affected by food availability. The eggs of starved specimens are smaller and develop as pelagic lecithotrophs, while the eggs of well-fed specimens develop into capsular metamorphic juveniles. This plasticity allows some progeny to survive in unstable estuarine environments (Chester, 1996). Similarly, a study of two congeneric species of the gastropod *Thais* (*Thais lamellosa* and *Thais emarginata*) showed a differential egg production response to food availability. *Thais lamellosa* spawns once a year and annual fecundity is proportional to body size. An increase in food availability results in an increase in size and a related increase in fecundity. On the contrary, *Thais emarginata* spawns several times a year. An increase in food availability does not result in growth in this species, but fecundity increases as a result of a greater number of spawning events (Spight and Emlen, 1976).

The opistobranch *Alderia modesta* exhibits poecilogony in the natural environment. This results in mixed populations with adults producing large eggs that develop into lecithotrophic larvae, small eggs that develop into planktotrophic larvae and mixed egg masses where eggs develop into both lecithotrophic and planktotrophic larvae. When a population of *A. modesta* was experimentally starved, the adults that previously produced lecithotrophic larvae switched to producing planktotrophic larvae or mixed egg masses. *A. modesta* is a specialist herbivore feeding exclusively on *Vaucheria* spp. The author suggests that the switch from lecithotrophy to planktotrophy is an adaptation to the disappearance of *Vaucheria* spp. in the natural environment, increasing larval dispersal away from an unfavourable habitat (Krug, 1998).

In echinoderms, Vadas (1977) showed that diet quality rather than quantity affected the production of reproductive material in the echinoids *Strongylocentrotus droebachiensis* and *Strongylocentrotus franciscanus*. A later study on the reproductive cycle of *S. droebachiensis* from differing habitats confirmed Vadas' results (Meidel and Scheibling, 1998). These authors found that the gonads were qualitatively similar (same temporal patterns of abundance of each type of ovarian cell) but that the gonadal mass of females differed between sites. The quantity of gut contents was similar between sites, but the quality of the gut contents was higher in individuals feeding on kelp beds and grazing fronts than in those from barren sites. These individuals also had a higher fecundity (Meidel and Scheibling, 1998). Similar observations of a higher reproductive output in populations feeding on higher quality or on a higher quantity of food have been found for the echinoids *Lytechinus variegatus* (Beddingfield and McClintock, 1998), *Arbacia lixula* and *Strongylocentrotus droebachiensis* (Bertram and Strathmann, 1998) and *Evechinus chloroticus* (Brewin *et al.*, 2000), and for the asteroid *Leptasterias epichlora* (George *et al.*, 1990; George, 1994a). Also, it has been suggested that brooding in some asteroids, such as *Leptasterias hexactis* or the Antarctic *Neosmilaster georgianus*, can compromise the ability to obtain food. During the long incubation phases, these brooding females starve, causing a depletion of energy reserves in the pyloric caeca. This reduction in the energy stored for future reproduction can delay oocyte development in the ovaries and ultimately limit the frequency of individual reproduction (Bosch and Slattery, 1999).

In deep-sea echinoderms, the most common life-history strategy involves the quasi-continuous production of a small number of large eggs with direct or lecithotrophic development (Shick *et al.*, 1981; Tyler *et al.*, 1982a, b; Gage and Tyler, 1991; Ramirez Llodra, 2000; Ramirez Llodra *et al.*, 2002). For example, the three porcellanasterid asteroids *Hyphalaster inermis*, *Styracaster horridus* and *Styracaster chuni* from the north-east Atlantic abyssal plains produce large eggs (~600 μm) in a quasi-continuous pattern. The possible effect of differing food availability on the reproductive output of these asteroids from different locations has been studied (Ramirez Llodra *et al.*, 2002). The data showed that adult growth and fecundity were higher in the specimens living in the richer environments (higher amount of food and better food quality).

In the deep sea there are also a few species that reproduce seasonally, producing a large number of small eggs that develop into planktotrophic larvae. It has been suggested that selection for seasonal reproduction with small eggs, high fecundity and planktotrophic development in the ophiuroid *Ophiura ljungmani*, the asteroids *Plutonaster bifrons* and *Dytaster grandis* and the echinoid *Echinus affinis* is related to a seasonal flux of phytodetritus to the sea bed. This input of organic matter may fuel gametogenesis and/or be used by the planktotrophic larvae in the water column (Tyler and Gage, 1980; Tyler and Pain, 1982; Tyler *et al.*, 1990; Campos-Creasey *et al.*, 1994).

6.2. Fecundity and latitude and depth

The variation of fecundity and egg size with changing depth and latitude is ultimately related to environmental factors, such as temperature and food quality and quantity that affect the adults and the larvae. There is a general trend towards the production of fewer but larger eggs in colder waters (Clarke and Gore, 1992). High latitudes or deep waters are characterized by low temperatures. In these environments, both embryonic and larval developments are slowed down by cold waters, and there is a trend to produce larger eggs (reducing fecundity) in order to provide the offspring with enough energy for a longer developmental time (see examples below). However, metabolic rates also decrease and therefore physiological processes are less energetically demanding. Additional information on the effects of temperature on development patterns is needed to understand the observed clines in egg sizes (Clarke and Gore, 1992).

In many crustacean species, the deeper water populations are composed of larger individuals that have long reproductive lifespans and produce larger eggs (Mauchline, 1988). King and Butler (1985) studied five pandalid shrimp from the Pacific with depth ranges between 200 and 800 m and found that specimens from deeper waters produced larger eggs. Nevertheless, the relative brood size and annual reproductive effort were not correlated with depth, indicating that the higher reproductive output of deep-water females is a consequence of their larger size.

In crustaceans, production of larger eggs results in the hatching of advanced larvae with a higher survival probability (Herring, 1974b; Omori, 1974; King and Butler, 1985). The larvae of many species that reproduce in deep waters migrate upwards through the water column to shallow depths. Deep-water larvae are exposed to fluctuations in food, temperature and other environmental factors and to predation during their extended periods in the water column, and the risk of larval mortality increases with depth (Thorson, 1950; King and Butler, 1985). In view of this risk of higher larval mortality in deep water, it appears to be more efficient to produce a few large eggs with a superior survival rate rather than a higher number of smaller eggs (King and Butler, 1985). However, Company and Sardà (1997) working on five deep-water pandalid shrimp from the western Mediterranean over a depth range between 150 and 1100 m, found no correlation between egg size and depth, but a decreasing brood size with increasing depth. These authors suggest that in the iso-thermal conditions of the Mediterranean Sea below 200 m, the pandalid shrimp do not need to compensate for larval mortality related to a tem-perature gradient with an increase of egg size.

A similar trend to that of deep-water species is found in decapod crustaceans from high latitudes, where females reach a larger size, have delayed maturity and produce larger eggs. The caridean shrimps *Chorismus antarcticus* and *Notocrangon antarcticus* from the Weddell Sea (high Antarctica) mature at a larger size and produce a smaller num-ber of larger eggs than their counterpart populations from South Georgia (subAntarctic) (Gorny *et al.*, 1992). Also, the snapping shrimp *Betaeus emarginatus* from Chile has a lower reproductive output than other alpheids from lower latitudes (Lardies and Wehrtmann, 1997). These high-latitude species follow a similar pattern to deep-water species, producing a smaller number of larger eggs. The reproductive output decreases with depth and latitude, but the larvae have a better energy provision and therefore survival probability (Clarke, 1979; Gorny *et al.*, 1992; Lardies and Wehrtmann, 1997). A similar cline has been found in the polar isopod *Ceratoserolis* sp. from the Weddell Sea, which produces larger eggs in colder water conditions (Clarke and Gore, 1992).

In the harpacticoid copepod *Scottolana canadensis* there are differences in growth and reproductive traits between latitudinally-separated populations, reflecting differences in environmental factors such as temperature (Lonsdale and Levinton, 1986, 1989). These authors suggested that *Scottolana canadensis* is adapted to the ambient temperature conditions of the habitat, and that evolution maximizes fitness through changes in feeding efficiency at different temperatures.

Relations between fecundity and egg size with depth have also been found in bivalves. The work on populations of *Placopecten magellanicus* living between 6 and 76 m depth, already noted (p. 145), shows that there can be a decrease in fecundity with increasing depth, related to temperature and food availability (MacDonald and Thompson, 1985b; Barber *et al.*, 1988). Specimens of the bivalve *Yoldia notabilis* from north-east Japan at a slightly deeper site (14 m) mature 1 year later than a shallower population (10 m), and age-specific reproductive effort was lower in the former. Nevertheless, size-specific reproductive effort did not show significant differences between populations, suggesting that egg production was related to size and not to age (Nakaoka, 1994). This author proposes that differences in food supply to the two populations affects reproductive effort through differences in growth rate.

While analysing the effects of maternal and larval nutrition on growth and form of planktotrophic larvae in the echinoid *Strongylocentrotus droebachiensis*, Bertram and Strathmann (1998) compared females from a shallow (6 m) and a deep (100 m) population. The authors showed that the growth of larvae is little affected by differences in maternal habitat, but that the ovaries of females in the deep population, where adult nutrition is poorer, were smaller and produced smaller eggs.

A study of energetic relationships and biogeographical location has shown differences in the fecundity, growth and reproduction of the reef coral *Pocillopora damicornis* (Richmond, 1987). This is an example of regional variability where biotic and environmental factors may be modulating the life history of the species. The colonies in the Indo-West Pacific have a slow growth rate (0.5–1.5 cm yr^{-1}) and reproduce throughout the year by brooding their embryos and releasing planulae over a period of several days each month. In contrast, colonies from the Eastern Pacific have a higher growth rate (3.6–$6\,\mathrm{cm\,yr^{-1}}$) and do not produce planulae. In this population asexual reproduction through fragmentation is important. The allocation of energy to biomass was similar in both populations, but the Indo-West Pacific colonies invest mainly in reproduction, while the Eastern Pacific colonies invest in growth. Richmond (1987) suggested two hypotheses to explain the lack of planulation in the Eastern Pacific population. Firstly, asexual reproduction by fragmentation would offer the advantage of a relatively large initial colony size. This is important in an environment with abundant predators and where competition for settling larvae is high because of the presence of fast-growing sponges and bryozoans. Secondly, the high temperature fluctuations of the Eastern Pacific could constrain gonad development. This would allow for a higher allocation of energy to somatic material, resulting in the higher growth rate observed in the Eastern Pacific population (Richmond, 1987).

7. SUMMARY

A thorough knowledge of the reproductive processes leading to the settlement of larvae and growth of juveniles is essential for understanding the dynamics of a population and the ecology of a community. These processes include the proliferation, growth and maturation of germ cells, spawning or copulation events, fertilization and embryo and larval development. Gametogenesis results in the production of mature eggs and sperm and plays a central role in reproduction. Gametogenesis is similar among invertebrates, but certain processes, such as vitellogenesis, are phylogenetically constrained. These are species-specific processes that determine patterns of egg production rate and constrain the reproductive response of the organism to environmental factors. Fecundity is the quantification of the number of offspring produced, and is therefore directly linked to the life-history strategy of the organism through bioenergetics and patterns of energy allocation.

The term fecundity refers to the number of offspring produced by a female in a certain time unit. This definition can be narrowed to adjust to the species studied. The basic unit quantified can be the oocyte, egg, embryo or larva, while the time over which fecundity is measured can vary from a day, a season, a year or a lifetime. Fecundity needs to be clearly defined for each study, in order to obtain the maximum information from the data analysed.

Because of the wide range of reproductive patterns among marine invertebrates, various methodologies have been developed to quantify fecundity. In brooding species, there is easy access to the embryos, allowing for precise quantification of fecundity through direct egg counts. In broadcast spawning species, ripe individuals can be induced to spawn their gametes under controlled laboratory conditions. Spawning can be induced through environmental shock such as thermal, light, osmotic or wave action shocks. Some species can also be induced to spawn through the injection of spawning-inducing chemicals, such as KCl in echinoids, 1-methyladenine in asteroids or serotonin in bivalves. A non-destructive method is used when the organisms need to be kept undamaged, by measuring the difference in weight before and after spawning. When organisms are not available alive, different, less accurate methods have to be used for the quantification of fecundity. In some species, maturing oocytes can be obtained by dissecting the ovaries. In species with small gonads or thin gonad walls, the oocytes may be counted directly by transmitted light under a microscope. When oocytes develop in well-organized ovaries with thick gonad walls, histological methods involving serial sectioning or the relationship between gonad volume and oocyte volume can be used for a broad estimation of fecundity.

The life-history strategy of an organism consists of a series of co-adapted traits evolved through natural selection. Life-history theory analyses the causes of observed differences in fitness. Life-history evolution models aim to predict the life-history strategy that an organism will have in a particular environment. The variables of a life-history strategy are growth rate, age and size at maturity, age- and size-specific fecundity, egg size, age- and size-specific reproductive effort, age- and size-specific mortality, larval developmental type and adult lifespan. Fecundity is one of the main variables in the life-history of a species and its direct link with energy allocation into reproduction plays a major role in modelling the covariation of reproductive traits.

Several models have been developed to explain the evolution of life-history strategies. The deterministic model (MacArthur and Wilson, 1967; Pianka, 1970) defines a continuum of strategies between two endpoints, the r selection and the K selection. In this model the population density in relation to environmental resources is identified as the main selective pressure. The r strategy is defined by a short lifespan, early maturity, small body size and high fecundity with the production of small eggs. This life-history strategy would be characteristic of organisms living under variable or unpredictable environments. The K strategy is defined by a long lifespan, delayed maturity, large body size and low fecundity but with the production of high quality offspring (large eggs). This life history strategy would be characteristic of species inhabiting constant or predictable environments. Criticisms of the deterministic model include the lack of reference to morpho-functional constraints and demographic considerations.

The stochastic models (Cohen, 1968; Murphy, 1968; Schaffer, 1974) predict similar life-history strategies to the deterministic model, but the evolutionary pressure here is caused by environmental effects on the differential mortality of juveniles and adults.

The demography theory includes the more advanced models in life-history theory. The demography models use age- and size-specific fecundity and mortality as the basis of variation in fitness (Stearns, 1992). The selection of optimal life-history strategies is explained as the trade-off between current reproductive output and residual reproductive value, the latter being the lifetime product of survivorship and fecundity. The criticisms of the demography theory are that it assumes a stable age distribution of the population and it ignores the effects caused by morpho-functional constraints on the evolution of life-history traits.

The Winemiller–Rose model (Winemiller, 1992; Winemiller and Rose, 1992) is a demographic model, which is based on an amplification of the deterministic model. This model proposes a continuum on a three-dimensional triangle where each vertex is a boundary life-history strategy. The endpoint strategies are defined by three important demographic parameters: age at maturity, fecundity and juvenile survivorship.

In a life-history strategy, fecundity is directly related to other reproductive traits. The most important relationships are the trade-offs between fecundity and egg size, the relationship between fecundity and female size and age, that of fecundity and age at first maturity, and that of fecundity and reproductive effort. One of the main trade-offs in reproductive biology is between number and size of eggs produced. Given a certain allocation of energy to reproduction, there is a negative correlation between fecundity and the size of the eggs produced. The egg size is related to larval developmental type and time, and in the last three decades several models have been proposed to explain the evolution of egg size and fecundity in invertebrates (Vance, 1973a, b; Smith and Fretwell, 1974; Levitan, 1993; Podolsky and Strathmann, 1996). There are also design barriers that need to be taken into consideration, because these can affect the evolution of egg size and fecundity in a species.

The relationship between fecundity and female size and age varies among species, but these two factors can affect the number of eggs produced. In most species, larger females have higher fecundities, although in some organisms, fecundity is more closely related to age than size.

Fecundity is also closely dependent on age at first maturity. The demography theory explains the evolution of fecundity traits based on the balance between costs and benefits in the size at first maturity.

Early-maturing species have a higher probability of reaching maturity but have a reduction in future fecundity, and species with delayed maturity live longer, allowing for a higher lifetime fecundity.

The reproductive output of a species is defined by two components, the number of eggs (fecundity) and the size of eggs (quality). The allocation of resources to growth and reproduction has to be optimized over the lifespan of the organism, and there is a trade-off between current reproductive output and subsequent adult reproductive success and survival. Ultimately, it is lifetime fecundity that is optimized in the evolution of life-history strategies.

Although the gametogenic patterns of a species are phylogenetically constrained, fecundity is not a conservative character and varies with environmental factors such as habitat stability or food quantity and quality. There is strong evidence in many species that a higher food availability or higher food quality enhances the production of more and/or higher quality eggs. Similarly, a starvation period or low food availability or quality can decrease or even stop the production of eggs.

The effects of environmental factors such as temperature or food availability and quality are reflected in the different fecundity patterns found in different environments. Generally, there is a decrease in fecundity with increasing latitude and increasing depth, often related to less favourable environments with lower energy availability. Some Antarctic and deep-water species produce larger eggs that hatch into more advanced larvae, allowing for a higher survival probability of the larvae in a less favourable environment.

8. CONCLUSIONS

The study of life-history strategies in marine invertebrates, and of the selective pressures and constraints shaping their evolution, are of great value in understanding the dynamics of an ecosystem. This is essential for an adequate management of resources and conservation of still poorly-known ecosystems such as the deep sea, where anthropogenic pressure (i.e. fishing, oil exploitation) is increasing rapidly. It is possible to cause a lethal disturbance to a species by disruption of its reproductive biology by sublethal factors. Even when the adult organism can survive adverse conditions, these conditions may affect the reproductive success of the individuals, and the long-term effect on the species could be extinction (Tyler and Ramirez Llodra, in press). Understanding the factors that affect the plasticity of reproductive traits and the evolution of life histories is therefore crucial.

ACKNOWLEDGEMENTS

I would like to thank Professor Paul Tyler, Dr David Billett and Dr Maria Baker for discussions and helpful suggestions on an earlier version of the manuscript. The comments of Dr M. Bentley, Alan and Eve Southward and an anonymous referee were much appreciated and helped to improve the manuscript. The author was supported by a European Marie Curie fellowship (ERB4001GT980157) and a University of Southampton studentship, which are gratefully acknowledged.

REFERENCES

Adiyodi, R. G. and Subramoniam, T. (1983). Arthropoda – Crustacea. *In* "Reproductive Biology of Invertebrates: Oogenesis, Oviposition and Oosorption" (K. G. Adiyodi and R. G. Adiyodi, eds), pp. 443–495. John Wiley & Sons, Chichester.

Anger, K. and Moreira, G. S. (1998). Morphometric and reproductive traits of tropical caridean shrimps. *Journal of Crustacean Biology* **18**, 823–838.

Arnold, J. M. and Williams-Arnold, L. (1977). Cephalopoda: Decapoda. Chapter 5. *In* "Reproduction of Marine Invertebrates, Vol. IV. Molluscs: Gastropods and Cephalopods" (A. C. Giese and J. S. Pearse, eds), pp. 243–290. Academic Press, New York.

Baker, M. C. (2001). Fertilisation kinetics in marine invertebrates. PhD thesis, School of Ocean and Earth Science, University of Southampton, Southampton, UK.

Baker, M. C. and Tyler, P. A. (2001). Fertilisation success in the commercial gastropod *Haliotis tuberculata*. *Marine Ecology Progress Series* **211**, 205–213.

Barber, B. J., Getchell, R., Shumway, S. and Schick, D. (1988). Reduced fecundity in a deep-water population of the giant scallop *Placopecten magellanicus* in the Gulf of Maine, USA. *Marine Ecology Progress Series* **42**, 207–212.

Barnes, H. and Barnes, M. (1968). Egg number, metabolic efficiency of egg production and fecundity, local and regional variations in a number of cirripeds. *Journal of Experimental Marine Biology and Ecology* **2**, 135–153.

Bayne, B. L. and Worrall, C. M. (1980). Growth and production of mussels *Mytilus edulis* from two populations. *Marine Ecology Progress Series* **3**, 317–328.

Bayne, B. L., Salkeld, P. N. and Worrall, C. M. (1983). Reproductive effort and value in different populations of the marine mussel *Mytilus edulis* L. *Oecologia* **59**, 19–26.

Beckevar, N. (1981). Cultivation, spawning and growth of the giant clams *Tridacna gigas*, *T. derasa*, *T. squamosa* in Palau, Caroline Islands. *Aquaculture* **24**, 21–30.

Beddingfield, S. D. and McClintock, J. B. (1998). Differential survivorship, reproduction, growth and nutrient allocation in the regular echinoid *Lytechinus variegatus* (Lamarck) fed natural diets. *Journal of Experimental Marine Biology and Ecology* **226**, 195–215.

Beerman, R. D. (1977). Gastropoda: Opistobranchia. *In* "Reproduction of Marine Invertebrates. Vol. IV. Molluscs: Gastropods and Cephalopoda" (A. C. Giese and J. S. Pearse, eds), pp. 115–180. Academic Press, New York.

Bell, G. (1980). The costs of reproduction and their consequences. *American Naturalist* **116**, 45–76.

Bell, M. C. and Fish, J. D. (1996). Fecundity and seasonal changes in reproductive output of females of the gravel beach amphipod, *Pectenogammarus planicrurus*. *Journal of the Marine Biological Association of the United Kingdom* **76**, 37–55.

Bentley, M. G. and Pacey, A. A. (1992). Physiological and environmental control of reproduction in polychaetes. *Oceanography and Marine Biology, an Annual Review* **30**, 443–481.

Bentley, M. G., Olive, P. J. W., Garwood, P. R. and Wright, N. H. (1984). The spawning and spawning mechanism of *Nephtys caeca* (Fabricius, 1780) and *Nephtys hombergii* Savigny, 1818 (Annelida: Polychaeta). *Sarsia* **69**, 63–68.

Benton, T. G. and Grant, A. (1996). How to keep fit in the real world: elasticity analyses and selection pressures on life histories in a variable environment. *American Naturalist* **147**, 115–139.

Berrill, N. J. (1975). Tunicata. *In* "Reproduction of Marine Invertebrates, Vol. II. Entoprocts and Lesser Coelomates" (A. C. Giese and J. S. Pearse, eds), pp. 241–282. Academic Press, New York.

Berry, A. J. (1977). Gastropoda: Pulmonata. *In* "Reproduction of Marine Invertebrates, Vol. IV. Molluscs: Gastropods and Cephalopods" (A. C. Giese and J. S. Pearse, eds), pp. 181–226. Academic Press, New York.

Bertram, D. F. and Strathmann, R. R. (1998). Effects of maternal and larval nutrition on growth and form of planktotrophic larvae. *Ecology* **79**, 315–327.

Bosch, I. (1989). Contrasting modes of reproduction in two Antarctic asteroids of the genus *Porania*, with a description of unusual feeding and non-feeding larval types. *Biological Bulletin* **177**, 77–82.

Bosch, I. and Slattery, M. (1999). Costs of extended brood protection in the Antarctic sea star, *Neosmilaster georgianus* (Echinodermata: Asteroidea). *Marine Biology* **134**, 449–459.

Brewin, P. E., Lamare, M. D., Keogh, J. A. and Mladenov, P. V. (2000). Reproductive variability over a four-year period in the sea urchin *Evechinus chloroticus* (Echinoidea: Echinodermata) from differing habitats in New Zealand. *Marine Biology* **137**, 543–557.

Brey, T., Pearse, J., Basch, L., McClintock, J. and Slattery, M. (1995). Growth and production of *Sterechinus neumayeri* (Echinoidea; Echinodermata) in McMurdo Sound, Antarctica. *Marine Biology* **124**, 279–292.

Bridges, T. S. (1996). Effects of organic additions to sediment, and maternal age and size, on patterns of offspring investment and performance in two opportunistic deposit-feeding polychaetes. *Marine Biology* **125**, 345–357.

Bridges, T. S., Levin, L. A., Cabrera, D. and Plaia, G. (1994). Effects of sediment amended with sewage, algae, or hydrocarbons on growth and reproduction in two opportunistic polychaetes. *Journal of Experimental Marine Biology and Ecology* **177**, 99–119.

Byrne, M. (1991). Reproduction, development and population biology of the Caribbean ophiuroid *Ophionereis olivacea*, a protandric hermaphrodite that broods its young. *Marine Biology* **111**, 387–399.

Calvo, M., Templado, J. and Penchaszadeh, P. E. (1998). Reproductive biology of the gregarious Mediterranean vermetid gastropod *Dendropoma petraeum*. *Journal of the Marine Biological Association of the United Kingdom* **78**, 525–549.

Campos-Creasey, L. S., Tyler, P. A., Gage, J. D. and John, A. W. G. (1994). Evidence for coupling the vertical flux of phytodetritus to the diet and seasonal life history of the deep-sea echinoid *Echinus affinis*. *Deep-Sea Research* **41**, 369–388.

Cassai, C. and Prevedelli, D. (1999). Fecundity and reproductive effort in *Ophryotrocha labronica* (Polychaeta: Dorvilleidae). *Marine Biology* **133**, 489–494.

Caswell, H. (1980a). On the equivalence of maximizing reproductive value and maximizing fitness. *Ecology* **61**, 19–24.

Caswell, H. (1980b). The evolution of "mixed" life histories in marine invertebrates and elsewhere. *American Naturalist* **117**, 529–536.

Caswell, H. (1983). Phenotypic plasticity in life-history traits: demographic effects and evolutionary consequences. *American Zoologist* **23**, 35–46.

Charlesworth, B., (1980). "Evolution in Age-structured Populations". Cambridge University Press, Cambridge.

Charlesworth, B. (1990). Life and times of the guppy. *Nature* **346**, 313–321.

Chester, C. M. (1996). The effect of adult nutrition on the reproduction and development of the estuarine nudibranch, *Tenellia adspersa* (Nordmann, 1845). *Journal of Experimental Marine Biology and Ecology* **198**, 113–130.

Cheung, S. G. and Lam, S. (1999). Effect of food availability on egg production and packaging in the intertidal scavenging gastropod *Nassarius festivus*. *Marine Biology* **135**, 281–287.

Chia, F.-S. and Walker, C. W. (1991). Echinodermata: Asteroidea. *In* "Reproduction of Marine Invertebrates. Vol. VI. Echinoderms and Lophophorates" (A. C. Giese, J. S. Pearse and V. B. Pearse, eds) pp. 301–353. Boxwood Press, California.

Chia, F.-S., Oguro, C. and Komatsu, M. (1993). Sea-star (Asteroid) development. *Oceanography and Marine Biology: Annual Review* **31**, 223–257.

Chia, F.-S., Gibson, G. and Qian, P.-Y. (1996). Poecilogony as a reproductive strategy of marine invertebrates. *Oceanologica Acta* **19**, 203–208.

Chornesky, E. A. and Peters, E. C. (1987). Sexual reproduction and colony growth in the scleractinian coral *Porites astreoides*. *Biological Bulletin* **172**, 161–177.

Christiansen, F. B. and Fenchel, T. M. (1979). Evolution of marine invertebrate reproductive patterns. *Theoretical Population Biology* **16**, 267–282.

Clark, S. (1988). A two phase photoperiodic response controlling the annual gametogenic cycle in *Harmothoe imbricata* (L.) (Polychaeta: Polyonidae). *Invertebrate Reproduction and Development* **14**, 245–266.

Clarke, A. (1979). On living in cold water: K-strategies in Antarctic Benthos. *Marine Biology* **55**, 111–119.

Clarke, A. (1987). Temperature, latitude and reproductive effort. *Marine Ecology Progress Series* **38**, 89–99.

Clarke, A. (1993a). Egg size and egg composition in polar shrimps (Caridea: Decapoda). *Journal of Experimental Marine Biology and Ecology* **168**, 189–203.

Clarke, A. (1993b). Reproductive trade-offs in caridean shrimps. *Functional Ecology* **7**, 411–419.

Clarke, A. and Gore, D. J. (1992). Egg size and composition in *Ceratoserolis* (Crustacea: Isopoda) from the Weddell Sea. *Polar Biology* **12**, 129–134.

Clarke, A., Hopkins, G. C. E. and Nilssen, E. M. (1991). Egg size and reproductive output in the deep-water prawn *Pandalus borealis* Kroyer, 1838. *Functional Ecology* **5**, 724–730.

Cohen, D. (1968). A general model of optimal reproduction in a randomly varying environment. *Journal of Ecology* **56**, 219–228.

Company, J. B. and Sardà, F. (1997). Reproductive patterns and population characteristics in five deep-water pandalid shrimps in the Western Mediterranean along a depth gradient (150–1100 m). *Marine Ecology Progress Series* **148**, 49–58.

Corey, S. (1987). Comparative fecundity of four crayfish in SW Ontario, Canada (Decapoda, Astacidea). *Crustaceana* **52**, 276–286.

Corey, S. and Reid, D. M. (1991). Comparative fecundity of decapod crustaceans. I. The fecundity of thirty-three species of nine families of Caridean shrimp. *Crustaceana* **60**, 270–294.

Crump, R. G. and Emson, R. H. (1978). Some aspects of the population dynamics of *Asterina gibbosa* (Asteroidea). *Journal of the Marine Biological Association of the United Kingdom* **58**, 451–466.

Cruz Rivera, E. and Hay, M. E. (2000). Can quantity replace quality? Food choice, compensatory feeding, and fitness of marine mesograzers. *Ecology* **81**, 201–219.

Culling, C. F. A. (1974). "Handbook of Histopathological and Biochemical Techniques". Butterworth, London.

Dyrynda, P. E. J. and Ryland, J. S. (1982). Reproductive strategies and life histories in the cheilostome marine bryozoans *Chartella papyracea* and *Bugula flabellata*. *Marine Biology* **71**, 241–256.

Ebert, T. A. (1996). The consequences of broadcasting, brooding and sexual reproduction in echinoderm metapopulations. *Oceanologica Acta* **19**, 217–226.

Eckelbarger, K. J. (1983). Evolutionary radiation in polychaete ovaries and vitellogenic mechanisms: their possible role in life-history patterns. *Canadian Journal of Zoology* **61**, 487–504.

Eckelbarger, K. J. (1986). Vitellogenic mechanisms and the allocation of energy to offspring in polychaetes. *Bulletin of Marine Science* **39**, 426–443.

Eckelbarger, K. J. (1994). Diversity of metazoan ovaries and vitellogenic mechanisms: implications for life history theory. *Proceedings of the Biological Society of Washington* **107**, 193–218.

Eckelbarger, K. J. and Larson, R. L. (1988). Ovarian morphology and oogenesis in *Aurelia aurita* (Scyphozoa: Semaeostomae): ultrastructural evidence of heterosynthetic yolk formation in a primitive metazoan. *Marine Biology* **100**, 103–115.

Eckelbarger, K. J. and Watling, L. (1995). Role of phylogenetic constraints in determining reproductive patterns in deep-sea invertebrates. *Invertebrate Biology* **114**, 256–269.

Einum, S. and Fleming, I. A. (2000). Highly fecund mothers sacrifice offspring survival to maximize fitness. *Nature* **405**, 565–567.

Emson, R. H. and Crump, R. G. (1979). Description of a new species of *Asterina* (Asteroidea) with an account of its ecology. *Journal of the Marine Biological Association of the United Kingdom* **59**, 77–94.

Fong, P. P. and Pearse, J. S. (1992a). Photoperiodic regulation of parturition in the self fertilising viviparous polychaete *Neanthes limnicola* from central California. *Marine Biology* **112**, 81–89.

Fong, P. P. and Pearse, J. S. (1992b). Evidence for a programmed circannual life cycle modulated by increasing daylengths in *Neanthes limnicola* (Polychaeta: Nereidae) from central California. *Biological Bulletin* **182**, 289–297.

Fong, P. P., Deguchi, R. and Kyozoka, K. (1996). Serotonergic ligands induce spawning but not oocyte maturation in the bivalve *Mactra chinensis* from Central Japan. *Biological Bulletin* **191**, 27–32.

Fonseca-Larios, A. E. and Briones-Fourzan, P. (1998). Fecundity of the spiny lobster *Panulirus argus* (Latreille, 1804) in the Caribbean coast of Mexico. *Bulletin of Marine Science* **63**, 21–32.

Gadgil, M. and Bossert, W. H. (1970). Life historical consequences of natural selection. *American Naturalist* **104**, 1–24.

Gage, J. D. and Tyler, P. A. (1991). "Deep-sea Biology. A Natural History of Organisms at the Deep-sea Floor". Cambridge University Press, Cambridge.

Garwood, P. R. (1980). The role of temperature and daylength in the control of the reproductive cycle of *Harmothoe imbricata* (L.) (Polychaeta: Polynoidae). *Journal of Experimental Marine Biology and Ecology* **47**, 35–53.

Garwood, P. R. and Olive, P. J. W. (1982). The influence of photoperiod on oocyte growth and its role in the control of the reproductive cycle in the polychaete *Harmothoe imbricata* (L.). *International Journal of Invertebrate Reproduction* **5**, 161–165.

George, S. B. (1994a). Population differences in maternal size and offspring quality for *Leptasterias epichlora* (Brandt) (Echinodermata: Asteroidea). *Journal of Experimental Marine Biology and Ecology* **175**, 121–131.

George, S. B. (1994b). The *Leptasterias* (Echinodermata: Asteroidea) species complex: variation in reproductive investment. *Marine Ecology Progress Series* **109**, 95–98.

George, S. B., Cellario, C. and Fénaux, L. (1990). Population differences in egg quality of *Arbacia lixula* (Echinodermata: Echinoidea): proximate composition of eggs and larval development. *Journal of Experimental Marine Biology and Ecology* **141**, 107–118.

Giangrande, A., Geraci, S. and Belmonte, G. (1994). Life-cycle and life-history diversity in marine invertebrates and the implications in community dynamics. *Oceanography and Marine Biology, an Annual Review* **32**, 305–333.

Giese, A. C. and Pearse, J. S. (1974). Introduction: general principles. *In* "Reproduction of Marine Invertebrates. Vol. I" (A. C. Giese and J. S. Pearse, eds), pp. 546. Academic Press, London.

Gorny, M., Arntz, W. E., Clarke, A. and Gore, D. J. (1992). Reproductive biology of caridean decapods from the Weddell Sea. *Polar Biology* **12**, 111–120.

Gremare, A., Marsh, A. G. and Tenore, K. R. (1988). Short-term reproductive responses of *Capitella* sp. I (Annelida: Polychaeta) fed on different diets. *Journal of Experimental Marine Biology and Ecology* **123**, 147–162.

Griffiths, C. L. and King, J. A. (1979). Energy expended on growth and gonad output in the ribbed mussel *Aulacomya ater*. *Marine Biology* **53**, 217–222.

Gutt, J. (1991). Investigations on brood protection in *Psolus dubiosus* (Echinodermata: Holothuroidea) from Antarctica in spring and autumn. *Marine Biology* **111**, 281–286.

Hadfield, M. G. and Strathmann, M. F. (1996). Variability, flexibility and plasticity in life histories of marine invertebrates. *Oceanologica Acta* **19**, 323–334.

Hamel, J.-F., Conand, Ch., Pawson, D. L. and Mercier A. (2001). The sea cucumber *Holothuria scabra* (Holothuroidea: Echinodermata): its biology and exploitation as Beche-de-Mer. *Advances in Marine Biology* **41**, 131–226.

Hannah, R. W., Jones, S. A. and Long, M. R. (1995). Fecundity of the ocean shrimp (*Pandalus jordani*). *Canadian Journal of Fisheries and Aquatic Sciences* **52**, 2098–2107.

Hansen, B. (1968). Brood-protection in a deep-sea holothurian, *Oneirophanta mutabilis* Theel. *Nature* **217**, 1062–1063.

Hardege, J. D. and Bentley, M. G. (1997). Spawning synchrony in *Arenicola marina*: evidence for sex pheromonal control. *Proceedings of the Royal Society of London* B **264**, 1041–1047.

Hardege, J. D., Bentley, M. G., Beckmann, M. and Müller, C. (1996). Sex pheromones in marine polychaetes: volatile organic substances (VOS) isolated from *Arenicola marina*. *Marine Ecology Progress Series* **139**, 157–166.

Hartnoll, R. G., Rice, A. L. and Attrill, M. J. (1992). Aspects of the biology of the galatheid genus *Munida* (Crustacea, Decapoda) from the Porcupine Seabight, Northeast Atlantic. *Sarsia* **76**, 231–246.

Havenhand, J. N. and Todd, C. D. (1988). Physiological ecology of *Adalaria proxima* (Alder et Hancock) and *Onchidoris muricata* (Müller) (Gastropoda: Nudibranchia). II. Reproduction. *Journal of Experimental Marine Biology and Ecology* **118**, 173–189.

Hayashi, I. (1980). Structure and growth of a shore population of the omer *Haliotis tuberculata*. *Journal of the Marine Biological Association of the United Kingdom* **60**, 431–437.

Hendler, G. (1991). Echinodermata: Ophiuroidea. *In* "Reproduction of Marine Invertebrates, Vol. VI. Echinoderms and Lophophorates" (A. C. Giese, J. S. Pearse and V. B. Pearse, eds), pp. 356–513. Boxwood Press, California.

Herring, P. (1967). Observations on the early larvae of three species of *Acanthephyra* (Crustacea, Decapoda, Caridea). *Deep-Sea Research* **14**, 325–329.

Herring, P. J. (1974a). Observations on the embryonic development of some deep-living decapod crustaceans, with particular reference to species of *Acanthephyra*. *Marine Biology* **25**, 25–33.

Herring, P. J. (1974b). Size, density and lipid content of some decapod eggs. *Deep-Sea Research* **21**, 91–94.

Hines, A. H. (1982). Allometric constraints and variables of reproductive effort in Brachyuran crabs. *Marine Biology* **69**, 309–320.

Hines, A. H. (1986). Larval problems and perspectives in life histories of marine invertebrates. *Bulletin of Marine Science* **39**, 506–525.

Hines, A. H. (1988). Fecundity and reproductive output in two species of deep-sea crabs, *Geryon fenneri* and *G. quinquedens* (Decapoda: Brachyura). *Journal of Crustacean Biology* **8**, 557–562.

Hirai, S., Kishimoto, T., Kadam, L., Kanatani, H. and Koide, S. S. (1988). Induction of spawning and oocyte maturation by 5-hydroxytryptamine in the surf clam. *Journal of Experimental Zoology* **245**, 318–321.

Hoegh-Guldberg, O. and Emlet, R. B. (1997). Energy use during the development of a lecithotrophic echinoid. *Biological Bulletin* **192**, 27–40.

Honkoop, P. J. C. and Van der Meer, J. (1997). Reproductive output of *Macoma balthica* populations in relation to winter-temperature and intertidal-height mediated changes of body mass. *Marine Ecology Progress Series* **149**, 155–162.

Honkoop, P. J. C., Van der Meer, J., Beukema, J. J. and Kwast, D. (1998). Does temperature-influenced egg production predict the recruitment in the bivalve *Macoma balthica*? *Marine Ecology Progress Series* **164**, 229–235.

Hopper, D. R., Hunter, C. L. and Richmond, R. H. (1998). Sexual reproduction of the tropical sea cucumber, *Actinopyga mauritiana* (Echinodermata: Holothuroidea), in Guam. *Bulletin of Marine Science* **63**, 1–9.

Ikegami, S., Honji, N. and Yoshida, M. (1978). Light-controlled production of spawning-inducing substance in jelly fish ovary. *Nature* **272**, 611–612.

Ivanova, M. B. and Vassilenko, S. V. (1987). Relationship between number of eggs, brood weight and female body weight in Crustacea. *International Revue der Gesamte Hydrobiologie* **72**, 147–169.

Jaeckle, W. B. (1995). Variation in the size, energy content and biochemical composition of invertebrate eggs: correlates to the mode of larval development. *In* "Ecology of Marine Invertebrate Larvae" (L. R. McEdward, ed.), pp. 49–77. CRC Press, Boca Raton, Florida.

Jónasdóttir, S. H. (1994). Effect of food quality on the reproductive success of *Acartia tonsa* and *Acartia hudsonica*: laboratory observations. *Marine Biology* **121**, 67–81.

Juneja, R. and Koide, S. S. (1996). Biochemical pathways involved in serotonin-regulated *Spisula* oocyte maturation and fertilisation. *Invertebrate Reproduction and Development* **30**, 47–53.

Kanatani, H. (1975). Maturation-inducing substances in asteroid and echinoid oocytes. *American Zoologist* **15**, 493–505.

Kautsky, N. (1982). Quantitative studies on gonad cycle, fecundity, reproductive output and recruitment in Baltic *Mytilus edulis* populations. *Marine Biology* **68**, 143–160.

King, M. G. and Butler, A. J. (1985). Relationship of life-history patterns in deep-water caridean shrimps (Crustacea: Natantia). *Marine Biology* **86**, 129–138.

Kleppel, G. S. and Hazzard, S. E. (2000). Diet and egg production of the copepod *Acartia tonsa* in Florida Bay. II. Role of the nutritional environment. *Marine Biology* **137**, 111–121.

Knight-Jones, P. and Bowden, N. (1984). Incubation and scissiparity in sabellidae (Polychaeta). *Journal of the Marine Biological Association of the United Kingdom* **64**, 809–818.

Kojis, B. L. and Quinn, N. J. (1981). Aspects of sexual reproduction and larval development in the shallow water hermatypic coral *Goniastrea australensis* (Edwards and Haine 1857). *Bulletin of Marine Science* **31**, 558–573.

Komatsu, M., Kano, Y. T., Yoshizacua, H., Akabane, S. and Oguro, C. (1979). Reproductive development of the hermaphroditic sea-star *Asterina minor* Hayashi. *Biological Bulletin* **157**, 258–274.

Kosobokova, K. N. (1992). Experimental study of the fecundity of the Antarctic copepod *Calanus propinquus*. *Oceanology* **32**, 89–93.

Krol, R. M., Hawkins, W. E. and Overstreet, R. M. (1992). Reproductive components. *In* "Microscopic Anatomy of Invertebrates, Vol. 10: Decapod Crustacea" (F. W. Harrison and A. G. Humes, eds), pp. 295–343. Wiley-Liss, New York.

Krug, P. J. (1998). Poecilogony in an estuarine opisthobranch: planktotrophy, lecithotrophy, and mixed clutches in a population of the ascoglossan *Alderia modesta*. *Marine Biology* **132**, 483–494.

Laabir, M., Poulet, S. A. and Ianora, A. (1995). Measuring production and viability of eggs in *Calanus helgolandicus*. *Journal of Plankton Research* **17**, 1125–1142.

Lalli, C. M. and Wells, F. E. Jr. (1973). Brood protection in an epipelagic thecosomatous pteropod, *Spiratella* ("*Limacina*") *inflata* (D'Orbigny). *Bulletin of Marine Science* **23**, 933–941.

Langton, R. W., Robinson, W. E. and Schick, D. (1987). Fecundity and reproductive effort of sea scallops *Placopecten magellanicus* from the Gulf of Maine. *Marine Ecology Progress Series* **37**, 19–25.

Lardies, M. A. and Wehrtmann, I. S. (1997). Egg production in *Betaeus emarginatus* (H. Milne Edwards, 1837) (Decapoda: Alpheidae): fecundity, reproductive output and chemical composition of eggs. *Ophelia* **46**, 165–174.

Last, K. and Olive, P. J. W. (1999). Photoperiodic control of growth and segment proliferation by *Nereis* (*Neanthes*) *virens* Sars in relation to real time and state of maturity. *Marine Biology* **134**, 191–200.

Levin, L. A. (1984). Multiple patterns of development in *Streblospio benedictii* Webster (Sipionidae) from three coasts of North America. *Biological Bulletin* **166**, 494–508.

Levin, L. A. and Bridges, T. S. (1994). Control and consequences of alternative developmental modes in a poecilogonous polychaete. *American Zoologist* **34**, 323–332.

Levin, L. A. and Bridges, T. S. (1995). Pattern and diversity in reproduction and development. *In* "Ecology of Marine Invertebrate Larvae" (L. R. McEdward, ed.), pp. 1–48. CRC Press, Boca Raton, Florida.

Levin, L. A. and Creed, E. L. (1986). Effect of temperature and food availability on reproductive responses of *Streblospio benedictii* (Polychaeta: Spionidae) with planktotrophic or lecithotrophic development. *Marine Biology* **92**, 103–113.

Levin, L. A., Caswell, H., DePatra, K. D. and Creed, E. L. (1987). Demographic consequences of larval development mode: planktotrophy vs. lecithotrophy in *Streblospio benedictii*. *Ecology* **68**, 1877–1886.

Levin, L. A., Zhu, J. and Creed, E. (1991). The genetic basis of life-history characters in a polychaete exhibiting planktotrophy and lecithotrophy. *Evolution* **45**, 380–397.

Levin, L. A., Plaia, G. R. and Huggett, C. L. (1994). The influence of natural organic enhancement on life histories and community structure of bathyal polychaetes. *In* "Reproduction, Larval Biology, and Recruitment of the Deep-sea Benthos" (C. M. Young and K. J. Eckelbarger, eds), pp. 261–283. Columbia University Press, New York.

Levin, L., Caswell, H., Bridges, T., DiBacco, C., Cabrera, D. and Plaia, G. (1996). Demographic responses of estuarine polychaetes to pollutants: life table response experiments. *Ecology Applied* **6**, 1295–1313.

Levitan, D. R. (1993). The importance of sperm limitation to the evolution of egg size in marine invertebrates. *American Naturalist* **141**, 517–536.

Levitan, D. R. (1996). Predicting optimal and unique egg sizes in free-spawning marine invertebrates. *American Naturalist* **48**, 174–188.

Linton, D. L. and Taghan, G. L. (2000). Feeding, growth and fecundity of *Capitella* sp. I in relation to sediment organic concentration. *Marine Ecology Progress Series* **205**, 229–240.

Lonsdale, D. J. and Levinton, J. S. (1986). Growth rate and reproductive differences in a widespread estuarine harpacticoid copepod (*Scottolana canadensis*). *Marine Biology* **91**, 231–237.

Lonsdale, D. J. and Levinton, J. S. (1989). Energy budgets of latitudinally separated *Scottolana canadensis* (Copepoda: Harpacticoida). *Limnology and Oceanography* **34**, 324–331.

MacArthur, R. H. and Wilson, E. O. (1967). "Theory of Island Biogeography". Princeton University Press, Princeton.

McCann, K. and Shuter, B. (1997). Bioenergetics of life history strategies and the comparative allometry of reproduction. *Canadian Journal of Fisheries and Aquatic Science* **54**, 1289–1298.

McClary, D. J. and Mladenov, P. V. (1989). Reproductive pattern in the brooding and broadcasting sea star *Pteraster militaris*. *Marine Biology* **103**, 531–540.

MacDonald, B. A. and Thompson, R. J. (1985a). Influence of temperature and food availability on the ecological energetics of the giant scallop *Placopecten magellanicus*. I. Growth rates of shell and somatic tissue. *Marine Ecology Progress Series* **25**, 279–294.

MacDonald, B. A. and Thompson, R. J. (1985b). Influence of temperature and food availability on the ecological energetics of the giant scallop *Placopecten magellanicus*. II. Reproductive output and total production. *Marine Ecology Progress Series* **25**, 295–303.

McEdward, L. R. (1997). Reproductive strategies of marine benthic invertebrates revisited: facultative feeding by planktotrophic larvae. *American Naturalist* **150**, 48–72.

McEdward, L. R. and Carson, S. F. (1987). Variation in egg organic content and its relationship with egg size in the starfish *Solaster stimpsoni*. *Marine Ecology Progress Series* **37**, 159–169.

McEdward, L. R. and Chia, F. S. (1991). Size and energy content of eggs from echinoderms with pelagic lecithotrophic development. *Journal of Experimental Marine Biology and Ecology* **147**, 95–102.

McEdward, L. R. and Colter, L. K. (1987). Egg volume and energy content are not correlated among sibling offspring of starfish: implications for life-history theory. *Evolution* **41**, 914–917.

McGinley, M. A., Temme, D. H. and Geber, M. A. (1987). Parental investment in offspring in variable environments: theoretical and empirical considerations. *American Naturalist* **130**, 370–398.

McHugh, D. (1993). A comparative study of reproduction and development in the polychaete family Terebellidae. *Biological Bulletin* **185**, 153–167.

McHugh, D. and Rouse, G. W. (1998). Life history evolution of marine invertebrates: new views from phylogenetic systematics. *TREE* **14**, 182–186.

Mauchline, J. (1988). Egg and brood sizes of oceanic pelagic crustaceans. *Marine Ecology Progress Series* **43**, 251–258.

Mauro, N. A. (1975). The premetamorphic developmental rate of *Phragmatopoma lapidosa* Kinberg, 1867, compared with that in temperate sabellariids (Polychaeta: Sabellariidae). *Bulletin of Marine Science* **25**, 387–392.

Medeiros-Bergen, D. E. and Ebert, T. A. (1995). Growth, fecundity and mortality rates of two intertidal brittlestars (Echinodermata: Ophiuroidea) with contrasting modes of development. *Journal of Experimental Marine Biology and Ecology* **189**, 47–64.

Meidel, S. K. and Scheibling, R. E. (1998). Annual reproductive cycle of the green sea urchin, *Strongylocentrotus droebachiensis*, in differing habitats in Nova Scotia, Canada. *Marine Biology* **131**, 461–478.

Mendez, N., Linke-Gamenick, I. and Forbes, V. E. (2000). Variability in reproductive mode and larval development within the *Capitella capitata* species complex. *Invertebrate Reproduction and Development* **38**, 131–142.

Menge, B. A. (1975). Brood or broadcast? The adaptive significance of different reproductive strategies in the two intertidal sea stars *Leptasterias hexactis* and *Pisaster ochraceus*. *Marine Biology* **31**, 87–100.

Monsalvo-Spencer, P., Maeda-Martinez, A. N. and Reynoso-Granados, T. (1997). Reproductive maturity and spawning induction in the Catarina scallop *Agropecten ventricosus* (= *circularis*) (Sowerby II, 1842). *Journal of Shellfish Research* **16**, 67–70.

Morgan, A. D. (2000). Induction of spawning in the sea cucumber *Holothuria scabra* (Echinodermata: Holothuroidea). *Journal of the World Aquaculture Society* **31**, 186–194.

Morse, A. (1977). Hydrogen peroxide induces spawning in mollusks, with activation of prostaglandin endoperoxide synthetase. *Science* **196**, 298–300.

Mortensen, T., (1927). "Handbook of the Echinoderms of the British Isles". Oxford University Press, Oxford.

Moss, G. A., Illingworth, J. and Tong, L. J. (1995). Comparing two simple methods to induce spawning in the New Zealand abalone (paua), *Haliotis iris. New Zealand Journal of Marine and Freshwater Research* **29**, 329–333.

Murphy, G. I. (1968). Pattern in life history and the environment. *American Naturalist* **102**, 391–403.

Nakaoka, M. (1994). Size-dependent reproductive traits of *Yoldia notabilis* (Bivalvia: Protobranchia). *Marine Ecology Progress Series* **114**, 129–137.

Nelson, K. (1991). Scheduling of reproduction in relation to moulting and growth in malacostracan crustaceans. *Crustacean Issues* **7**, 77–113.

Nichols, D., Bishop, G. M. and Sime, A. T. T. (1985). Reproductive and nutritional periodicities in populations of the European sea urchin *Echinus esculentus* (Echinodermata: Echinoidea) from the English Channel. *Journal of the Marine Biological Association of the United Kingdom* **65**, 203–220.

Ohtomi, J. (1997). Reproductive biology and growth of the deep-water pandalid shrimp *Plesionika semilaevis* (Decapoda: Caridea). *Journal of Crustacean Biology* **17**, 81–89.

Olive, P. J. W. (1980). Control of the reproductive cycle in female *Eulalia viridis* (Polychaeta: Phyllodocidae). *Journal of the Marine Biological Association of the United Kingdom* **61**, 941–958.

Olive, P. J. W. (1985). Physiological adaptations and the concepts of optimal reproductive strategy and physiological constraint in marine invertebrates. *In* "Physiological Adaptation of Marine Invertebrates" (M. S. Laverack, ed.), pp. 268–300. Society for Experimental Biology, Cambridge.

Olive, P. J. W. and Bentley, M. G. (1980). Hormonal control of oogenesis, ovulation and spawning in the annual reproductive cycle of the polychaete *Nephtys hombergi* Sav. (Nephtyidae). *International Journal of Invertebrate Reproduction* **2**, 205–221.

Olive, P. J. W. and Morgan, P. J. (1991). The reproductive cycles of four British intertidal *Nephtys* species in relation to their geographical distribution (Polychaeta: Nephtyidae). *Ophelia* **5**, 351–361.

Olive, P. J. W., Porter, J. S., Sandeman, N. J., Wright, N. H. and Bentley, M. G. (1997). Variable spawning success of *Nephtys hombergi* (Annelida: Polychaeta) in response to environmental variation: a life history homeostasis? *Journal of Experimental Marine Biology and Ecology* **215**, 247–268.

Olive, P. J. W., Rees, S. W. and Djunaedi, A. (1998). The influence of photoperiod and temperature on oocyte growth in the semelparous polychaete *Nereis* (*Neanthes*) *virens* Sars. *Marine Ecology Progress Series* **172**, 169–183.

Olive, P. J. W., Lewis, C. and Beardall, V. (2000). Fitness components of seasonal reproduction: an analysis using *Nereis virens* as a life history model. *Oceanologica Acta* **23**, 377–389.

Omori, M. (1974). The biology of pelagic shrimps in the ocean. *Advances in Marine Biology* **12**, 233–324.

Orton, J. H. (1920). Sea temperature, breeding and distribution in marine animals. *Journal of the Marine Biological Association of the United Kingdom* **12**, 339–366.

Pain, S. L., Tyler, P. A. and Gage, J. D. (1982a). The reproductive biology of the deep-sea asteroids *Benthopecten simplex* (Perrier), *Pectinaster filholi* Perrier, and *Pontaster tenuispinus* Duben & Koren (Phanerozonia: Benthopectinidae) from the Rockall Trough. *Journal of Experimental Marine Biology and Ecology* **65**, 195–211.

Pain, S. L., Tyler, P. A. and Gage, J. D. (1982b). The reproductive biology of *Hymenaster membranaceus* from the Rockall Trough, North-East Atlantic Ocean, with notes on *H. gennaeus*. *Marine Biology* **70**, 41–50.

Painter, S. D., Chong, M. G., Wong, M. A., Gray, A., Cormier, J. G. and Nagle, G. T. (1991). Relative contributions of the egg layer and egg cordon to pheromonal attraction and the induction of mating and egg-laying behavior in *Aplysia*. *Biological Bulletin* **181**, 81–94.

Painter, S. D., Clough, B., Garden, R. W., Sweedler, J. V. and Nagle, G. T. (1998). Characterization of *Aplysia* attractin, the first water-borne peptide pheromone in invertebrates. *Biological Bulletin* **194**, 120–131.

Parker, N. R., Mladenov, P. V. and Grange, K. R. (1997). Reproductive biology of the antipatharian black coral *Antipathes fiordensis* in Doubtful Sound, Fiordland, New Zealand. *Marine Biology* **130**, 11–22.

Partridge, J. K. (1977). Studies on *Tapes decussatus* (L.) in Ireland. PhD thesis, Department of Zoology, University College of Galway and Shellfish Research Laboratory, Galway.

Paulet, Y. M. and Boucher, J. (1991). Is reproduction mainly regulated by temperature or photoperiod in *Pecten maximus*? *Invertebrate Reproduction and Development* **19**, 61–70.

Pearse, A. G. E. (1961). "Histochemistry: Theoretical and Applied". Churchill and Churchill, London.

Pearse, J. S. (1979). Polyplacophora. *In* "Reproduction of Marine Invertebrates". Vol. V (A. C. Giese and J. S. Pearse, eds), pp. 25–85. Academic Press, New York.

Pearse, J. S. and Cameron, A. R. (1991). Echinodermata: Echinoidea. *In* "Reproduction of Marine Invertebrates. Vol. VI. Echinoderms and Lophophorates" (A. C. Giese, J. S. Pearse and V. B. Pearse, eds), pp. 514–663. Boxwood Press, California.

Pianka, E. R. (1970). On "r" and "K" selection. *American Naturalist* **104**, 592–597.

Podolsky, R. D. and Strathmann, R. R. (1996). Evolution of egg size in free-spawners: consequences of the ferilization-fecundity trade-off. *American Naturalist* **148**, 160–173.

Pond, D., Harris, R., Head, R. and Harbour, D. (1996). Environmental and nutritional factors determining seasonal variability in the fecundity and egg viability of *Calanus helgolandicus* in coastal waters off Plymouth, UK. *Marine Ecology Progress Series* **143**, 45–63.

Poulin, E. and Feral, J.-P. (1995). Pattern of spatial distribution of a brood-protecting schizasterid echinoid, *Abatus cordatus*, endemic to the Kerguelen Islands. *Marine Ecology Progress Series* **118**, 179–186.

Prevedelli, D. and Simonini, R. (2000). Effects of salinity and two food regimes on survival, fecundity and sex ratio in two groups of *Dinophilus gyrociliatus* (Polychaeta: Dinophilidae). *Marine Biology* **137**, 23–29.

Prevedelli, D. and Zunarelli Vandini, R. (1998). Effects of diet on reproductive characteristics of *Ophryotrocha labronica* (Polychaeta: Dorvilleidae). *Marine Biology* **132**, 163–170.

Prevedelli, D. and Zunarelli Vandini, R. (1999). Survival, fecundity and sex ratio of *Dinophilus gyrociliatus* (Polychaeta: Dinophilidae) under different dietary conditions. *Marine Biology* **133**, 231–236.

Qian, P.-Y. and Chia, F.-S. (1991). Fecundity and egg size are mediated by food quality in the polychaete worm *Capitella* sp. *Journal of Experimental Marine Biology and Ecology* **148**, 11–25.

Qian, P.-Y. and Chia, F.-S. (1992a). Effect of ageing on reproduction in a marine polychaete *Capitella* sp. *Journal of Experimental Marine Biology and Ecology* **156**, 23–38.

Qian, P.-Y. and Chia, F.-S. (1992b). Effects of diet type on the demographics of *Capitella* sp. (Annelida: Polychaeta): lecithotrophic development vs. planktotrophic development. *Journal of Experimental Marine Biology and Ecology* **157**, 159–179.

Qian, P.-Y. and Chia, F.-S. (1994). *In situ* measurements of recruitment, mortality, growth, and fecundity of *Capitella* sp. (Annelida: Polychaeta). *Marine Ecology Progress Series* **111**, 53–62.

Ram, J. L., Crawford, G. W., Walker, J. U., Mojares, J. J., Patel, N., Fong, P. P. and Kyozuka, K. (1993). Spawning in the zebra mussel (*Dreissensia polymorpha*): activation by internal or external application of serotonin. *Journal of Experimental Zoology* **265**, 587–598.

Ramirez Llodra, E. (2000). Reproductive patterns of deep-sea invertebrates related to energy availability and phylogeny. PhD thesis, School of Ocean and Earth Science, University of Southampton, Southampton, UK.

Ramirez Llodra, E., Tyler, P. A. and Copley, J. T. P. (2000). Reproductive biology of three caridean shrimp, *Rimicaris exoculata*, *Chorocaris chacei* and *Mirocaris fortunata* (Caridea: Decapoda), from hydrothermal vents. *Journal of the Marine Biological Association of the United Kingdom* **80**, 473–484.

Ramirez Llodra, E., Tyler, P. A. and Billett, D. S. M. (2002). Reproductive biology of porcellansterid asteroids from three abyssal sites in the Northeast Atlantic with contrasting food input. *Marine Biology*, published online 8th February.

Razouls, S., Razouls, C. and Huntley, M. (1991). Development and expression of sexual maturity in female *Calanus pacificus* (Copepoda: Clanoidea) in relation to food quality. *Marine Biology* **110**, 65–74.

Reed, C. (1991). Bryozoan. Chapter 3. *In* "Reproduction of Marine Invertebrates, Vol. VI. Echinoderms and Lophophorates" (A. C. Giese, J. S. Pearse and V. B. Pearse, eds), pp. 86–246. Boxwood Press, California.

Reid, D. M. and Corey, S. (1991). Comparative fecundity of decapod crustaceans. II. The fecundity of fifteen species of Anomuran and Brachyuran crabs. *Crustaceana* **61**, 175–189.

Richmond, R. H. (1987). Energetic relationships and biogeographical differences among fecundity, growth and reproduction in the reef coral *Pocillopora damicornis*. *Bulletin of Marine Science* **41**, 594–604.

Richmond, R. H. and Hunter, C. L. (1990). Reproduction and recruitment of corals: comparisons among the Caribbean, the Tropical Pacific, and the Red Sea. *Marine Ecology Progress Series* **60**, 185–203.

Rinkevich, B. and Loya, Y. (1979a). The reproduction of the Red Sea coral *Stylophora pistillata*. I. Gonads and planulae. *Marine Ecology Progress Series* **1**, 133–144.

Rinkevich, B. and Loya, Y. (1979b). The reproduction of the Red Sea coral *Stylophora pistillata*. II. Synchronization in breeding and seasonality of planulae shedding. *Marine Ecology Progress Series* **1**, 145–152.

Rodhouse, P. G. (1978). Energy transformation by the oyster *Ostrea edulis* L. in a temperate estuary. *Journal of Experimental Marine Biology and Ecology* **34**, 1–22.

Rose, M. (1983). Theories of life-history evolution. *American Zoologist* **23**, 15–23.

Rupert, E. E. and Barnes, R. D. (1994). "Invertebrate Zoology". Saunders College Publishing, New York.

Sakai, K. (1998). Effect of colony size, polyp size and budding mode on egg production in a colonial coral. *Biological Bulletin* **195**, 319–325.

Sastry, A. N. (1979). Pelecypoda (excluding Ostreidae). In "Reproduction of Marine Invertebrates", Vol. V (A. C. Giese and J. S. Pearse, eds), pp. 113–292. Academic Press, New York.

Schaffer, W. M. (1974). Optimal reproductive effort in fluctuating environments. *American Naturalist* **108**, 781–790.

Schatt, P. and Feral, J.-P. (1996). Completely direct development of *Abatus coratus*, a brooding schizasterid (Echiodermata: Echinoidea) from Kerguelen, with description of perigastrulation, a hypothetical new mode of gastrulation. *Biological Bulletin* **190**, 24–44.

Scheltema, R. S. (1994). Adaptations for reproduction among deep-sea benthic molluscs: an appraisal of the existing evidence. In "Reproduction, Larval Biology, and Recruitment of the Deep-sea Benthos" (C. M. Young and K. J. Eckelbarger, eds), pp. 44–75. Columbia University Press, New York.

Scherle, W. (1970). A simple method for volumetry of organs in quantitative stereology. *Mikroskopic Bd.* **26**, 57–60.

Schoenmakers, H. J. N., Colenbrander, P. H. J. M., Peute, J. and Van Oordt, P. G. W. J. (1981). Anatomy of the ovaries of the starfish *Asterias rubens* (Echinodermata). A histological and ultrastructural study. *Cell and Tissue Research* **217**, 577–597.

Sewell, M. A. (1994). Birth, recruitment and juvenile growth in the intraovarian brooding sea cucumber *Leptosynapta clarki*. *Marine Ecology Progress Series* **114**, 149–156.

Sewell, M. A. and Chia, F.-S. (1994). Reproduction on the intraovarian brooding apodid *Leptosynapta clarki* (Echinodermata: Holothuroidea) in British Columbia. *Marine Biology* **121**, 285–300.

Sewell, M. A. and Young, C. M. (1997). Are echinoderm egg size distributions bimodal? *Biological Bulletin* **193**, 297–305.

Shick, J. M. (1991). Reproduction and population structure. In "A Functional Biology of Sea Anemones" (P. Calow, ed.), pp. 228–277. Chapman & Hall, London.

Shick, J. M., Taylor, W. F. and Lamb, A. N. (1981). Reproduction and genetic variation in the deposit-feeding sea star *Ctenodiscus crispatus*. *Marine Biology* **63**, 51–66.

Shilling, F. M. and Manahan, D. T. (1994). Energy metabolism and amino acid transport during early development of Antarctic and temperate echinoderms. *Biological Bulletin* **187**, 398–407.

Smiley, S., McEuen, F. S., Chafee, C. and Krishan, S. (1991). Echinodermata: Holothuroidea. In "Reproduction of Marine invertebrates, Vol. VI. Echinoderms and Lophophorates" (A. C. Giese, J. S. Pearse and V. B. Pearse, eds), pp. 664–750. Boxwood Press, California.

Smith, C. C. and Fretwell, S. D. (1974). The optimal balance between size and number of offspring. *American Naturalist* **108**, 499–506.

Smith, L. D. and Hughes, T. P. (1999). An experimental assessment of survival, re-attachment and fecundity of coral fragments. *Journal of Experimental Marine Biology and Ecology* **235**, 147–164.

Sokal, R. R. and Rohlf, J. F. (1995). "Biometry: The Principles and Practice of Statistics in Biological Research", 3rd edn. W. H. Freeman, New York.

Somers, K. M. (1991). Characterizing size-specific fecundity in crustaceans. *Crustacean Issues* **7**, 357–378.

Southwood, T. R. E. (1988). Tactics, strategies and templets. *Oikos* **52**, 3–18.

Spight, T. M. and Emlen, J. (1976). Clutch sizes of two marine snails with a changing food supply. *Ecology* **57**, 1162–1178.

Stearns, S. C. (1977). The evolution of life history traits: a critique of the theory and a review of the data. *Annual Review of Ecology and Systematics* **8**, 145–171.

Stearns, S. C. (1980). A new view of life history evolution. *Oikos* **35**, 266–281.

Stearns, S. C. (1992). "The Evolution of Life Histories". Oxford University Press, Oxford.

Stella, V., Lopez, L. S. and Rodriguez, E. M. (1996). Fecundity and brood biomass investment in the estuarine crab *Chasmagnathus granulatus* Dana, 1851 (Decapoda, Brachyura, Grapsidae). *Crustaceana* **69**, 306–312.

Strathmann, R. R. and Strathmann, M. F. (1982). The relationship between adult size and brooding in marine invertebrates. *American Naturalist* **119**, 91–101.

Strathmann, R. R., Strathmann, M. F. and Emson, R. H. (1984). Does limited brood capacity link adult size, brooding, and simultaneous hermaphroditism? A test with the starfish *Asterina phylactica*. *American Naturalist* **123**, 796–818.

Styan, C. A. (1998). Polyspermy, egg size and fertilization kinetics of free spawning marine invertebrates. *American Naturalist* **152**, 290–297.

Sumida, P. Y. G., Tyler, P. A., Lampitt, R. S. and Gage, J. D. (2000). Reproduction, dispersal and settlement of the bathyal ophiuroid *Ophiocten gracilis* in the NE Atlantic. *Marine Biology* **137**, 623–630.

Tester, P. A. and Turner, J. T. (1990). How long does it take copepods to make eggs? *Journal of Experimental Biology and Ecology* **141**, 169–182.

Thessalou-Legaki, M. and Kiortsis, V. (1997). Estimation of the reproductive output of the burrowing shrimp *Callianassa thirhena*: a comparison of three different biometrical approaches. *Marine Biology* **127**, 435–442.

Thompson, R. J. (1979). Fecundity and reproductive effort of the blue mussel (*Mytilus edulis*), the sea urchin (*Strongylocentrotus droebachiensis*), and the snow crab (*Chionoecetes opilio*) from populations in Nova Scotia and Newfoundland. *Journal of the Fisheries Research Board of Canada* **36**, 955–964.

Thorson, G. (1950). Reproduction and larval development of marine bottom invertebrates. *Biological Reviews* **25**, 1–45.

Todd, C. D. and Havenhand, J. N. (1988). Physiological ecology of *Adalaria proxima* (Alder et Hancock) and *Onchidoris muricata* (Müller) (Gastropoda: Nudibranchia). III. Energy budgets. *Journal of Experimental Marine Biology and Ecology* **118**, 191–205.

Tuomi, J., Hakala, T. and Haukioja, E. (1983). Alternative concepts of reproductive effort, cost of reproduction and selection in life-history evolution. *American Zoologist* **23**, 25–34.

Tyler, P. A. and Billett, D. S. M. (1987). The reproductive ecology of elasipodid holothurians from the N.E. Atlantic. *Biological Oceanography* **5**, 273–296.

Tyler, P. A. and Gage, J. D. (1980). Reproduction and growth of the deep-sea brittlestar *Ophiura ljungmani* (Lyman). *Oceanologica Acta* **3**, 177–185.

Tyler, P. A. and Gage, J. D. (1983). The reproductive biology of *Ypsilothuria talismani* (Holothroidea: Dendrochirota) from the North-East Atlantic. *Journal of the Marine Biological Association of the United Kingdom* **63**, 609–616.

Tyler, P. A. and Pain, S. L. (1982). The reproductive biology of *Plutonaster bifrons, Dytaster insignis* and *Psilaster andromeda* (Asteroidea: Astropectinidae) from the Rockall Trough. *Journal of the Marine Biological Association of the United Kingdom* **62**, 869–887.

Tyler, P. A. and Ramirez Llodra, E. (in press). Larval and reproductive strategies on European continental margins. *In* "Ocean Margin Systems". Springer Verlag.

Tyler, P. A., Pain, S. L. and Gage, J. D. (1982a). Gametogenic cycles in deep-sea phanerozoan asteroids from the N.E. Atlantic. *In* "Proceedings of the International Echinoderms Conference, Tampa Bay, 14–17.09.81" (J. M. Lawrence, ed.), pp. 431–435. A. A. Balkema, Rotterdam.

Tyler, P. A., Pain, S. L. and Gage, J. D. (1982b). The reproductive biology of the deep-sea asteroid *Bathybiaster vexillifer*. *Journal of the Marine Biological Association of the United Kingdom* **62**, 57–69.

Tyler, P. A., Gage, J. D. and Billett, D. S. M. (1985). Life-history biology of *Peniagone azorica* and *P. diaphana* (Echinoderma: Holothuroidea) from the north-east Atlantic Ocean. *Marine Biology* **89**, 71–81.

Tyler, P. A., Billett, D. S. M. and Gage, J. D. (1987). The ecology and reproductive biology of *Cherbonniera utriculus* and *Molpadia blakei* from the N.E. Atlantic. *Journal of the Marine Biological Association of the United Kingdom* **67**, 385–397.

Tyler, P. A., Billett, D. S. M. and Gage, J. D. (1990). Seasonal reproduction in the seastar *Dytaster grandis* from 4000m in the North-East Atlantic Ocean. *Journal of the Marine Biological Association of the United Kingdom* **70**, 173–180.

Tyler, P. A., Gage, J. D., Paterson, G. J. J. and Rice, A. L. (1993). Dietary constraints on reproductive periodicity in two sympatric deep-sea astropectinid seastars. *Marine Biology* **115**, 267–277.

Vadas, R. L. (1977). Preferential feeding: an optimization strategy in sea urchins. *Ecological Monographs* **47**, 337–371.

Van Dover, C. L. and Williams, A. (1991). Egg size in squat lobsters (Galatheoidea): constraint and freedom. *Crustacean Issues* **7**, 143–156.

Van Dover, C. L., Factor, J. R., Williams, A. B. and Berg, C. J. J. (1985). Reproductive patterns of decapod crustaceans from hydrothermal vents. *Biological Society of Washington Bulletin* **6**, 223–227.

Van Veghel, M. L. J. and Bak, R. P. M. (1994). Reproductive characteristics of the polymorphic Caribbean reef building coral *Monastrea annularis*. III. Reproduction in damaged and regenerating colonies. *Marine Ecology Progress Series* **109**, 229–233.

Van Veghel, M. L. J. and Kahmann, M. E. H. (1994). Reproductive characteristics of the polymorphic Caribbean reef building coral *Monastrea annularis*. II. Fecundity and colony structure. *Marine Ecology Progress Series* **109**, 221–227.

Vance, R. R. (1973a). On reproductive strategies in marine benthic invertebrates. *American Naturalist* **107**, 352–361.

Vance, R. R. (1973b). More on reproductive strategies in marine benthic invertebrates. *American Naturalist* **107**, 353–361.

Vianey-Liaud, M. and Lancaster, F. (1994). Effects of intermittent starvation on mature *Biomphalaria glabrata* (Gastropoda: Planorbidae). *Hydrobiologia* **291**, 125–130.

Walker, C. W. (1980). Spermatogenic columns, somatic cells and the microenvironment of germinal cells in the testes of asteroids. *Journal of Morphology* **166**, 81–107.

Wallace, C. C. (1985). Reproduction, recruitment and fragmentation in sympatric species of the coral genus *Acropora*. *Marine Biology* **88**, 217–233.

Watson, G. J. and Bentley, M. G. (1997). Evidence for a coelomic maturation factor controlling oocyte maturation in the polychaete *Arenicola marina* (L.). *Invertebrate Reproduction and Development* **31**, 297–305.

Watson, G. J. and Bentley, M. G. (1998). Action of CMF (coelomic maturation factor) on oocytes of the polychaete *Arenicola marina* (L.). *Journal of Experimental Zoology* **281**, 65–71.

Watson, G. J., Williams, M. E. and Bentley, M. G. (2000a). Can synchronous spawning be predicted from environmental parameters? A case study of the lugworm *Arenicola marina*. *Marine Biology* **136**, 1003–1017.

Watson, G. J., Langford, F. M., Gaudron, S. M. and Bentley, M. G. (2000b). Factors influencing spawning and pairing in the scale worm *Harmothoe imbricata* (Annelida: Polychaeta). *Biological Bulletin* **199**, 50–58.

Webber, H. H. (1977). Gastropoda: Prosobranchia. *In* "Reproduction of Marine Invertebrates, Vol. IV. Molluscs: Gastropods and Cephalopods" (A. C. Giese and J. S. Pearse, eds), pp. 1–98. Academic Press, New York.

Wehrtmann, I. S. and Andrade, G. (1998). Egg production in *Heterocarpus reedi* from northern Chile, with a comparison between iced and living females. *Ophelia* **49**, 71–82.

Wells, M. J. and Wells, J. (1977). Cephalopoda: Octopoda. *In* "Reproduction of Marine Invertebrates, Vol. IV. Molluscs: Gastropods and Cephalopods" (A. C. Giese and J. S. Pearse, eds), pp. 291–336. Academic Press, New York.

Wendt, D. E. (1998). Effect of larval swimming duration on growth and reproduction of *Bugula neritina* (Bryozoa) under field conditions. *Biological Bulletin* **195**, 126–135.

Whittaker, R. H. and Goodman, D. (1979). Classifying species according to their demographic strategy. I. Population fluctuations and environmental heterogeneity. *American Naturalist* **113**, 185–200.

Wilbur, H. M., Tinke, D. W. and Collins, J. P. (1974). Environmental certainty, trophic level and resource availability in life history evolution. *American Naturalist* **108**, 805–817.

Williams, G. C. (1966). Natural selection, the costs of reproduction and a refinement of Lack's principle. *American Naturalist* **100**, 687–690.

Williams, T. D. and Jones, M. B. (1999). Effect of temperature and food quantity on the reproduction of *Tisbe battagliai* (Copepoda: Harpacticoida). *Journal of Experimental Marine Biology and Ecology* **236**, 273–290.

Wilson, W. H. (1991). Sexual reproductive modes in Polychaetes: classification and diversity. *Bulletin of Marine Science* **48**, 500–516.

Winemiller, K. and Rose, K. (1992). Patterns of life history diversification in North American fishes: implications for population regulation. *Canadian Journal of Fisheries and Aquatic Science* **49**, 2196–2218.

Winemiller, K. O. (1992). Life-history strategies and the effectiveness of sexual selection. *Oikos* **63**, 318–327.

Yamashita, M., Mita, K, Yoshida, N. and Kondo, T. (2000). Molecular mechanisms of the initiation of oocyte maturation: general and species specific aspects. *Progress in Cell Cycle Research* **4**, 115–129 (L. Meijer, A. Jezquel and B. Decommun, eds). Kluwer Academic/Plenum Publishers.

Young, C. M. and Tyler, P. A. (1993). Embryos of the deep-sea echinoid *Echinus affinis* require high pressure for development. *Limnology and Oceanography* **38**, 178–181.

Young, C. M., Tyler, P. A. and Fénaux, L. (1997). Potential for deep sea invasion by Mediterranean shallow water echinoids: pressure and temperature as stage-specific dispersal barriers. *Marine Ecology Progress Series* **154**, 197–209.

Young, C. M., Ekaratne, S. U. K. and Cameron, L. J. (1998). Thermal tolerances of embryos and planktotrophic larvae of *Archaeopneustes hystrix* (A. Agassiz) (Spatangoidea) and *Stylocidaris lineata* (Mortensen) (Cidaroidea), bathyal echinoids from the Bahamian Slope. *Journal of Experimental Marine Biology and Ecology* **223**, 65–76.

Zajac, R. N. (1986). The effects of intra-specific density and food supply on growth and reproduction in an infaunal polychaete, *Polydora ligni* Webster. *Journal of Marine Research* **44**, 339–359.

Zal, F., Jollivet, D., Chevaldonné, P. and Desbruyères, D. (1995). Reproductive biology and population structure of the deep-sea hydrothermal vent worm *Paralvinella grasslei* (Polychaeta: Alvinellidae) at 13°N at the East Pacific Rise. *Marine Biology* **122**, 637–648.

Zeeck, E., Harder, T., Beckmann, M. and Müller, C. T. (1996). Marine gamete-release pheromones. *Nature, London* **382**, 214.

Ecology of Southern Ocean Pack Ice

Andrew S. Brierley[1] and David N. Thomas[2]

[1]*Gatty Marine Laboratory, School of Biology, University of St Andrews, Fife, KY16 8LB, UK*
[2]*School of Ocean Sciences, University of Wales-Bangor, Menai Bridge, Anglesey, LL59 5EY, UK*
e-mail: andrew.brierley@st-andrews.ac.uk, d.thomas@bangor.ac.uk

Around Antarctica the annual five-fold growth and decay of sea ice is the most prominent physical process and has a profound impact on marine life there. In winter the pack ice canopy extends to cover almost 20 million square kilometres – some 8% of the southern hemisphere and an area larger than the Antarctic continent itself (13.2 million square kilometres)

ADVANCES IN MARINE BIOLOGY VOL. 43
ISBN 0-12-026143-X

– and is one of the largest, most dynamic ecosystems on earth. Biological activity is associated with all physical components of the sea-ice system: the sea-ice surface; the internal sea-ice matrix and brine channel system; the underside of sea ice and the waters in the vicinity of sea ice that are modified by the presence of sea ice. Microbial and microalgal communities proliferate on and within sea ice and are grazed by a wide range of proto- and macrozooplankton that inhabit the sea ice in large concentrations. Grazing organisms also exploit biogenic material released from the sea ice at ice break-up or melt. Although rates of primary production in the underlying water column are often low because of shading by sea-ice cover, sea ice itself forms a substratum that provides standing stocks of bacteria, algae and grazers significantly higher than those in ice-free areas. Decay of sea ice in summer releases particulate and dissolved organic matter to the water column, playing a major role in biogeochemical cycling as well as seeding water column phytoplankton blooms. Numerous zooplankton species graze sea-ice algae, benefiting additionally because the overlying sea-ice ceiling provides a refuge from surface predators. Sea ice is an important nursery habitat for Antarctic krill, the pivotal species in the Southern Ocean marine ecosystem. Some deep-water fish migrate to shallow depths beneath sea ice to exploit the elevated concentrations of some zooplankton there. The increased secondary production associated with pack ice and the sea-ice edge is exploited by many higher predators, with seals, seabirds and whales aggregating there. As a result, much of the Southern Ocean pelagic whaling was concentrated at the edge of the marginal ice zone. The extent and duration of sea ice fluctuate periodically under the influence of global climatic phenomena including the El Niño Southern Oscillation. Life cycles of some associated species may reflect this periodicity. With evidence for climatic warming in some regions of Antarctica, there is concern that ecosystem change may be induced by changes in sea-ice extent. The relative abundance of krill and salps appears to change interannually with sea-ice extent, and in warm years, when salps proliferate, krill are scarce and dependent predators suffer severely. Further research on the Southern Ocean sea-ice system is required, not only to further our basic understanding of the ecology, but also to provide ecosystem managers with the information necessary for the development of strategies in response to short- and medium-term environmental changes in Antarctica.

Technological advances are delivering new sampling platforms such as autonomous underwater vehicles that are improving vastly our ability to sample the Antarctic under sea-ice environment. Data from such platforms will enhance greatly our understanding of the globally important Southern Ocean sea-ice ecosystem.

1. GENERAL INTRODUCTION

"The sea, as we advanced, became covered more thickly with small fragments of ice and then sheets of broken pack appeared ahead and on the starboard bow. At first these floes were just two or three hundred yards across with a mile or so of clear water in between. Then they became larger and larger till we were pushing our way through belts of broken pack over a mile wide and stretching to the right as far as the eye could see. Each mass of ice is made up of closely packed pieces. It is a strange sight: the dull grey sky above, the almost black water, and the dazzling white ice rising and falling on the gentle swell. The water appears like ink because it receives so little of the light which is nearly all reflected by the ice; it has the darkness of a cave."

(Sir Alistair Hardy, 1967)

The annual growth and decay of sea ice around the Antarctic continent is one of the most prominent physical processes on Earth. In winter the sea ice extends northwards of the geographic, continental and political bounds of Antarctica. Its influence pervades the Southern Ocean as a whole, and, through interactions with thermohaline circulation and climate, far beyond. The Southern Ocean sea-ice realm is thus of global significance. The periodic advance and retreat of the sea ice, from a minimum extent of about 3.5 million square kilometres in February to a maximum of about 19 million square kilometres in September (Gloersen *et al.*, 1992; Parkinson and Gloersen, 1993; Parkinson, 1998) (also see Plate 1), creates "one of the largest and most dynamic ecosystems on earth" (Arrigo and Lizotte, 1998). This ecosystem is remote and difficult to study and, because for many years all but the periphery was inaccessible, it remained less well known than other components of the global environment. Antarctica was outside the theatre of interest during the Cold War and, as a consequence, basic information on sea-ice thickness as gathered by military submarines in the Arctic is lacking for the Southern Ocean. With the advent of increasingly powerful ice-breaking research vessels, drifting ice camps and, very recently, autonomous underwater vehicles, sea-ice research has developed significantly in the past 20 years. There is now a substantial body of literature covering various aspects pertinent to Southern Ocean sea-ice ecology. Focusing mainly on developments in the past 15 years, we present an up-to-date overview of available knowledge on ecological processes within, and associated with, the Southern Ocean sea-ice realm.

Previous reviews have considered physical controls on the development of Antarctic sea-ice communities (Ackley and Sullivan, 1994) and the role that sea ice plays in structuring ecosystems (Eicken, 1992). Horner (1985b) edited a book on sea-ice biota that, although including a chapter on

Arctic sea-ice fauna (Carey, 1985), focused predominantly on sea-ice algae and bacteria and included little on higher trophic levels. These works were advanced by the reviews of Garrison (1991), Horner *et al.* (1992), Legendre *et al.* (1992), Palmisano and Garrison (1993), Kirst and Wiencke (1995), Garrison and Mathot (1996), Nichols *et al.* (1995, 1999a), Lizotte and Arrigo (1998), Staley and Gosink (1999), Lizotte (2001) and Thomas and Dieckmann (2002a, b). However, these reviews tend to focus on specific aspects, processes or groups of organisms of the sea-ice system. Although there are a number of broadly ranging, non-specialist, introductory discussions that touch on the ecology of sea ice (e.g. Garrison *et al.* 1986; Nicol and de la Mare, 1993; Thomas and Dieckmann, 1994; Stevens, 1995a, b; 1996; Thomas 1996; Nicol and Allison, 1997; Brierley and Reid, 1999; Copley, 2000), there has been little attempt to synthesize in the primary literature the available information on the diverse elements that make up the ecology of the pack ice zone. New data have heightened awareness of the importance and vulnerability of the Antarctic sea-ice environment. Significant advances include discovery of a potential major mid-twentieth century decline in sea-ice extent (de la Mare, 1997); periodic oscillations in sea-ice extent (Murphy *et al.*, 1995); the Antarctic Circumpolar Wave climatic phenomenon (White and Peterson, 1996), driven in part by the El Niño Southern Oscillation; the link between sea-ice extent and Antarctic krill abundance, and the well-being of krill predators (Loeb *et al.*, 1997; Brierley *et al.*, 1999a; Reid and Croxall, 2001); and the first continuous survey of krill beneath sea ice, and of sea-ice thickness (Brierley *et al.*, 2002). Concentrating on pack ice (oceanic ice as opposed to the more permanent sea ice that is *fast* to land, which has been studied extensively from research stations ashore) and the link to the pelagic environment, this review seeks to incorporate these new findings within a broad description of the interactions of plant and animal life across all trophic levels within the physical environment of the seasonal Southern Ocean sea ice. The review draws together frazils of information into what is hopefully a consolidated pack of information. We use these observations as a basis for estimation of possible future impacts of climate change on the sea-ice environment.

Antarctic sea ice has been the focus of study by Working Groups of the Scientific Committee on Antarctic Research (SCAR), including the Antarctic Sea Ice Processes and Climate group (ASPeCt) and Coastal and Shelf Ecology of the Antarctic Sea-Ice Zone group (CS-EASIZ). With the recent launch of several major new international research initiatives contributing to the Southern Ocean component of GLOBEC (Global Ocean Ecosystems Dynamics), a review of the present state of knowledge of the ecology of Southern Ocean sea ice is timely.

2. THE PHYSICAL ENVIRONMENT

In winter, sea ice in the Southern Ocean extends to form a continuous girdle around the Antarctic continent. This distribution is in marked contrast to that in the Arctic Ocean, where an ice-covered basin is surrounded by land (see Stonehouse, 1989). This and other differences in the biology of sea ice in the Arctic and Southern Oceans have been tabulated by Spindler (1990), Horner *et al.* (1992), Legendre *et al.* (1992) and Kirst and Wiencke (1995). Throughout this review we endeavour to comply with terminology defined by Horner and others in the SCOR (Scientific Committee on Oceanic Research) Working Group on the ecology of sea-ice biota.

2.1. Seasonality, sea-ice formation and melting

Southern Ocean sea ice reaches a maximum extent of about 19 million km^2 in September (Plate 1A and Figure 1) and consists typically of an area of 15 million km^2 of ice and 4 million km^2 of open water in the form of leads or polynyas. The transition between open and ice-covered water, the sea-ice edge, is often defined by the 15% concentration line (15% of the sea

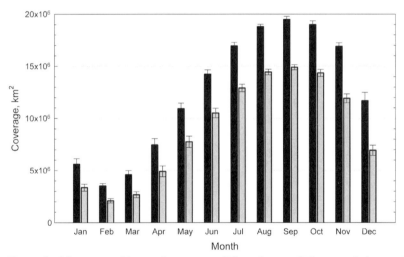

Figure 1 Mean monthly sea-ice extent (■) and area (▧) around Antarctica (based on 15% sea-ice concentration) during the period from January 1979 to December 1996 as determined by satellite observations and interpretation using the NASA2 algorithm (see Cavalieri *et al.*, 1984). Error bars are 1 standard deviation. Data were provided by S. Harangozo (British Antarctic Survey) and colleagues in the EU-funded PELICON project (see www-iup.physik.uni-bremen.de/iuppage/pelicon_e.html).

surface being ice covered, e.g. Stammerjohn and Smith, 1997). At the height of summer, in February, this line bounds some 3.5 million km² of ocean. As in winter, the ocean will not be completely ice-covered and the actual area of sea ice may be only 2 million km². Mean monthly sea-ice areas and extents are shown in Figure 1. The temporal progression from open to ice-covered water, and *vice versa*, can occur extremely rapidly. For example, in the Weddell Sea (335°E), sea ice receded on average 10.5°E of latitude (1167 km) between December and January over the period 1979–1993 (Plate 1). In the same sector sea ice extended on average 5.8°E (645 km) between April and May. These rates of sea-ice decay and growth are equivalent to movements of the sea-ice edge at 1.6 and 0.9 km h^{-1} respectively: with the exception of tides over mudflats, the passage from day to night or the spread of forest fires, it is difficult to conceive of a faster changing biological environment.

The interface between the seasonal pack ice and open water is not a clear-cut boundary, but a complex belt with distinct characteristics. This marginal ice zone (MIZ), which may be 100–200 km wide, is influenced by the penetration of waves and swell that break the pack into numerous smaller floes. Floe size increases rapidly with distance from the sea-ice edge (Maykut, 1985) because the destructive power of swells is damped out by the sea ice itself (e.g. Squire *et al.*, 1995). Broken floes in the early stages of melting form a patchy and dynamic wind-blown matrix (Turner and Owens, 1995), and introduction of fresh meltwater at the surface can lead to pronounced vertical stratification of the underlying water column (Smith and Nelson, 1986) (see Figure 2). However, this is not always the case and development of stratification is highly dependent on prevailing winds and the degree of surface mixing (see section 5). Fronts can develop at the sea-ice edge where the stratified waters abut mixed waters beyond the influence of sea-ice melt (Niebauer and Alexander, 1985; Murphy *et al.*, 1998a). These fronts track the receding sea-ice edge, resulting in a temporally and spatially highly patchy environment. Conversely, as sea ice forms, high salinity brine is rejected from the developing ice crystals (Wadhams, 2000). The brine gathers first between the sea-ice crystals, in the pockets and long, narrow channels of the developing sea-ice matrix. The brine inclusions, or pores (Cottier *et al.*, 1999; Eicken *et al.*, 2000), vary tremendously in size, distribution and degree of connectivity. Large brine structures extend vertically downwards through the developing ice sheet. Brine migrates, primarily under the influence of gravity, down these structures that are known commonly as brine channels (see Figures 2 and 3), before draining into the underlying water column. Introduction of high density brine destabilizes the upper water column, resulting in convective mixing and the eventual formation of deep Antarctic Bottom Water. Sea-ice growth and melting thus has a significant influence on the circulation of

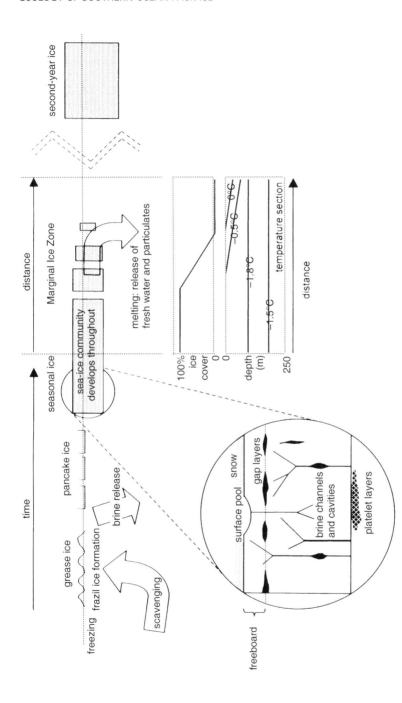

Figure 2 Schematic representation of sea-ice structure, showing ice formation in autumn, melt at the sea-ice edge in spring, and the underlying water column. Adapted from data in Eicken (1992), Horner *et al.* (1992), Ackley and Sullivan (1994) and Priddle *et al.* (1996).

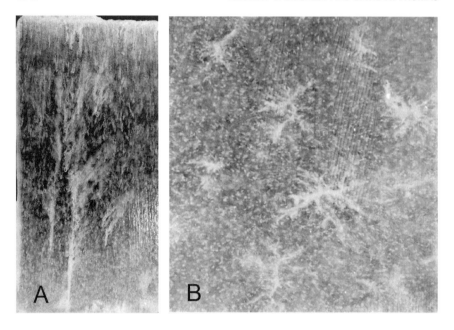

Figure 3 Sections of sea ice, 2 cm thick, grown in the Hamburg Ship Model Basin (HSVA) ice tank in January 1997, showing brine channel structure. (A) A horizontal section 20 × 22 cm and (B) a vertical section 22 × 10 cm from the sea-ice/air interface to the sea-ice/water interface. The "star burst" shapes in the horizontal section correspond to the brine channel side-branches that feed the main vertical channels. Initial sea-ice growth rate was about 2 mm h^{-1}, leading to lateral channel separation of about 6.5 channels per 100 cm^2. Channel separation is similar to that found in naturally grown sea ice (cf. Wakatsuchi and Saito, 1985). More details are given in Cottier (1999) and Cottier *et al.* (1999). (Photographs courtesy of F. Cottier.)

the oceans and, through heat dissipation, global climate (Yuang and Martinson, 2000).

The physical structure of sea ice has been described in detail by Maykut (1985). Further information can be found in Weeks and Ackley (1982), Jefferies (1998), Leppäranta (1998) and Wadhams (2000). Since this physical structure (see Figures 1 and 2) has such a major bearing on the biological processes within and adjacent to sea ice, some of the most pertinent details are included here. For any understanding of sea-ice ecology, a basic appreciation of how sea ice is formed and the processes governing its growth and ultimate decay is a prerequisite.

With the onset of autumn, fast ice grows out from the land to meet pack ice forming at sea. As surface seawater temperature falls (to about −1.9°C at salinity 35; T_{freezing}°C ≈ −0.055 · s) ice crystals 3–4 mm in diameter begin to form. These frazil ice crystals, which may also form at depth

(Penrose *et al.*, 1994), accumulate into a thickening, amorphous layer of grease ice (Horner *et al.*, 1992). This layer is broken readily by waves or swell but, if cold persists, the crystals continue to grow in a loose matrix that develops the upturned edges that are characteristic of pancake ice. The pancakes subsequently fuse into a solid ice sheet that may be 10 cm thick (Maykut, 1985; Lange *et al.*, 1989). Once a solid sheet has formed, ice growth slows and frazil crystals cease forming. Columnar crystals grow on the underside of the sheet creating congelation ice (Lange *et al.*, 1989). Brine channels develop in columnar ice and have a characteristic branching, tree-like structure (Weissenberger, 1992; Cottier, 1999) (see Figure 3). Brine channels initially occur in great numbers (Wakatsuchi and Saito, 1985; Cole and Shapiro, 1998) but the smaller, less well-developed channels may wither as the larger channels draw away their supplies of brine. Brine channels flaw the physical integrity of sea ice (Cottier *et al.*, 1999), and the porosity of the ice is the key variable in controlling ice floe strength (Eicken *et al.*, 1991a). The solid ice sheet may be split by wind, waves and swell. The separated floes can be forced over or under one another in the process of rafting, and frazil ice may become sandwiched between congelation ice, adding further to the overall thickness. In the vicinity of ice shelves, sea ice may also thicken further by the collection of platelet ice beneath the congelation ice (Palmisano and Sullivan, 1985b; Dieckmann *et al.*, 1986). In the Southern Ocean, sea-ice thickness is believed seldom to exceed 2 m since 90% of sea ice is first-year ice that melts in the following summer (Wadhams, 2000). By contrast, Arctic sea ice may persist for tens of years and reach in excess of 5 m thick (Maykut, 1985). However, where rafting, pressure ridge formation and deformation of ice floes is extensive, sea ice >10 m thick can be found, especially in fields of multi-year ice (Wadhams, 2000). Only in embayments and regions where fast ice fails to break out in summer, or where pack ice accumulates in gyres such as the Weddell Sea, does the sea ice persist to any great extent beyond a single season to form multi-year ice. Despite the importance of sea-ice formation to global ocean circulation, little is known of ice thickness (and hence of ice volume) at large scales in the Southern Ocean. The waters around Antarctica did not have the same strategic importance as the Arctic Ocean during the Cold War years (Copley, 2000) and cannot boast the extensive series of observations from military submarines that were collected in the northern hemisphere (Wadhams *et al.*, 1985; Wadhams, 1995). In the Antarctic, sea-ice thickness measurements have been restricted largely to point observations from drilled holes or estimates made from ice-breaking ships on passage (Wadhams *et al.*, 1987), and knowledge generally of Antarctic sea-ice processes is substantially less than in the Arctic (Wadhams, 2000). Recently electromagnetic induction measurements that make use of the electrical conductivity of sea ice have been developed to measure sea-ice thickness (Haas, 1997, 1998; Haas *et al.*, 1997). Instruments exploiting this technique have been deployed at the bows of ice-breaking ships, or towed by helicopter (Kovacs *et al.*, 1987, 1995), thereby enabling sea-ice thickness to be measured along long transects. Another recent development for large-scale measurement of sea-ice thickness comprises acoustic observations from the *Autosub-2* autonomous underwater vehicle (AUV) (Millard *et al.*, 1998), which gathered over 210 km of along-transect estimates of sea-ice thickness in the northern Weddell Sea in 2001 (Brierley *et al.*, 2002).

Sea-ice floes may be separated by wind and current, leaving leads of open water. Although generally quite narrow (tens to hundreds of metres), leads may extend for many kilometres. Other non-linear-shaped openings enclosed by sea ice are called polynyas. Close to the shore, winds (katabatic or depressional) may blow ice offshore leaving areas of open water called coastal polynyas. These can extend for many tens of kilometres (they are useful for navigation) and often assume linear configuration. Coastal polynyas have been shown to occupy relatively small areas of the ice-covered part of the Weddell Sea (about 0.2% in winter), although they produce between 2.5 and 5% of the total Weddell Sea ice volume (Markus *et al.*, 1998). Polynyas may also form offshore if sufficient heat enters a region from the ocean to melt ice. Notable examples include polynyas in the Weddell Sea in the mid-1970s (Gordon and Comiso, 1988) and in Terra Nova Bay (Budillon and Spezie, 2000). The appearance, position, size and shape of the Weddell Sea polynya have been explained by variations in the large-scale oceanic flow past the Maud Rise Seamount resulting in a horizontal cyclonic eddy being shed from its north-east flank. The eddy transmits a divergent Ekman stress, resulting in an opening in the pack ice that is enhanced by atmospheric thermodynamic interactions inducing oceanic convection (Holland, 2001). Leads and polynyas may host accumulations of birds and mammals that exploit the feeding opportunities presented by an area of open water in an otherwise capped ocean (e.g. Gill and Thiele, 1997).

2.2. Synopsis of sea-ice habitats

Although there have been many advances in understanding of the ecology of sea ice over the past 15 years, the classic text of Horner (1985b) remains a valuable descriptor of how sea ice provides a range of habitats for a diverse array of biological processes. Her comprehensive study is now augmented by the works of Eicken (1992), Horner *et al.* (1992), Palmisano and Garrison (1993), Ackley and Sullivan (1994) and Garrison and Mathot (1996). Gutt (2001) has reviewed the physical impact of ice on benthic marine communities. These works document biological activity throughout sea ice, from the surfaces of floes, in deformation/melt ponds, in the surface layers of sea ice underlying snow, inside internal sea-ice interstices and beneath sea ice. Interior biological assemblages, which are most common higher in the sea ice, may be associated with faults in the sea-ice structure, internal melt regions or the brine channel system (see Figure 2).

Sea-ice ecology has now reached a stage where a closely linked collaboration between glaciologists and physicists concerned with the physical processes controlling sea ice, and biologists, is required for further development of the field. Although much has been learnt about the organisms that inhabit sea ice, we still know relatively little about the physical/biological interactions that structure the ecology of the ice and the adjacent water (but see Eicken, 1992; Ackley and Sullivan, 1994). In particular, developments in the understanding of small-scale variability of salinity, porosity and brine channel structures are important for sea-ice biologists (Eicken *et al.*, 2000; Krembs *et al.*, 2000, 2001b). Sea ice is far from homogenous. Cottier *et al.* (1999) showed the presence of steep salinity gradients and highly variable brine distribution, with areas of high bulk salinity in brine channels that are surrounded by brine-depleted ice. These rapid, small-scale changes in salinity will present considerable osmotic challenges to life within the ice. Brine is redistributed throughout sea ice during warming and melting phases by migration through inter-pore connections. Eicken *et al.* (2000) developed non-destructive tomographic magnetic resonance imaging (MRI) techniques to study directly the microstructure and thermal evolution of brine inclusions within sea ice. Their studies looked at ice warming from $-21°C$ to $-6°C$ and showed how pore size increased with increasing temperature, while pore numbers decreased as pores merged and coalesced. Such studies are vital for improving our understanding of the critical physical determinants of sea-ice porosity. Fine-scale variations in sea-ice texture, and in properties such as salinity, are the basis of the larger-scale variation in the distribution of sea-ice organisms that presents such a problem for comprehensive sea-ice sampling (Eicken *et al.*, 1991b; Swadling *et al.*, 1997). This has been highlighted by Krembs *et al.* (2002b) who demonstrated that boundary-layer flow along irregular under-ice topography alters the flux of porewater across the ice–water interface, as well as affecting significantly the spatial distribution of ice algae on the under-surfaces of the ice. They propose that changes in the under-ice topography brought about by the localized formation and melting of ice can explain, at least partially, the high variability in small-scale distribution of sea-ice organisms.

For biologists, knowledge of the shape and volume of the brine channel system (Figure 3) is a fundamental requirement. This knowledge was lacking until Weissenberger *et al.* (1992) devised a resin casting technique with a resolution from just a few micrometres to >30 mm that enabled hardened casts of internal sea-ice structure to be produced. These casts could subsequently be analysed by scanning electron microscopy. Although a rather laborious method for routine analyses, their technique enables up to 80% of the volume and extent of habitable space in different types of sea ice to be visualized. Another technique for quantifying surface area of interconnected brine channels is the rhodamine chloride absorption technique of Krembs *et al.* (2000), although measurements using this technique are restricted practically to laboratory conditions since the rhodamine has to be added before freezing. Krembs *et al.* (2000) also considered measurements of brine-channel frequency distribution and determined that the total internal surface area within sea ice ranged from 0.6 to 7.0 $m^2 kg^{-1}$ ice at $-2.5°C$, decreasing with decreasing temperatures. Between 6 and 41% of this area can be covered by microorganisms and, as temperatures decrease, this percentage cover increases in proportion to the decrease in surface area.

Although the resin visualizations of internal sea-ice structure produced by Weissenberger *et al.* (1992) did much to help envision the spatial distribution of brine channels and pockets within sea ice, they do not enable the study of spatial distribution of organisms, or the spatial relationships between them, within brine channels. Junge *et al.* (2001) used epifluorescence microscopy and image analysis to observe microorganisms in the pore spaces of sea ice. These methods enabled examination of bacteria and their associations with various sea-ice surfaces and with sea-ice algae (Krembs *et al.*, in press). The continued development of such techniques, especially when combined with other non-destructive methods such as MRI (Eicken *et al.*, 2000), will enhance greatly the information that can be obtained from more traditional methods. This enhancement is vital because it is at the level of the interaction of organisms' distribution within the brine channel and pocket system that the biological processes within sea ice are controlled.

Krembs *et al.* (2000) used glass capillaries ranging from 12 to 1420 μm in diameter as model substrates to mimic the brine channel habitat. With cultures of various taxa, they were able to investigate the flexibility of organisms to move inside the narrow geometric confines of brine channels, and make comparisons between the labyrinthine brine channel system and soil/sediment structure. There is possibly much that sea-ice ecologists could learn by looking at analogies in these systems that, at least superficially, are similar to sea ice. In a later work Krembs *et al.* (2002b) extended this comparison, hypothesizing that the exchange processes at the ice–water interface are mediated by hydrodynamic processes similar to the way that roughness-related advective porewater exchange plays a key role in the distribution of benthic organisms and local organic matter accumulation at the sediment–water interface.

The best studied sea-ice habitats are the underside and ice platelet layers of fast ice (e.g. Horner, 1985a; Dieckmann *et al.*, 1992; Arrigo *et al.*, 1995; McMinn *et al.*, 1999, 2000; Trenerry *et al.*, 2001). This is a result of the relative ease of access to these locations. In the pack ice, although flora and fauna have been observed on the peripheries of floes, and platelet ice is known to house biological activity in some regions (Dieckmann *et al.*, 1986; Smetacek *et al.*, 1992; Grossmann *et al.*, 1996), it is the upper ice surfaces that are best described. In recent years the high productivity of assemblages at the upper surface and in gap layers close to the sea-ice surface have led to the identification of these habitats as regions of extensive biological activity, especially during the summer (Garrison and Buck, 1991; Fritsen *et al.*, 1994, 1998; Thomas *et al.*, 1998). These sites receive high irradiance and often are associated with rotten, porous ice. In rotting ice ready exchange of inorganic nutrients can take place in the interior of floes and enables extensive algal growth (Thomas *et al.*, 1998; Kennedy *et al.*, submitted); nutrient exchange is constrained in solid sea ice. There is evidence that microenvironmental sites formed within the pack in summer from disintegrating sea ice may be important sites of high productivity, with quite different characteristics from the sea ice at large (Gleitz *et al.*, 1996a). Crack pools within disintegrating floes can range in size from tens of centimetres to a few metres across and, as is the case for cracks in fast ice (Günther *et al.*, 1999a, b), are sites where platelet ice can accumulate (Grossmann *et al.*, 1996). Gleitz *et al.* (1996a) showed that such sites are important for metazooplankton grazing. The cracks have direct contact with the underlying water and, rather like platelet layers, do not form the physical barrier to grazing organisms that sea ice does. In the Arctic, surface melt/flood ponds are a widespread phenomenon. Generally speaking, pools are less prevalent on Antarctic summer ice floes (Horner *et al.*, 1992; Ackley and Sullivan, 1994; Wadhams, 2000; Haas *et al.*, 2001), although in localized areas their density and size make them an important feature of the ecology (Garrison and Buck, 1991; Schnack-Schiel *et al.*, 2001b). Despite the potential importance of these features, particularly in summer which is the easiest time for sampling, relatively little work has been done on summer pack ice (see synopsis of Dieckmann *et al.*, 1998).

Snow loading, and the formation of snow ice and superimposed ice, leads to the formation of semi-continuous gap layers at or below the freeboard of ice floes (Eicken *et al.*, 1994; Haas *et al.*, 2001; Massom *et al.*, 2001) (see Figure 2). These gap layers are correctly referred to as freeboard layers, although often may be confused with, and described as, infiltration layers (Garrison and Buck, 1991). The terminology of these layers needs clarification by sea-ice glaciologists in light of this existing confusion, particularly given recent proposals for how the layers are formed (Ackley and Sullivan, 1994; Fritsen *et al.*, 1998, 2001; Haas *et al.*, 2001). Freeboard layers are increasingly being identified as biologically important. The high light transmission to these sites leads to high primary productivity (Kristiansen *et al.*, 1992, 1998; Fritsen *et al.*, 1994, 1998, 2001; Haas *et al.*, 2001) and, at times, grazing activity (Garrison and Buck, 1989; Thomas *et al.*, 1998; Schnack-Schiel *et al.*, 2001b). Assemblages formed high in floes support dense algal standing stocks ($>300\,\mu$g chl-$\alpha\,l^{-1}$) that are comparable to those found on the underside of fast ice and in platelet assemblages, and are among the highest concentrations recorded for Antarctic pack ice (Syvertsen and Kristiansen, 1993; Garrison and Mathot, 1996; Thomas *et al.*, 1998). It has been proposed that algal growth may reach such high levels in these sites that absorption of solar radiation by the algae might contribute to further ice melt (Ackley and Sullivan, 1994). However, Haas *et al.* (2001) have presented a model for the development of these surface gaps based purely on physical determinations, the predictions of which are substantiated when the results of a bio-optical model coupled to a thermodynamic model are considered (Zeebe *et al.*, 1996). Significant enhancement of biologically induced melting will only take place at algal standing stocks higher than 150 mg chl-$a\,$m^{-2} and with snow depths <5 cm. Standing stocks of freeboard and infiltration layers rarely meet these criteria.

3. SAMPLING THE SEA-ICE ENVIRONMENT

3.1. Sea-ice extent

Sporadic records of the extent of sea ice in the Southern Ocean are available from the earliest days of Antarctic exploration. Measures of ice extent and duration were first determined by charting the positions of the sea-ice edge as reported by various vessels throughout the year, or by observations of local freeze and thaw at land bases (Murphy *et al.*, 1995). These observations were usually restricted in either space or time. Continuous and consistent widespread records of sea ice have been available only since 1978 when the first of a series of passive microwave-detecting satellites was launched. These satellites provide data on the geographic extent, surface properties and concentration of sea ice quasi-synoptically throughout the Southern Ocean (Cavalieri *et al.*, 1984; Comiso, 1991, 2000; Comiso *et al.*, 1992; Comiso and Gordon, 1998; Drinkwater, 1998; Morris *et al.*, 1998; Markus *et al.* 1998; Haas *et al.*, 1999b; Haas, 2001). Sensors operating at various frequencies including visible, infrared, microwave, radar and lidar are able to detect sea ice. Passive microwave sensors are particularly appropriate because they work in both light and dark conditions, and the influence of atmospheric conditions, even at low sea-ice concentrations, is small. Depending upon the orbit of the satellite, complete global coverage can be achieved daily. Satellite passive microwave imagers such as the Scanning Multichannel Microwave Radiometer (SMMR) on board Nimbus-7 (1978–1987), the Special Sensor Microwave/Imager (SSM/I) on the DMSP satellites (from 1987 onwards) and active microwave instruments on board the European Space Agency's ERS-1 and ERS-2 satellites (from 1991) have provided continuous data for much of the past 20 years, enabling trends in regional and circum-Antarctic extent and area to be studied (e.g. Stammerjohn and Smith, 1997).

In addition to the spectacular seasonal changes in sea-ice extent described in section 2.1, time-series analyses of satellite-derived ice extent data show that around Antarctica ice extent is increasing by 0.9% per decade (Stammerjohn and Smith, 1997). The increase is not uniform around the continent though, and Parkinson (1998) concluded that for the Southern Ocean as a whole there is no clear indication of an overall shortening or lengthening of the ice season. Trends through 7 years of data show that ice seasons in the eastern Ross Sea, Amundsen Sea, far western Weddell Sea, far eastern Weddell Sea and the coastal regions off East Antarctica between 40° and 80°E have shortened. In contrast, sea-ice seasons have lengthened in the western Ross Sea, Bellingshausen Sea, central Weddell Sea and the 80°–135°E sector off the coast of East Antarctica.

Analyses of satellite data for the summer minimum extent of sea ice in the Weddell Sea from 1979 to 1995 have shown that correlations of summer ice characteristics with the subsequent winter are weak. However, low summer minimum extents are preceded by high winter maxima (Comiso and Gordon, 1998). Rather than temperature, it is evident that wind characteristics determine this variability. Oscillatory variations of sea-ice cover of the Amundsen and Bellingshausen Seas in relation to the El Niño Southern Oscillation (ENSO) with periods of 2.4 and 4.2 years also show a correlation with oscillations in scalar near-surface winds, implying a relationship between the winds and sea-ice distribution (Gloersen and Mernicky, 1998).

It has been suggested that there was a very large (up to 25%) reduction in sea-ice extent in the middle of the last century (de la Mare, 1997). This suggestion is based upon whaling records. Pelagic whaling focused on productive ice-edge regions and there was an international requirement to record the position of whale catches. However, in the absence of other corroborating data, this suggestion has not met with universal acceptance.

There are also marked interannual and regional variations in sea-ice extent. Interannual variability in regional sea-ice condition was first reported by Murphy *et al.* (1995) from records of fast ice duration at Signy Island in the South Orkneys. Between the mid-1960s and mid-1990s there was a pronounced subdecadal cyclicity in fast ice duration. This phenomenon, as manifested around Antarctica generally, has become known subsequently as the Antarctic Circumpolar Wave (White and Peterson, 1996). The wave appears to propagate as a dipole around Antarctica with an approximate 7-year periodicity, such that every 3 or 4 years any given longitude will experience a maximum in ice extent, whilst at the same time locations 90 degrees of longitude to the east or west will be experiencing lows. The strong seasonality and interannual variability of sea-ice extent will have a potential impact at all trophic levels within the Southern Ocean marine ecosystem, from primary production to higher predators. Indices of sea-ice condition have been developed, and these provide ecologists with a quantitative means to address questions relating to sea ice and ecosystem function (Smith *et al.*, 1998). Possible biological implications of changing sea-ice extent are considered further in section 6.2.

3.2. Sea-ice composition

3.2.1. In situ *sampling*

The stage has now been reached where sea-ice sampling can be conducted at all times of the year (Dieckmann *et al.*, 1998). Fast ice can be sampled year-round from bases ashore (Günther and Dieckmann, 1999, 2001) but winter remains the least-studied period because of the logistic difficulties posed to winter expeditions. Pack ice is less accessible than fast ice and generally requires an ice-strengthened vessel or helicopter for access. It is important that means are found to support activities collecting information about the ecology of sea ice during winter. Year-round fast ice studies have provided valuable information about physiochemical processes that may be driving species succession and temporal changes in grazing activity (Stoecker *et al.*, 1998; Günther and Dieckmann, 1999, 2001). Successional patterns found in fast ice environments show that similar initiatives are required in the pack ice if we are to understand the full implications of seasonal change for the ecology of Southern Ocean sea ice.

The majority of ice samples from the interior of Antarctic sea ice have been obtained using cores (see Plate 2A). Cores of various diameters, typically 10 cm, can be taken mechanically or by hand (Horner, 1990; Horner *et al.*, 1992) and can provide material from throughout the depth horizon of the sea ice. Samples so obtained are vulnerable to contamination with sea water and brine loss, especially from the deeper sections, and the consequent under-sampling of what is arguably the most important part of the sea ice biologically (Horner *et al.*, 1992). Cores can be extracted from beneath sea ice by SCUBA divers, and these cores may be less prone to leakage (Horner, 1990).

One of the greatest problems hampering understanding of sea-ice ecology is the difficulty in obtaining sufficient samples from which to identify important large-scale spatial trends. Patchiness in sea ice can be extreme. Differences of an order of magnitude in sea-ice algal standing stock and inorganic nutrient concentrations have been recorded among cores spaced just tens of centimetres to a few metres apart on the same floe (Eicken *et al.*, 1991b). The same study has shown that these parameters can vary by up to two orders of magnitude vertically within a single core. With such high degrees of spatial heterogeneity, it is evident that extrapolation of findings from point samples to areal estimates is fraught with difficulties. Furthermore, spatial heterogeneity may increase following rafting or the formation of pressure ridges. Large sample sizes are needed from such heterogeneous environments. Dieckmann *et al.* (1998) compiled sea-ice algal standing stock data from 448 cores collected during 13 campaigns spanning a decade of activity. The range of values obtained can be utilized for refining numerical models of the Antarctic pack ice, and for investigating spatial and temporal patterns of primary production (Arrigo *et al.*, 1997, 1998a, b).

The high degree of variation in distribution of algal biomass in sea ice (see section 4.1), which is related to the distribution of large-scale secondary pores (Eicken *et al.*, 1991b), has consequences for the distribution of algal grazers within the sea ice. However, Swadling *et al.* (1997) have shown that the horizontal distribution of metazoans, although not strongly correlated to chlorophyll distributions, "could vary as much at scales of less than one metre as it can on scales of several kilometres". On most ship-based expeditions it is not plausible to take sea-ice cores with the required frequency and spatial resolution to study such heterogeneity in detail. Many campaigns are reliant on short (a matter of hours) sea-ice stations where coring is but one part of a more extensive suite of measurements. The nature of most expeditions is that only rarely do they remain on station long enough for sufficient cores to be taken to address temporal changes in the sea ice. Given the high degree of within and between floes variation, detection of temporal developments is likely to remain problematical.

Longer-term observations of pack ice can be made from drifting ice stations (Fritsen *et al.*, 1994; Melnikov and Spiridonov, 1996). A notable example was the Ice Station Weddell-1 (ISW-1) deployed between

February and June 1992. This enabled the study of the dynamics of microbial communities in relation to changing physicochemical conditions on a 2–3 km^2 multi-year ice floe over 4 months (Fritsen *et al.*, 1994; Fritsen and Sullivan, 1997; Voronina *et al.*, 2001). The work conducted during that campaign showed clearly how vital such time-series are for our understanding of the seasonal development of sea-ice assemblages and the controlling factors governing them. Drift station campaigns are difficult to arrange logistically, but offer a realistic way of obtaining longer-term temporal data.

One of the most difficult elements of sea ice to sample is the bottom layer, which often is characterized by a fragile skeletal layer of dendritic crystals (Eicken, 1998; McMinn *et al.*, 1999, 2000; Wadhams, 2000; Trenerry *et al.*, 2001). Traditional coring techniques may result in this part of the core being severely disrupted or lost altogether (Horner, 1990), although Archer *et al.* (1996) were able to make video observations showing that careful coring can yield samples within which the bottom sections of the sea ice remain intact. The community living at the sea-ice/water interface is studied rarely, since this complex boundary is disturbed by coring activity (McMinn and Ashworth, 1998; McMinn *et al.*, 1999, 2000; Krembs *et al.*, 2001b; Trenerry *et al.*, 2001). The most obvious solution is to core or sample from beneath the sea ice using SCUBA (Horner, 1990; Lizotte and Sullivan, 1992; Syvertsen and Kristiansen, 1993; Menshenina and Melnikov, 1995; Robinson *et al.*, 1995a; Melnikov and Spiridonov, 1996), although the logistic and safety constraints are severe. Surface-demand diving techniques or rebreathing equipment must be used to avoid disturbance from divers' exhaust air bubbles.

Divers and remotely operated vehicles (ROVs) have made direct observations of the under-ice environment (Marschall, 1987, 1988; O'Brien, 1987; Hamner *et al.*, 1989; Bergström *et al.*, 1990; Quetin *et al.*, 1996). The horizontal range of these observations is limited by safety considerations and cable lengths (200 m for ROVs, 30 m for SCUBA) (Siegel *et al.*, 1990), and these techniques provide only limited windows of observation. The resulting data are seldom quantitative (Stretch *et al.*, 1988; Bergström *et al.*, 1990). Despite operational limitations associated with ROV umbilical cables, further development of ROV technology may provide an increasingly useful tool for sampling beneath sea ice (Gutt, 1995). Autonomous underwater vehicles (AUVs) are not hindered by cables but are as yet incapable of collecting samples from the underside of ice. Navigational limitations and worries about collision with the irregularities under ice floes, in particular keels or pressure ridges, preclude approach to within less than about 10 m of the lower surface. Macaulay (1994) pointed out the possibility of using AUVs as platforms for acoustic sampling under sea ice. AUV technology has since matured (Millard *et al.*,

1998), and this type of sampling has been accomplished recently (Brierley *et al.*, 2002). Continuous underway acoustic sampling by the *Autosub-2* AUV (Plate 2B) running at around 100 m depth along defined line transects from open to ice-covered waters has revealed the importance of the sea-ice edge as a zone of enhanced biological activity (Plate 1B). Bottom-moored upward-looking acoustic instruments have been used to make observations of the pelagic environment beneath sea ice (Kaufmann *et al.*, 1995). As with coring, point acoustic samples of water column biomass suffer because they are unable to resolve the sometimes intense spatial patchiness. The continuous sampling now possible with AUVs overcomes this problem.

Despite the potential damage caused by coring, a core hole is routinely the only access point for sampling organisms living in the sea-ice/water boundary layer. Various techniques have been utilized for sampling through core holes. The simplest is to pump large volumes of water from the sea-ice/water interface using a tube that can be positioned immediately under the sea ice and that can be extended several metres from the core hole. Schnack-Schiel *et al.* (1995, 1998) pumped up to 250 litres through 55 μm mesh nets for copepod enumeration. A finer resolution sampling can be achieved with the ADONIS sampler described by Dieckmann *et al.* (1992) (Figure 4). Originally designed for sampling the interstitial water in

Figure 4 Sampling the under-ice platelet layer and water column underlying sea ice using ADONIS (Dieckmann *et al.*, 1992). The sampling tubes (shown close-up in the inset) are positioned using the vertical pipe and samples are obtained with a vacuum pump. The vertical pipe is heated electrically to prevent samples from freezing.

platelet layers and the water column underlying fast ice, this sampling device is designed to profile at 0.12 m intervals to a depth of about 1 m below the under-surface of sea ice. The device is ideal for sampling water and organisms of the size up to protozooplankton, but is not suitable for sampling metazoans. Dieckmann *et al.* (1992) questioned the suitability of ADONIS samples for the measurement of dissolved gases in the water, but tests by Günther *et al.* (1999b) showed them to be suited ideally for profiling dissolved gases in water and platelet layers. Smetacek *et al.* (1992) used a similar apparatus involving an L-shaped support. They were able to direct tubing 1 m away from a core hole in order to collect water from the platelet layer and underlying water at 0.25 m intervals, down to a depth of 1.5 m below the sea ice. Herman *et al.* (1993) have described a multi-purpose L-shaped structure that can be adapted as a plankton sampler for the sea-ice/water interface, for incubating sea-ice cores with radio tracers for primary production estimates, and for measuring algal concentrations and profiles at subsea-ice surfaces. The NIPR-1 sampler (Fukuchi *et al.*, 1979) is a dedicated zooplankton sampling device that uses a propeller to push water through a net. This sampler enables high-resolution (0.25 m intervals) profiles to be made beneath sea ice.

One of the most difficult problems facing sea-ice researchers is sampling organisms living within the ice with as little disturbance and damage as possible. Even with the most stringent procedures, brine drainage occurs as soon as a core is removed from sea ice. In order to minimize loss of material, and contamination with brine from other horizons, it is routine practice to section cores on the ice and store individual sections for subsequent melting and analyses (Horner *et al.*, 1992). The best method for subsequent treatment of cores is the subject of debate. Melting the core sections, even in the dark at low temperatures, may damage delicate organisms as a result of the osmotic stresses associated with large reductions of salinity (Garrison and Buck, 1986). Damage may not be excessive for samples dominated by diatoms. Hypo-osmotic stress might lead to the release of dissolved organic solutes and/or ions, affecting measurements of inorganic nutrients and dissolved organic matter (DOM). Thomas *et al.* (1998, 2001a) present evidence indicating that this is not the case. Gentle melting of core sections in large volumes of filtered sea water is an accepted method for collecting organisms from the ice (Garrison and Buck, 1986). The method has been modified by melting ice in dialysis bags under running sea water at $-1°C$ (Syvertsen and Kristiansen, 1993; Giesenhagen *et al.*, 1999). This has the advantage of concentrating the biota in a relatively small volume quickly and with minimal damage.

Whatever the method of sea-ice melt, it is difficult to retrieve organisms for use in grazing rate experiments or productivity incubations. These parameters are best measured in the brine in which the organisms are living (Gleitz and Thomas, 1993; Gleitz *et al.*, 1995; Grossmann, 1994). Various methods are available for collecting sea-ice brines. The most versatile method is that of "sackhole" sampling, in which incomplete core holes are drilled into the sea ice and the brine that drains into the hole is collected (Garrison and Buck, 1986; Stoecker *et al.* 1992, 1993; Gleitz *et al.*, 1995). Care must be taken to avoid contamination by underlying sea water, which may penetrate from the ocean below and fill the sampling hole. The porosity and temperature of sea ice is crucial to the success of brine collection. In rotten, porous sea ice it is difficult to collect brine that is not contaminated by sea water, whereas in cold sea ice at temperatures lower than $-10°C$ it is difficult to collect brines in any quantity in a reasonable time because of the lack of brine in ice at these temperatures.

Direct sampling has enabled the estimation of brine gas concentrations for the first time (Gleitz *et al.*, 1995), and a better characterization of brine chemistry (see Thomas and Dieckmann, 2002b). However, any sampling method that relies on collection of the sample over a period of time where brine–atmosphere exchange can take place cannot give a precise measure of actual brine gas conditions. A second method that is used routinely to collect brine from sea ice is to centrifuge core sections at *in situ* temperatures (Weissenberger, 1992; Krembs *et al.*, 2000, 2001a, b). Centrifugation has the advantage that there is better control over the source of the brine; sackhole brine will come from the walls and bottom of the core hole. Experiments measuring DOM and inorganic nutrients (Giannelli *et al.*, 2001) have shown that centrifugation removes dissolved organic and inorganic constituents of brine to the same degree. A major disadvantage of both centrifugation and sackhole sampling is the uncertainty as to what percentage of particulate matter is retained within the ice. As centrifugation may continue for up to 20 minutes, brine collected in this way is unsuitable for accurate gas analyses.

Microelectrode technology for determining fine-scale gas fluxes and irradiance measurements in algal mats and sediment samples has been modified to measure the primary productivity of assemblages on the underside of the ice (McMinn and Ashworth, 1998; McMinn *et al.*, 1999, 2000; Kühl *et al.*, 2001; Rysgaard *et al.*, 2001; Trenerry *et al.*, 2001). It is likely that brine physicochemical properties will be more clearly understood only when such technology can be deployed readily in the field. Initially, microelectrodes were used on core sections that had been retrieved from ice floes (McMinn and Ashworth, 1998). In later studies the methods were adapted to allow under-ice deployment and *in situ* determination of photosynthetic activity under natural irradiance conditions, and a range of grazing pressures, nutrient conditions and under-ice current velocities (McMinn *et al.*, 2000; Kühl *et al.*, 2001; Rysgaard *et al.*, 2001; Trenerry *et al.*,

2001). A major limitation with this method is the difficulty of using fragile electrodes in hard ice. The challenge will be to devise electrodes or sensors that can be used to study processes within sea ice.

Many workers have extracted sea-ice brines to obtain material with which to make primary production and bacterial production estimates (Stoecker *et al.*, 1992; Gleitz and Thomas, 1993; Grossmann, 1994). Mock and Gradinger (1999) and Mock (2002) have adopted a different approach. They obtained whole cores, sectioned them, and put the sections into sealed chambers which they then placed back into the core hole for a period of incubation. This method seems to be a step towards gaining realistic estimates of *in situ* biological activity within the sea ice. Most previous attempts at *in situ* incubation, using radiotracers to measure rates of primary production, have been restricted to bottom communities (Mock and Gradinger, 1999). Other measurements of bacterial and algal productivity have been restricted to traditional incubations under controlled light and temperature regimes in the laboratory. As in the controversy between *in situ* and incubator methods for measuring water column primary production, there is a great deal of uncertainty as to how comparable, realistic or relevant rates obtained from sea ice are with either approach. Smith and Herman (1991) compared *in situ* and incubator methods in sea-ice primary productivity studies and showed that there were large discrepancies between the results from the two approaches. The differences were attributed to a suite of causes, and Smith and Herman (1991) could not arrive at a clear preference of one method over the other. As they stress, any valid production method must at a minimum account for production reflected by biomass increase. Considering the large-scale heterogeneity of biological assemblages within sea ice, any productivity measurements based on the techniques used to date can at best only be viewed as good estimates since even the biomass-increase criterion of success is hard to evaluate.

3.2.2. *Laboratory sampling*

There is increasing use of laboratory simulations of sea-ice conditions for research with sea-ice organisms because abiotic conditions can be controlled carefully. This trend is generally found in marine microbial ecology, where laboratory experiments account for >30% of the effort (Duarte *et al.*, 1997). Laboratory-based sea-ice studies range from incubations in the liquid phase, where brine salinities are altered in line with the temperature changes (Gleitz and Thomas, 1992; Gleitz *et al.*, 1996b), through to experiments where sea ice is actually produced under controlled conditions in what are simply mesocosm experiments with a sea-ice cover. The advantages of such an experimental approach are obvious considering the logistic difficulties of working for long periods in the field. Laboratory studies show striking species-specific differences in response to various temperature, salinity and nutrient conditions (Aletsee and Jahnke, 1992; Grossmann and Gleitz, 1993; Thomas and Gleitz, 1993; Mock and Gradinger, 2000). Such differences are not discernible, or rarely so, in field samples.

In the simplest form, laboratory simulation sea-ice experiments can be conducted in small water volumes of the order of tens of litres (e.g. Grossmann and Gleitz, 1993). Using somewhat simple technology (cool box, modified domestic freezer and magnetic stirrer) they were able to produce realistic sea-ice pancakes over 3 days, and were able to measure the development of associations between algal and bacterial cells contained in the brine that was collected by centrifuging the pancakes. On a larger scale, Weissenberger (1998) was able to maintain a living sea-ice assemblage including bacteria, algae, protozoans and metazoans in a 220 litre mesocosm over a 19-week period. In this experiment the sea ice was formed from sea water containing a natural diversity of organisms, and it was possible to follow the succession of organisms within the sea ice as it grew. Weissenberger (1998) states that the tank size was at the lower end of what is acceptable for sampling, and points out that the reduced ice thickness and limited ice/water ratio obtained may affect transport processes by brine motion and exchange with the underlying water. However, these detractions are minor in comparison with the wealth of information that can be obtained from such studies. Eicken et al. (1998) and Krembs et al. (2002) have taken sea-ice tank experiments one step further by adapting model-ship testing facilities (Figures 5 and 6; see the Hamburg Ship Model Basin (HSVA) www.hsva.de) and used a $180\,m^3$ synthetic sea water-filled basin to conduct mesocosm experiments. This mesocosm proved to be suitable for establishing a sea-ice algal assemblage, and it was possible to relate the succession of algal species to the changing physical properties of the sea ice (also see Krembs et al., 2000, 2001b). One limitation of such large facilities is the limited potential to measure "natural" chemical changes within developing sea ice and underlying water, because contamination from tank facilities is difficult to avoid. Experiments begun in the same facility as used by Krembs et al. (2000), but for which chemical contamination was a critical factor, employed smaller enclosures within the main basin (Haas et al., 1999a; Giannelli et al., 2001). The facilities used by Krembs et al. (2001a) had the advantage that currents and wave motion could be created, an important feature when trying to simulate natural conditions. In 2400 litre tank experiments, Weissenberger and Grossmann (1998) used different patterns of water movement to demonstrate how important such features are for determining the enrichment of algae and bacteria into a growing sheet of sea

Figure 5 The environmental ice tank of the Hamburg Ship Model Testing Basin (http://www.hsva.de). In this configuration, mesocosm experiments were being conducted within the three containers which had a surface area of $4\,m^2$. The tank can be used without such compartments, giving a total ice area of approximately $180\,m^2$, where surface wind, under-ice water current and wave machines can be employed. (Photograph courtesy of M. Granskog.)

Figure 6 The ice tank at Hamburg. Approximately 10-cm thick ice being collected from the tank, representing 10 days growth at an ambient temperature of $-10°$C. (Photograph courtesy of M. Granskog.)

ice. The ease with which physical and chemical conditions can be regulated in ice tank facilities such as these are likely to take our understanding of sea-ice biology and ecology much further than would be possible by field campaigns alone. The opportunity to produce and grow sea ice under strictly defined conditions will also expedite our understanding of the physical and chemical control of species selection and succession within sea ice. Another advantage of tank facilities is the opportunity they offer for testing new sensors, sampling equipment and technologies for future use in the field. However, these developments should also be structured, heeding the warnings of Duarte *et al.* (1997) who effectively argued for "greater efforts towards field experiments designed to interplay with observational and comparative approaches, leading to conclusive tests of key hypotheses and paradigms through carefully designed, crucial eco-system experiments". A good example of how this desire can be fulfilled is demonstrated by the success of the recent iron fertilization experiments in the Southern Ocean, SOIREE (Boyd *et al.*, 2000; Boyd and Laws, 2001) and EISENEX (V. Smetacek, personal communication).

As an alternative to artificially generating sea ice, a number of investigators have collected natural sea-ice biota for use in microcosm experiments. These experiments have, for example, investigated sea-ice microbial food web structure and studied the effect of melting ice on water column microbial dynamics. The approach commonly taken is to collect natural assemblages, add them to larger volumes of water, and measure the changes in assemblage dynamics and development under various controlled abiotic conditions. These experiments have been used to investigate the influence of the chemical constituents of sea ice, such as high dissolved organic matter concentrations, on the biota of water collected in various regions of the Southern Ocean (Kuosa *et al.*, 1992; Giesenhagen *et al.*, 1999), or even on uni-algal isolates derived from Antarctic waters (Brandini and Baumann, 1997). A more novel approach than "simple" tank microcosms was that employed by Riebesell *et al.* (1991) who used 10 litre plexiglass cylinders that were rotated, and demonstrated the high potential for aggregate formation in sea-ice algal assemblages.

4. TROPHIC PROCESSES

The previous sections of this review have described how, almost without exception, every component of the physical Southern Ocean sea-ice realm is capable of accommodating some form of life at some stage. The next section describes the full spectrum of trophic processes, ranging from primary production to higher-level predation, associated with Southern Ocean sea ice. Sea-ice biota refers to organisms at all trophic levels that live in, on or associated with sea ice for some or all of their life cycle (Horner *et al.*, 1992). Such organisms, whether benthic or pelagic, are termed sympagic ("with ice"). Autochthonous organisms live out most of their lives in association with the ice. Allochthonous organisms are associated with ice only temporarily (Lønne and Gullicksen, 1991). A comprehensive list of terms associated with ice-related biology and ecology is given in Horner *et al.* (1992).

4.1 Primary production

A major task still to be accomplished in Antarctic sea-ice research is to determine the magnitude of microalgal production, and how this relates to the overall estimates of primary production in the Southern Ocean (Legendre *et al.*, 1992; Arrigo *et al.*, 1997, 1998a, b; Lizotte, 2001). The collation of large datasets, including satellite data from the coastal zone colour scanner (CZCS), has enabled a more global picture to be formed of the annual variation in algal standing stocks in sea ice (Dieckmann *et al.*, 1998; Lizotte, 2001). Recent estimates for total annual primary production of Antarctic sea-ice assemblages range from 63 to $70 \, \text{Tg} \, \text{C} \, \text{yr}^{-1}$, which equate to about 5% of the estimated total primary production of the sea-ice zone (Lizotte, 2001). However, as Lizotte (2001) stresses, the relative importance of primary production in sea ice is probably not just a simple function of the contribution it makes to the annual total. The timing and geographic location of the ice-based contribution are likely to be critical factors determining the importance of sea-ice primary production.

Kirst and Wiencke (1995) and Lizotte (2001) have reviewed most of the work to date on primary production within Antarctic sea ice. We now know much about the types of photosynthetic organisms found within sea ice, and less about the chemical and physical constraints acting upon them, but have an increasing body of information about adaptive metabolism/physiologies that enable them to function within the sea-ice matrix. Unfortunately studies of seasonal or complete annual cycles of primary production in the sea-ice zone are lacking. Such data are required for prediction of carbon sequestration in the Southern Ocean, especially in relation to climatic change (Priddle et al., 1992; Lizotte, 2001).

Phytoplankton blooms at the sea-ice edge can be between four and eight times as productive as open-water blooms (Smith and Nelson, 1986). The introduction of low salinity waters at the ocean surface when sea ice melts stabilizes the upper water column, reducing the likelihood of algae being mixed downwards into light-limited depths. In the Southern Ocean, the spring phytoplankton bloom tracks polewards, beginning in October, in the wake of the melting ice edge. The warming, increasingly photo-irradiated upper waters, are seeded by algae released from the decaying sea ice.

Primary production also occurs within the sea ice, with the most conspicuous sign of algae being the often dense brownish discolouration on the underside of sea ice revealed when ice breakers pass through the pack. Hooker (1847) was the first to identify diatoms in Antarctic sea ice. One hundred and forty years later Horner (1985c) reported that "organisms other than diatoms in the size range 2 to $20 \, \mu\text{m}$ are just being recognized from the ice". It is now clear that most of the algal groups in the Southern Ocean plankton are also found in sea ice (Palmisano and Garrison, 1993), although it is true that sea-ice algal assemblages are generally dominated by pennate diatoms. Ackley and Sullivan (1994) have summarized the mechanism by which algae may become incorporated within sea ice (also see Melnikov, 1995). This mechanism, sea-ice scavenging, is relatively unselective and explains the similarity in species composition between the water column and the sea ice. When frazil ice forms, phytoplankton cells adhere to the ice crystals and subsequently become incorporated within the sea-ice matrix as the frazil ice coagulates (Garrison et al., 1983, 1989). This so-called scavenging of algae is augmented by a filtering and trapping of algal cells within grease ice that is driven by the pumping effect of Langmuir circulation, as well as that induced by progressing waves (Weissenberger and Grossmann, 1998). Given the lack of selectivity of ice scavenging, it is likely that the relative importance of nano- and microflagellates to the autotrophic sea-ice community has been (and still is) underestimated. This is most likely because of the difficulties associated with the techniques required to preserve flagellate samples for enumeration (Garrison and Buck, 1986). Despite the high diversity of autotrophs within sea ice, that also includes autotrophic dinoflagellates and ciliates, two small pennate diatoms, *Fragilariopsis cylindrus* (Grunow) Krieger and *F. curta* (Van Heurck) Hustedt, and the prymnesiophyte *Phaeocystis antarctica* Karsten are the dominant species in blooms in the sea-ice zone (Lizotte, 2001). Gleitz *et al.* (1998) found that at high diatom standing stocks species diversity decreases. This has also been reported by Gleitz and Thomas (1993), who showed that as first-year sea ice grew and high algal standing stocks established, the assemblages were dominated by only a very few small diatom species. Taking into consideration the findings of other studies, Gleitz and Thomas (1993) suggested that pore and channel size was the major factor in the preferential accumulation of a few smaller species within sea ice. However, Gleitz *et al.* (1998) subsequently concluded that it was the physiological capacity of these species to maintain high growth rates in the spring and summer, in connection with their life history cycles, that may be the key to the prominence of so few diatom species in the ice.

Phaeocystis species are more usually found in sea-ice habitats not constrained by the brine channel systems, such as surface ponds, rotten summer sea ice or freeboard/infiltration layers. Especially in the latter, these are situations where the contraints of salinity, temperature and low light do not inhibit primary production as they do in interior ice assemblages, thereby enabling high standing stocks (including diatoms) to accumulate (Garrison and Buck, 1991; Fritsen *et al.*, 1994, 1998; Fritsen and Sullivan, 1997; Thomas *et al.*, 1998; Kennedy *et al.*, submitted). This high growth capacity in these assemblages is further enhanced by the possibility of inorganic nutrient exchange, as is the case for sea-ice bottom assemblages. Although much is discussed about the constraints of space, salinity and low temperature within sea ice, and the effects that this has on biology, it is these surface and ice-bottom assemblages in which the greatest standing stocks of algae ($<400\,\mu g\,chl\text{-}a\,l^{-1}$) have been measured in the pack ice. In terms of overall primary production budgets especially, more attention must be given to understanding factors controlling the establishment, initiation, development and fate of these assemblages (Arrigo *et al.*, 1998a; Fritsen *et al.*, 1998). It is not just surface assemblages associated with porous freeboard/infiltration layers, however, that are so productive. Dense dinoflagellate and chrysophyte assemblages can develop in the upper sea-ice interior, and high rates of primary production have been measured at these sites, especially in spring when the upper sea-ice temperature is low and brine salinities are high (Stoecker *et al.*, 2000). These algal assemblages are often poorly defined, but they may make an important additional contribution to total sea-ice primary production.

Within the sea ice, light, salinity, temperature and inorganic nutrient limitations will all combine to reduce rates of primary production (Eicken, 1992; Palmisano and Garrison, 1993; Kirst and Wiencke, 1995). Attenuation of irradiance by sea ice is the major factor limiting algal growth in the ice itself and in the water column beneath. Irradiance within sea ice has been measured, but light levels beneath floes are less well described and the dependence of light transmission on ice thickness is likely to be decidedly non-linear (Priddle *et al.*, 1996). Knowledge of the available photosynthetically active radiation (PAR) levels beneath sea ice are essential to model the contribution of primary production in ice-covered seas to total primary productivity (cf. Arrigo *et al.*, 1997, 1998a, b). Underway measures of PAR beneath pack ice have recently been made from the *Autosub-2* AUV in conjunction with underway measures of sea-ice thickness (Brierley *et al.*, 2002).

Irradiance, salinity and temperature levels within the sea-ice matrix can vary considerably. Each of these variables may have a major impact on primary production. Much of the work on the effects of irradiance on sea-ice algae is summarized by Kirst and Wiencke (1995). Carbon assimilation and growth has been measured at photon flux densities as low as 0.2–$2.9\,\mu$mol photons m^{-2} s^{-1}. Photosynthetic characteristics are generally those of low light-adapted algae, with high photosynthetic efficiency, and saturation and photoinhibition at low irradiance (Lizotte *et al.*, 1998). Rates of photoacclimation in sea-ice algae are similar to those of phytoplankton. There is some evidence that at low light levels the contribution of β-carboxylation reactions to total carbon assimilation may increase (Robinson *et al.*, 1995b; Mitchell and Beardall, 1996). Little attention has been given to respiration in sea-ice algae, although there is some evidence that acclimation in respiratory metabolism may enhance the carbon metabolism efficiency under conditions of low irradiances and low temperatures (Thomas *et al.*, 1992).

The effects of salinity on sea-ice algal growth are independent of those induced by temperature or light, and there is a multiplicative relationship between salinity and light or temperature (Arrigo and Sullivan, 1992). Available data suggest that the ranges of temperature and salinity over which carbon assimilation and growth of sea-ice algae can take place may vary greatly. Carbon assimilation was measured by Arrigo and Sullivan (1992) in salinities ranging from 3 to 90, and by Gleitz and Kirst (1991) up to salinities of 110. Aletsee and Jahnke (1992) measured the minimum growth limits of Arctic *Nitzschia frigida* Grunow to be $-8°C$ at a corresponding salinity of 145, although the generation time of 60 days was exceptionally long. In contrast, the generation times were 50–70 hours at $-4°C$ and a salinity of 73. These data together point to the existence of a highly tolerant algal community that is able to function over the majority of the range of conditions likely to be encountered in sea ice.

A consequence of active algal growth and high photosynthetic activity within the sea-ice matrix is a much altered chemistry within the matrix compared with outside: depletion of dissolved inorganic carbon and major inorganic nutrients, high pH values and strong oxygen supersaturation prevail (Gleitz *et al.*, 1995; Günther *et al.*, 1999b). The depletion of inorganic carbon and high oxygen concentrations can also coincide with extraordinarily high concentrations of ammonia and phosphorus (Dieckmann *et al.*, 1991a; Arrigo *et al.*, 1995; Thomas *et al.*, 1998), which is indicative of efficient regeneration and accumulation of regenerated nutrients. Günther *et al.* (1999b) have proposed a hypothetical relationship describing various stages of the interaction of primary production, dissolved gases, DOM and inorganic nutrients during the development of sea-ice algal blooms. As in many other aspects of sea-ice biology discussed in this review, it is likely that future time-series measurements may provide the key to our understanding of the complex geochemical dynamics underway in the semi-enclosed sea-ice matrix. Complete sets of data on concentrations of particulates and dissolved organic and inorganic constituents are required before schemes such as those of Günther *et al.* (1999b) can be verified. It is beyond the scope of this review to discuss sea-ice core biogeochemistry, but that topic is discussed in detail in a related review by Thomas and Dieckmann (2002b). Although nutrient regeneration rates can be high within the ice, and closed systems can develop at times, the highest standing stocks $(0.5–1\,\mathrm{mg\,chl}\text{-}a\,\mathrm{l}^{-1})$ of sea-ice algae are produced where resupply of nutrients from sea water is possible (Arrigo *et al.*, 1995; Thomas *et al.*, 1998; Kennedy *et al.*, submitted).

An obvious characteristic of pack ice is that, during winter, organisms in the ice will experience periods of very low irradiance. Irradiance may often not exceed compensation levels, and extended periods of full darkness may be experienced, especially if floes are covered in snow. Winter studies are rare, but there are indications that heterotrophic components of sea-ice assemblages may dominate at these times and that, as a result, detritus may accumulate in sea ice by the end of winter (Garrison and Close, 1993; Kivi and Kuosa, 1994). Of course it is not just light levels that are at a minimum in winter, since temperature and salinity constraints will also be most severe in winter months. However, Melnikov (1998) has shown that winter production in the ice may not be as minimal as is sometimes assumed, especially under thin first-year ice with little snow cover.

The possibility of a switch to heterotrophy, in particular uptake of exogenous amino acids and glucose in the DOM pool, has been postulated as a means for winter survival by sea-ice algae (Palmisano and Sullivan, 1985a; Rivkin and Putt, 1987). As reviewed by Zaslavskaia *et al.* (2001), all the necessary activities for glucose metabolism exist in diatom cells, and the potential to transform metabolism from obligate photoautotrophy to full heterotrophy can be achieved through metabolic engineering. Whether such switches to heterotrophy by sea-ice algae are really possible is still open to much speculation. However, dark survival by algae is well documented, especially in diatom species: many species have been shown to produce resting cells that are morphologically similar to the vegetative cells, but are physiologically dormant. Such resting stages are also known from nutrient limited/starved algae (reviewed by Peters and Thomas, 1996a, b). Although cyst and spore formation is well documented for sea-ice dinoflagellates and chrysophytes (Garrison and Close, 1993; Stoecker *et al.*, 1997, 1998, 2000), it is not so common among ice diatoms (Garrison and Mathot, 1996). Physiological resting stages are likely to be used by several algal species, including diatoms, within the ice during adverse winter conditions. Stoecker *et al.* (1997, 1998) showed evidence as to how dinoflagellates and chrysophytes in sea ice encyst just before ice melt. Cyst formation is evidently an adaptation for dispersal in the water column during summer and autumn, and also for overwintering in upper cold, hypersaline ice. The cysts are released into the surface waters following melt and are incorporated into new ice formation during autumn. Some excystment events in early autumn may contribute to brief autumn blooms, with cells then re-encysting before winter (Stoecker *et al.*, 1997). Stoecker *et al.* (2000) state that cyst formation in the dinoflagellate *Polarella glacialis* Montresor (see Montresor *et al.*, 1999) is most likely triggered by nutrient depletion, in particular nitrogen depletion. In contrast statocyst formation in the chrysophytes at the same site was not triggered by nutrient depletion. These organisms often exhibit density-dependent control of sexuality and statocyst formation.

Excess lipid formation has often been reported as a feature of sea-ice algal assemblages, and this is generally attributed to light and/or nutrient stress. However, as well as increased lipid production, metabolic changes may result in different classes of lipid being produced at different times (Nichols *et al.*, 1989; Priscu *et al.*, 1990; Fahl and Kattner, 1993; Gleitz *et al.*, 1996a; McMinn *et al.*, 1999; Mock and Gradinger, 2000). There are very obvious species differences in this response (Lizotte and Sullivan, 1992; Thomas and Gleitz, 1993), and increased lipids in sea-ice algal assemblages are not universal. It appears that often the switching of allocation of assimilated carbon between the pools of low molecular weight metabolites and polymeric carbohydrates is the major response to changing nutrient, temperature and light conditions in sea ice (Gleitz and Kirst, 1991; Gleitz and Thomas, 1992, 1993; Thomas and Gleitz, 1993). Overall increased lipid production under low light conditions is often associated with glycolipid production, the main class of lipid in chloroplasts and, in particular, the thylakoid membranes that are highly developed in shade-adapted sea-ice algae (Mock and Gradinger, 2000). Mock and Kroon (personal communication) showed that the lipid composition of thylakoid membranes of sea-ice algae is altered significantly in response to low light, nitrogen depletion and low temperatures, and that the regulation of fatty acid desaturation of thylakoid membrane lipids is probably the key for efficient electron transport under these conditions. Similar processes are key to the thermosaline regulation of many psychrophilic bacteria isolated from sea ice (Nichols *et al.*, 1995, 1999a, 1999b, 2000; Staley and Gosink, 1999; Sheriden and Brenchley, 2000; Thomas and Dieckmann, 2002a, 2002b).

4.2. Microbial activity in sea ice

One of the areas of sea-ice ecological research that has made the greatest progress over the past 10 years is the study of bacterial activity within sea ice. Aspects of bacterial biodiversity, physiology and biochemistry have been reviewed by Helmke and Weyland (1995), Nichols *et al.* (1995, 1999a), Staley and Gosink (1999) and Pomeroy and Wiebe (2001). There is a rich diversity of bacteria in Antarctic sea ice, falling into a wide variety of phylogenetic groups. Archaea have also been found in sea ice (DeLong, 1998), which until recently was surprising considering the thermophilic characteristics of the Archaea. However, as pointed out by Smith (2001) molecular analysis indicates that Archaea inhabit a wide range of non-hyperthermal habitats. Another intriguing finding is the formation of gas vacuoles by bacteria in sea ice, a phenomenon not reported from any other marine habitat (Staley and Gosink, 1999). The formation of gas vacuoles has been hypothesized to be a mechanism enabling bacteria to rise in the water column, bringing them into contact with sea-ice algae or even causing them to rise within the ice itself. An alternative hypothesis is that the bacteria produce gas vacuoles as a dispersal mechanism.

Sea-ice scavenging, the mechanism by which algal cells are incorporated into sea ice (see section 4.1), does not apply to bacterial cells. Instead it appears that bacteria are incorporated into new sea ice primarily by physical adhesion to concentrations of algal cells and aggregates. This has been shown in field investigations (Grossmann, 1994; Grossmann and Dieckmann, 1994) and laboratory studies, with the amount of bacteria becoming incorporated into sea ice being highly dependent on the algal species being scavenged into the grease ice layer (Grossmann and Gleitz, 1993). These relationships between bacterial colonization and algal species composition appear to be a result of differences in algal cell morphology and the subsequent differences in cell surface area available for the bacteria to inhabit. Some epiphytic bacteria have modifications enabling them to attach to diatom surfaces, and exopolymeric layers of bacteria have even been shown to penetrate diatom hosts (Sullivan and Palmisano, 1984). Archer *et al.* (1996) observed that epiphytic bacteria on a diatom (*Entomoneis* sp.) contributed up to 93% of the total bacterial biomass, but that increases in free-living bacteria occurred between November and December. The close association of bacteria with algal cells and particles that is evidently important for the inoculation of new sea ice is a feature that has often been described for older sea-ice assemblages (Palmisano and Garrison, 1993; Grossmann, 1994; Delille and Rosiers, 1996).

Grossmann and Dieckmann (1994) recorded an apparent preferential incorporation of large bacteria into sea ice and proposed that, for larger cells (volumes close to $1 \mu m^3$), physical enrichment was important for their incorporation. Another strategy that has been invoked for the incorporation of bacteria into new sea ice is ice nucleation. Here the surfaces of bacteria themselves act as catalysts promoting the formation of ice crystals. A number of bacterial strains have strong ice-nucleating capabilities (Sullivan, 1985; Nichols *et al.*, 1995), some of which are abundant in sea ice. If ice nucleation does occur, then bacteria in surface waters involved in the process must have sophisticated osmotic stress- and freeze-avoidance strategies enabling them to cope with the extreme physicochemical stresses that predominate in the liquid phase during sea-ice formation (Nichols *et al.*, 1995).

There is a gradual transition in bacterial community composition from the dominance of psychrotolerant to obligate psychrophilic bacteria as sea ice progresses from grease to pancake ice and through to a consolidated ice sheet (Delille, 1992; Helmke and Weyland, 1995). It is well established, however, that low temperature alone does not account for the selective enrichment of psychrophiles (Helmke and Weyland, 1995; Pomeroy and Wiebe, 2001). Rather, it has been proposed that a bacterial species' ability to sequester substrates at increasingly low temperatures may play a major role (Nedwell, 1999; Reay *et al.*, 1999; Pomeroy and Wiebe, 2001). In direct contrast to Helmke and Weyland (1995), Nichols *et al.* (1995) consider salinity to be a primary factor controlling bacterial growth and survival in sea-ice brines. They suggest that the frequency, magnitude and rate of salinity variation may be a selective factor in the control of psychrophilic bacterial populations. Selection may also be brought about by physical means (Nichols et al., 1995) with elevated levels of organic matter in the sea ice providing solid surfaces for bacterial attachment and a stable matrix for the close association of bacteria and nutrient source. Whatever the selective mechanism, extremely high measurements of DOM in sea ice, at times several orders of magnitude higher than that of sea water (Thomas et al., 1998, 2001a, b; Herborg et al., 2001; Carlson and Hansell, in press), indicate either a possible lack of ability by heterotrophic bacteria to utilize available substrates or a reduced substrate affinity. As Pomeroy and Wiebe (2001) point out, the lower substrate affinity of heterotrophic bacteria in sea ice may result in the one situation in nature where DOM may accumulate, at least in the short term. As a consequence, upon ice melt high pulses of DOM may be released into the water column (Brandini and Baumann, 1997; Kahler et al., 1997; Scott et al., 2000) and may possibly contribute to the initiation of phytoplankton blooms. However, unless there is a very stable water column and quiescent conditions below the sea ice, it would seem more likely that rapid dilution of such pulses will occur. Their significance is more likely to be greatest in the skeletal layers at the underside of ice floes and in the boundary layers of the ice–water interface (Krembs and Engel, 2001; Krembs et al., 2002).

The suppression of some bacterial populations recruited from the water column and the development of a community adapted to the sea ice is apparent from gross bacterial activity measurements (Kottmeier and Sullivan, 1990). Immediately following incorporation into new sea ice, there is a strong reduction of bacterial metabolic activity as water column bacteria decline (Grossmann and Gleitz, 1993; Grossmann, 1994). As the ice grows this gross metabolic reduction is reversed as the psychrophilic population becomes established. As sea ice ages, an accumulation of bacterial biomass is recorded that is a result of both bacterial division and an increase in the volume of the individual bacterial cells. Grossmann and Dieckmann (1994) summarize the discussion about the common observation of large bacteria in sea ice. Various explanations for the phenomenon have been put forward, including a response to low temperature and high nutrient conditions. It seems probable, however, that the large bacterial cell sizes are more likely to result from low bacterial mortality, with individual bacteria being largely protected from bacterivoral organisms and consequently enjoying a longer life, even though metazoan and protozoan grazers have a remarkable capacity for moving in small channels (Krembs et al., 2000). Reduced grazing pressure resulting from restricted consumer access is a widely cited reason for the observed build up of algal and bacterial biomass in wide ranging sea-ice habitats although, as Grossmann et al. (1996) state, this is contrary to the concept of the existence of a highly developed microbial food web in sea ice. However, there are some observations of situations within sea ice where there are large algal standing stocks, abundant grazers and high levels of DOM coupled with evidently high bacterial activity (Thomas *et al.*, 1998).

The coupling between primary production by sea-ice algae and bacterial production is not straightforward. Estimates vary from bacterial production levels at around 10% of sea-ice algal production, to instances where bacterial consumption of organic carbon exceeds production by algae (Grossmann and Dieckmann, 1994 and references therein). In autumn and winter the metabolic activity of bacteria in established, consolidated sea ice contributes substantially to overall heterotrophy of sea-ice microbial communities, whereas in spring and summer sea-ice bacterial production contributes just a small percentage of the total microalgal production. Because of the high rates of primary production in newly formed sea ice, and the inhibition of total bacterial activity discussed above, there is initially a high ratio of primary production to bacterial production (Grossmann and Dieckmann, 1994). The occurrences at times of very high concentrations of DOM in sea ice are not always well correlated with primary production (Thomas *et al.*, 2001a). The nature of this organic matter remains largely unknown, although recent evidence has shown that much of it may be in the form of mucopolysaccharide gels produced by algae and/or bacteria (Herborg *et al.*, 2001; Krembs *et al.*, in press).

Large quantities of extracellular polymeric substances (EPS), such as mucopolysaccharides, within the brine channel system may greatly alter brine fluid viscosity (Krembs *et al.*, in press). If it does transpire that EPS is present in large quantities, then we will have to rethink our impression of the nature of the brine channel system. Rather than a liquid-filled labyrinth it is possibly more like a weak gel-filled system, with obvious consequences for the movement of organisms and transport of solutes (see section 4.3.1). The findings of Krembs *et al.* (in press) clearly show that diatoms may actually lie in a matrix of exopolymeric substances that may anchor the cells in the brine channel system and act as a non-crystallized area around the cells. These circumstances may also pertain to some bacteria. The consequences for bacteria–algae interactions, as well as the effects of such matrices for algal and bacteria grazers, will be profound. The production of EPS, coupled with the narrow channel space, might well be the reason for the apparent breakdown of the classic microbial loop in the sea-ice matrix and the subsequent accumulation there of DOM. EPS production serves various functions that enhance the competitive success of microbes, including preventing efficient uptake by grazers and maybe acting after ingestion as a barrier to slow down the penetration of digestive enzymes (Jürgens and Güde, 1994). This is an intriguing new facet of sea-ice ecology and, as research develops, it is likely that much will be learnt here from the general concepts already known from biofilm studies in aquatic microbiology, and the effect of EPS on the permeability and solute transport in other porous media such as rocks, soils, sediments and aquifers (Meyer-Reil, 1994; Decho, 2000).

The concept that EPS secretions result in microenvironments within sea ice in which attachment to surfaces is enabled, activity of exoenzymes is enhanced, nutrients are sequestered and protection against grazers and toxins is gained (Decho, 1990), is one that is also suggested by the release of ice-active molecules by all species of diatom found in sea ice (Raymond et al., 1994; Raymond, 2000). It has been suggested that these ice-active substances (IASs) may be glycoproteins that bind preferentially to ice crystals, causing pitting. This pitting may in turn alter the optical properties of sea ice and may also help to maintain fine pore space structure. However, Raymond (2000) points to the possibility that ice pitting might not actually serve any function. Rather it might be that some interaction of the IASs on the surface of cells increases the ability of cells to stick to ice surfaces, or is involved in protecting cells from freeze-thaw damage. Since IASs appear to be ubiquitous for sea-ice diatoms, it is also possible that they may also be secreted by the psychrophilic bacteria that are favoured as sea ice grows. An interesting feature of these IASs is that they seem to be produced mainly while sea ice is growing; in established sea ice their production is much reduced (Raymond, 2000).

Kirst et al. (1991) and DiTullio et al. (1998) have shown that sea-ice algae have the potential to produce significant quantities of dimethyl-sulphoniopropionate (DMSP) in response to low temperatures and high salinities in the brines. DMSP is cleaved into equimolar concentrations of dimethyl sulphide (DMS) and acrylic acid by enzyme activity and/or a result of grazing activity or virus lysis (Malin and Kirst, 1997). DMSP is a potential bacterial substrate, as is DMS. The role of acrylic acid has received relatively little attention, which is surprising since it has long been speculated that acrylic acid can retard bacterial growth. However, it is known that some marine bacteria can metabolize acrylic acid (Noordkamp et al., 1998 and references therein), and Slezak et al. (1994) have shown that it is only when acrylic acid concentrations are very high ($>10 \mu M$) that inhibition of bacterial growth actually takes place. Considering the high levels of DMSP that have been detected in sea ice (e.g. Turner et al., 1995), there seems to be a potential for inhibitory acrylic acid concentrations to develop. At very high sea-ice algal biomass levels ($>500 \mu g \, chl \, \alpha \, l^{-1}$), bactericidal and/or bacteriostatic compounds from the algae are thought to explain decreases in bacterial growth such as are seen in first-year bottom sea-ice assemblages (Monfort et al., 2000). However, this decrease might not be related to DMSP or acrylic acid production. Guglielmo et al. (2000) refer to bacterial activity being inhibited, although in this case they consider the possible inhibition was coupled with extremely high aminopeptidase and β-glucosidase activity in the sea ice. Wolfe *et al.* (1997) have shown that acrylic acid produced from the breakdown of DMSP may be central as a deterrent against the protozoan herbivory that is initiated through a grazing-activated mechanism. However, the significance of the DMSP-DMS-acrylic acid system may be even more profound in sea-ice systems. Wolfe (2000) summarizes how these chemicals have far-reaching implications – acting as kairomones at many spatial scales, serving a variety of purposes from attracting bacteria, acting as alarm cues for metazoans and aiding procellariiform seabirds to locate food patches (Nevitt *et al.*, 1995).

Very little information has been published about viruses or fungi in Antarctic sea ice, though they are common in Antarctic water and sediment samples, and marine viruses are known to have profound implications for biogeochemical cycling and ecology in the world's oceans (Fuhrman, 1999). Riemann and Schaumann (1993) found thraustochytrids in the mucilage tubes of diatoms that were also colonized by large numbers of bacteria. Besides this report, however, and two others cited by the authors that mention fungi, viruses or fungi have not been recorded as being significant components of Antarctic sea-ice assemblages. Marchant *et al.* (2000) report a greatly enhanced virus to bacteria ratio (VBR) in sea-ice zones, and the only major study of viruses in sea ice (Maranger *et al.*, 1994) found VBRs to be amongst the highest reported for any natural samples. Viruses were concentrated in sea ice by the same factors as bacteria when compared with the underlying water column, and occurred in greatest abundance in those parts of the sea ice where bacteria were most active. The VBR ratio was found to decrease with time, which Maranger *et al.* (1994) interpreted as showing the possible proliferation of phage-resistant bacteria with time, or that viral lytic activity increased in conjunction with higher bacterial growth rates. As the study of the microbial ecology of sea ice develops it is clear that the role of viruses will have to be given a far greater emphasis.

4.3. Secondary production

4.3.1. *Grazers within sea ice*

There are some situations in the sea-ice environment where large numbers of grazers are found in association with high algal standing stocks (e.g. Garrison and Buck, 1991; Stoecker *et al.*, 1993; Archer *et al.*, 1996; Schnack-Schiel *et al.*, 1998, 2001a, b; Thomas *et al.*, 1998; Gradinger, 1999). These include surface melt ponds, gap layers and bottom ice assemblages where the mobility of grazers is unconstrained. Although the sea-ice environment generally is characterized by low temperatures and high salinities, lack of space is the greatest restriction for protozoan and metazoan grazers attempting to exploit the high standing stocks of algae and bacteria that accumulate within the sea-ice matrix. Since the brine channel system is space-limited, grazing pressure is not as high there as the algal stocks could otherwise support. The pioneering work of Krembs *et al.* (2000) has done much to enable us to envision how effective a refuge from grazers the brine channel system can be. Using glass capillaries ranging in size from 12 to 1420 μm in diameter as model substrates to mimic the brine channel habitat, they were able to monitor the movement and colonization of brine channel proxies by turbellarians, rotifers, nematodes, harpacticoid copepods, flagellates, amoebae, diatoms and bacteria. Only rotifers and turbellaria were able to traverse "channels" significantly smaller than their body diameter. Turbellarians apparently changed their body dimensions in response to salinity changes. Rotifers traversed channels just 57% the width of their body diameter. Larger amphipods avoided narrow passages and indeed most of the other organisms tested simply congregated in the narrowest of tubes into which they could physically fit. Krembs *et al.* (2000) concluded that pore spaces within sea ice $\leq 200\ \mu$m in diameter are refugia in which bacteria, pennate diatoms, flagellates and small protozoans benefit from very much reduced grazing pressure.

Large quantities of extracellular polymeric substances (EPS, see section 4.2) in the brine channels of sea ice (Krembs and Engel, 2001; Krembs *et al.*, in press) could further restrict movement and thereby reduce further grazing by both protozoans and metazoans. In addition to physical hampering, exopolymers can also depress feeding (Passow and Alldredge, 1999), although conversely they can themselves be ingested and used as a food source by some protozoans and metazoans. Passow and Alldredge (1999) found that aggregates of EPS formed with nano-sized particles that should normally be too small for the filtering apparatus of *Euphausia pacifica* (Hansen), actually served as an alternative food source for this species, thereby enhancing rather than depressing grazing. The full role of EPS in sea-ice secondary production will only become clear once the nature and distribution of EPS within the brine channel system is clarified. If, for example, the findings of Passow and Alldredge (1999) for *E. pacifica* hold for *Euphausia superba* (Dana) (Antarctic krill), then accumulations of EPS at the sea-ice/ocean interface described by Krembs and Engel (2001) could greatly enhance krill feeding there, and could explain some observations of increased krill abundance at this interface (see section 4.3.3).

Schnack-Schiel *et al.* (2001a) produced a synopsis of the main contributers to the meiofauna of Weddell Sea sea ice compiled from various research campaigns. This extended the synopsis presented by Gradinger (1999). Foraminiferans dominated in terms of abundance (48%), followed by turbellarians (23%), harpacticoid copepods (14%) and calanoid copepods (7%). However, expressed in terms of carbon biomass, turbellarians accounted for 53%, the two orders of copepod contributed 20% each and foraminferans made up only 6% of the total. Gradinger (1999) calculated that sea-ice meiofauna as a whole is not food limited and probably does not constrain the accumulation of sea-ice algae in summer. Gradinger's (1999) estimate of total meiofaunal ingestion of $1.1\,g\,C\,m^{-2}\,yr^{-1}$ equates to only 16% of Antarctic sympagic algal production, and this does not consider bacteria or heterotrophic flagellates as potential food sources. These are, however, very much general figures and several studies have shown that protozoan and metazoan heterotrophs can contribute up to 90% of the total sea-ice biota (Garrison and Buck, 1991). As Gradinger (1999) concludes, the next step is to obtain reliable *in situ* grazing rates for sea-ice grazers, which hopefully will become available with the further development of methods such as those used by Krembs *et al.* (2000).

Foraminiferans, although represented by just the single species *Neogloboquadrina pachyderma* (Ehrenberg) are very conspicuous in Antarctic sea ice. Dieckmann *et al.* (1991b) showed clearly that this species is incorporated into sea ice when it is formed under dynamic conditions, and that a substantial part of the population in the ice-covered Southern Ocean is forced to spend part of the life cycle within the sea ice. Although the distribution of foraminiferans in sea ice is difficult to correlate with algal biomass, in samples of high algal standing stocks foraminiferans were present in very large numbers indeed (Spindler and Dieckmann, 1986; Dieckmann *et al.*, 1991b; Thomas *et al.*, 1998). They clearly feed and grow within the ice, and are the only planktic foraminifera known to be able to survive salinities up to the 82 found in sea-ice brine.

Although turbellarians are evidently important contributors to the sea-ice meiofauna (Schnack-Schiel *et al.*, 2001a), detailed investigations of their biology in Antarctic sea ice are limited (Janssen and Gradinger, 1999). Acoel turbellarians were unknown in Antarctic waters up to 1990, but were found by Kurbjeweit *et al.* (1993) and Janssen and Gradinger (1999) in large numbers in summer sea ice, feeding on ice diatoms, as part of well-developed assemblages that also included copepods. Turbellarians isolated from sea ice have wide salinity tolerance, up to salinities of 80, and they utilize the potential conveyed by such osmotic adaptability to change body size in narrow, hypersaline channels (Gradinger and Schnack-Schiel, 1998).

Among the best documented grazers on sea-ice assemblages are the crustaceans often observed in large numbers on the undersides of overturned ice floes. Amphipods and euphausiids are particularly conspicuous in this respect, but are seldom found within the sea ice itself and exploit predominantly ice surfaces on the periphery of ice floes. They are, however, occasionally reported in surface ponds and gap water layers. The ecology of these organisms is discussed in greater detail in section 4.3.2. Copepods, on the other hand, can be found within sea ice at concentrations of over 400 individuals per litre, and particularly high abundances of copepods can accumulate in the dense algal assemblages characteristic of porous summer sea ice (Garrison and Buck, 1991; Schnack-Schiel et al., 1998; 2001b; Thomas et al., 1998). Life history strategies may result in distinct differences in population structures (see section 4.3.2.1) between copepods within sea ice and the same species in open water. Although cyclopoid and poecilostomoid species have been recorded within sea ice, calanoid and harpacticoid species are most common (Schnack-Schiel et al., 1998; Gunther et al., 1999a), with three species dominating the records made to date. These are the calanoids *Stephos longipes* (Giesbrecht) and *Paralabidocera antarctica* (I. C. Thompson) and the harpacticoid *Drescheriella glacialis* (Dahms and Dieckmann) (Tanimura et al., 1984a, b, 1996; Hopkins and Torres, 1988; Dahms et al., 1990; Schnack-Schiel et al., 1995, 1998, 2001b; Swadling et al., 1997, 2000a; Günther et al., 1999a). Both *P. antarctica* and *D. glacialis* have been shown to have broad salinity tolerances (Dahms et al., 1990; Swadling et al., 2000a). This is in marked contrast to two other copepod species that are common in Antarctic waters and the pack ice zone, *Calanus propinquus* (Brady) and *Metridia gerlachei* (Giesbrecht), which were found to be incapable of surviving even slightly elevated salinities (Gradinger and Schnack-Schiel, 1998). *C. propinquus* is often found in surface waters underlying sea ice (Kurbjewit et al., 1993), and has been shown to feed on sea-ice diatoms released from the ice. This species has never been observed in direct contact with sea ice (Gradinger and Schnack-Schiel, 1998), however, and its poor salinity tolerance may explain why. Furthermore, whilst sea-ice melt, through its initiation of the spring bloom, may precipitate conditions favourable for some copepod species, freezing may not. The elevated salinities beneath forming ice (>34) have been shown to be lethal for both *C. propinquus* and *M. gerlachei* (Gradinger and Schnack-Schiel, 1998), and neither of these species would survive incorporation within forming ice.

A strong degree of spatial patchiness is inherent in the distributions of all the sea-ice copepod studies cited above. Although sometimes good correlation may be found between copepod abundance and standing stocks of algae, there is often very little correlation. Different species may also show very different biogeographic distributions (Swadling et al., 2000a). *Drescheriella glacialis* probably has a circum-Antarctic distribution, whereas *Paralabidocera antarctica* is found mainly along the east Antarctic coast, occurring only in low numbers elsewhere. In contrast, *Stephos longipes* is very abundant in the west but occurs only in low numbers in eastern waters. Günther *et al.* (1999b) found some evidence that *S. longipes* and *P. antarctica* may exclude each other, with high abundance of one species always being associated with the absence, or only limited presence, of the other. However, *Stephos longipes*, *Paralabidocera antarctica* and *Drescheriella glacialis* have been found together in platelet layers in the Weddell Sea (Günther *et al.*, 1999b), dominating a fauna that was particularly diverse in copepod species as compared to the (low) diversity normally found in sea ice (Swadling *et al.*, 2000a). *P. antarctica* has also been found in a maritime lake near Davis Station, East Antarctica, that became isolated from the ocean 5000 years BP (Swadling *et al.*, 2000b), and its intense grazing there plays a key role in controlling microbial biomass and community structure (Swadling and Gibson, 2000). *Stephos longipes* and other copepod species recorded from sea ice have been found in a tide crack in the McMurdo ice shelf, and tide cracks such as this are considered to be important nursery grounds for the nauplii and copepodite stages of these species (Knox *et al.*, 1996). Among the dominant copepod species found in sea ice, there appear to be very different life history strategies that work in different ways to maintain populations in the ice. Schnack-Schiel *et al.* (1995) proposed that *Stephos longipes* has a 1-year life cycle synchronized to the annual pack ice formation. Adults and nauplii are found in the ice in winter when few individuals are in the water column. Even though high numbers may remain in rotten summer sea ice (Schnack-Schiel *et al.*, 1998, 2001b), *S. longipes* is most abundant in surface waters in the summer, where development from copepodite stages CI to CIII takes place rapidly. In autumn *S. longipes* is most abundant in mid waters, although the species does populate new ice, maybe through the mediation of highly sticky eggs being laid on rising frazil ice crystals (Kurbjeweit *et al.*, 1993). A young population therefore overwinters in the ice, whereas an older population (CIV) migrates into deeper waters or possibly to the sea floor.

Dreschiella glacialis has a very different association with ice to that of *Stephos longipes*. This species can complete its full life cycle in the ice, with reproduction taking place year-round where ice persists (Dahms *et al.*, 1990; Schnack-Schiel *et al.*, 1998). However, a pelagic or benthic phase in the life cycle is usually necessary in most of the Antarctic pack where ice does not last for more than one season. Although the adults are good swimmers, nauplii and copepodids do not swim (Dahms *et al.*, 1990), and Schnack-Schiel *et al.* (2001b) propose that the dynamics of seasonal melting and refreezing within ice floes, and the effects this has on the entrained copepods, may be fundamental for maintaining copepod populations within the ice where recruitment from the benthos is not possible. How the species recolonizes sea ice in deep waters when no ice persists throughout summer is still unresolved. In contrast *Paralabidocera antarctica* has a 1-year life cycle inhabiting the ice–water interface throughout its lifespan (Hoshiai *et al.*, 1996; Tanimura *et al.*, 1996). With the onset of ice growth in autumn, nauplii enter sea ice and develop to copepodid stage CIII before developing into a pelagic phase in spring. Even during this pelagic phase this species remains largely in the sea-ice/water interface where ice persists.

Copepods feeding extensively on diatoms may suffer an impaired egg-hatching success as a result of aldehydes within the diatoms that can arrest embryonic development (Miralto *et al.*, 1999). The hypothesis is a controversial one, as discussed by Tang and Dam (2001). However, if there is a reduced fecundity in copepods as a result of a high diatom diet, then the consequences for copepods grazing within sea ice may be great since diatoms form the major part of these grazers' diet. It remains to be seen, however, if sea-ice diatoms contain these detrimental aldehydes. If they do, then sea ice would be an interesting system in which to further test the associations proposed by Miralto *et al.* (1999), owing to the close coupling of copepod grazing activity with diatom distribution in the ice.

The contribution of ciliates to the overall heterotrophic biomass within sea-ice assemblages varies greatly between studies, from <10% (Archer *et al.*, 1996) to >70% (see citations in Scott *et al.*, 2001). This apparent difference may be a result, at least in part, of the absence in young sea ice of the large numbers of ciliates that are found in older ice. This in turn has been ascribed to rapid ciliate population growth (generation times of hours to days) within sea ice (Petz *et al.*, 1995).

There seems to be a high degree of endemism amongst Southern Ocean ciliates, and many of these Antarctic specialists are found in sea ice, at times in very high abundance (Garrison and Buck, 1991; Stoecker *et al.*, 1993; Petz *et al.*, 1995; Song and Wilbert, 2000). In the most detailed study to date on ciliate distribution in sea ice, Petz *et al.* (1995) recorded 68 ciliate species, 55 of which were found exclusively in sea ice, six solely in the pelagic and seven in both sea ice and water. Twenty of the sea-ice ciliates were new to science. Corliss and Snyder (1986) identified 26 taxa in sea-ice samples, most of which were not common in the plankton. Whilst it is true that a few species found in the sea ice are also common in the plankton (Garrison and Gowing, 1993; Palmisano and Garrison, 1993), there may be a very large group of ciliate species that are found only in sea ice. There are, however, difficulties associated with preservation methods that confound comparisons of ciliates from sea-ice and pelagic habitats (Stoecker *et al.*, 1994; Petz *et al.*, 1995).

Ciliates ingest particles over a wide size range, from bacteria and detritus to microalgae and other ciliates, with the size of particle ingested being determined largely by the size of the grazer. Scott *et al.* (2001) showed that the non-loricate ciliate *Pseudocohnilembus* (Evans and Thompson) isolated from sea ice grazes principally on bacteria-sized particles, although it can ingest organisms as large as nanoplankton and particles as small as femtoplankton and colloids.

Other heterotrophic protozoans, including amoebae, nanoflagellates, euglenoids and dinoflagellates, have all been found in sea ice in various degrees of abundance. Choanoflagellates have been reported to be highly abundant in some studies (Garrison and Buck, 1991; Garrison and Close, 1993), but of little importance to the overall heterotrophic community in others (Archer *et al.*, 1996). Heterotrophic euglenoids have seldom been viewed as a key component of the heterotrophic assemblages in sea ice, although they did contribute >50% of the total heterotrophic biomass in the study of Archer *et al.* (1996). Heterotrophic dinoflagellates including phagotrophic species are often found in sea ice (Buck *et al.*, 1990). Several groups of dinoflagellates and flagellates engulf prey the same size as themselves. Others have feeding behaviour and apparatus enabling them to feed on prey considerably bigger than themselves (see discussions by Klass, 1997; Smetacek, 1999b, 2001); sea-ice diatoms and equal-sized organisms will thus be readily consumed. Protozoa–protozoa predation is also very likely to occur in the sea ice.

If size limitations do not prevent this host of heterotrophic organisms from penetrating the labyrinth of brine channels to graze on sea-ice bacteria, algae and EPS, then an active microbial food web/network may exist within sea ice (Palmisano and Garrison, 1993). However, there may be a large proportion of the sea-ice matrix where this potential trophic link cannot play a role, and usual "microbial loop" control processes will break down. The regions of sea ice where such grazing activities are more likely to take place are "open plan" bottom ice layers, platelet layers, surface ponds and porous gap layers where brine geometry restrictions are not prevalent.

Garrison and Buck (1991) calculated that grazing by ciliates, dino-flagellates and nanoflagellates alone is potentially highly significant in controlling the development of, and succession within, the sea-ice community. There have been suggestions of general trends in successional sequences in sea ice, with high autotrophic biomass in the early spring giving way to increasing heterotrophic components in late spring and summer (Garrison and Mathot, 1996). Stoecker *et al.* (1993) discuss how such successions are a result of a combination of *in situ* growth, migration, predation and losses due to brine drainage and salinity changes. Marine microbial interactions are highly complex and this complexity is probably enhanced in sea ice compared with open-water situations. Studies such as those of Archer *et al.* (1996) that follow the changes in microbial carbon biomass within sea ice over time, and studies to infer carbon flow, are vital for extending our understanding of microbial interactions. Interpretation of such studies is hampered by the high spatial and temporal variability within sea ice (Stoecker *et al.*, 1993) that is seldom resolved by short field campaigns.

Grazing by both metazoans and protozoans within sea ice will result in the production of large quantities of faecal pellets (Garrison and Buck, 1989; Buck *et al.*, 1990). Detritus may reach high concentrations in older ice (Garrison and Close, 1993) which, on ice ablation and break-up, will contribute a pulse of quickly-sedimenting material to the underlying water column and sediments (see section 5). Metazoan grazing may account for the high levels of ammonia found in many sea-ice cores (Thomas and Dieckmann, 2002a, b). Metazoan feeding behaviour also leads to release of DOM and hastens nutrient regeneration in sea ice (Thomas *et al.*, 1998; Günther *et al.*, 1999b).

Although the Antarctic sea-ice grazing community is diverse and complex, there is a striking absence from the Antarctic sea-ice ecology literature of reports of nematodes or rotifers. To date there is only one report of nematodes (Blome and Riemann, 1999), and that was from a single sample (from first-year ice in the Bellingshausen Sea in a water depth of 388 m). This paucity of nematodes is in stark contrast to Arctic sea ice, where free-living species belonging to the superfamily Monhysteroida are found in abundance (Tchesunov and Riemann, 1995; Riemann and Sime-Ngando, 1997). Gut content analyses show that these nematodes feed on other organisms, including other nematodes, but there is also some evidence that uptake of DOM is a major source of nutrition (Tchesunov and Riemann, 1995). Suitable conditions for nematodes should therefore prevail in Antarctic sea ice. No rotifers have been found in Antarctic sea ice either, even though these too are common in Arctic samples (Gradinger, 1999). The reasons for this Arctic/Antarctic difference are unclear, and it remains possible that they are simply sampling artifacts. It may just be a matter of time before more comprehensive sampling reveals a more complete faunal record for Antarctic sea ice. Foraminiferans, which are very abundant in Antarctic sea ice, had for many years remained unknown from Arctic sea ice but have now been found there, albeit only in a few samples (Dieckmann *et al.*, 1991b; Gradinger, 1999). Alternatively, it may be that suitable vectors for colonization of sea ice are not present in the Antarctic.

Whereas many planktic organisms are simply caught up in the sea-ice matrix and have an enforced seasonal inclusion, some organisms evidently have life histories that require an ice phase, or at least have phases that exploit the ice in ways that other similar species do not. For several of these sympagic fauna, the mechanisms by which they are incorporated into sea ice is unclear. Many of these organisms do not have an obvious planktic life. The most obvious vector is that residual multi-year ice contains populations of organisms that act as inoculum for newly forming ice. However, in the Antarctic only a very small percentage of the sea ice lasts for more than one season, so this seems unlikely. For most planktic organisms it is widely accepted that colonization of the ice happens during new ice formation and the scavenging (see section 4.1) of organisms into grease ice slicks that are further inoculated by the pumping of water through these effective filters (Garrison *et al.*, 1983, 1989; Weissenberger and Grossmann, 1998). For some, however, the vectors or mechanisms leading to the ice phase of the organisms' life remain unknown. For coastal regions with shallow water depths it is not difficult to imagine colonization of the sea ice from the benthos by larval stages, even in species with poor swimming capabilities. The number of marine invertebrate larvae is considerably higher than had been expected under Thorson's rule (Stanwell-Smith *et al.*, 1999) and larval dispersal may be widespread. Another commonly-cited vector in shallow water is lifting of organisms from the benthos attached to anchor ice (Reimnitz *et al.*, 1987; Schnack-Schiel *et al.*, 1995). However, most of the pack ice formation overlies water several thousand metres deep and here mechanisms of colonization by non-planktic organisms remain enigmatic. Ice platelets can be formed at great depths (Dieckmann *et al.*, 1986; Penrose *et al.*, 1994) in large quantities potentially acting as a vector for lifting organisms to overlying waters. However, this phenomenon will be limited to localized patches, and is hardly a widespread process that could explain how organisms colonize sea ice overlying deep waters.

Turbellarians have been shown to spawn in sea ice (Janssen and Gradinger, 1999) in January and February. Eggs, juveniles and adults will be released into the water column upon ice melt but, although sea-ice turbellarian species can swim, none have been reported in the plankton and it is presumed that they sink to the sea floor. Janssen and Gradinger (1999) suggest that the species of turbellarians in sea ice may have an adhesive disk that allows them to attach to crustaceans before being released from the ice. Swimming crustaceans, including amphipods that migrate from the sea floor to the ice peripheries, common ice copepods and also other species such as *Calanus propinquus* that do not themselves enter the ice but spend periods of time just beneath it (Kurbjeweit *et al.*, 1993; Gradinger and Schnack-Schiel, 1998), may act as vectors for transferral to different ice floes either in the long or shorter term.

Many ciliate species have been described from the ice, with no planktic form found. Although Petz *et al.* (1995) referred to the colonization taking place via resting spores, there is no direct evidence for this. There is a similar conundrum for Arctic nematodes. Riemann and Sime-Ngando (1997) discussed a number of ways by which nematodes with no planktic form may inoculate ice and similar speculation can be made for Antarctic turbellarians and the "non planktic" ciliate species. Other animals may also play a colonization role: some nematodes known from sea ice have been found in whale baleen plates (Lorenzen, 1986).

4.3.2. The community beneath sea ice

The sub-ice habitat – the sea water directly beneath sea ice – accommodates numerous allochthonous and autochthonoubetes sympagic zooplankton. Sea-ice cover also influences the abundance of other fauna that, although not dependent upon the ice *per se*, respond to the presence of associated sympagic life. The annual advance and retreat of sea ice, for example, is often discussed as a factor affecting food availability for macrozooplankton (e.g. Ross and Quetin, 1986; Smetacek *et al.*, 1990), either via the direct supply of food materials from sea ice or indirectly by providing conditions favourable for phytoplankton blooms.

Much of our knowledge on the distribution and abundance of zoo-plankton and nekton beneath Antarctic sea ice has arisen from two major ship-based research programmes to the Weddell and Scotia Seas in the 1980s; the European *Polarstern* Study (EPOS; see for example, Eicken, 1992), and the American AMERIEZ (Antarctic Marine Ecosystem Research at the Ice Edge Zone; see for example, Daly, 1998) programme. Data from these cruises continue to yield new insights into the ecology of sea ice.

4.3.2.1. *Copepods* As a component of the AMERIEZ programme, Hopkins, Lancraft and colleagues (e.g. Hopkins and Torres, 1988; Lancraft *et al.*, 1989, 1991) collected zooplankton with a variety of different-sized nets from open- and ice-covered waters in summer and winter in the Weddell and Scotia Seas. Their, and other data enable zooplankton communities in sea-ice-covered and adjacent open waters to be compared. In the vicinity of the sea-ice edge, in the <1 mm size range, they found that copepods of the genera *Oithona*, *Oncaea*, *Ctenocalanus* and *Mirocalanus* were dominant. Species composition was generally similar in open and ice-covered waters, although a member of the harpacticoid family Tisbidae was found frequently under ice but seldom in open water. Microzooplankton abundance was sparse in the upper 50 m beneath ice, and Hopkins and Torres (1988) suggested that this may have been a result of the presence of cooler waters there. In the 1–20 mm size range the calanoid copepods *Metridia gerlachei*, *Calanus propinquus* and *Calanoides acutus* (Giesbrecht) contributed more than half of the total water column biomass. These species, and *Rhincalanus gigas* (Brady), *Euchaeta antarctica* (Giesbrecht) and *Oithona similis* (Claus) are generally common throughout the Southern Ocean (see Ward *et al.*, 1996; Atkinson *et al.*, 1997; Voronina, 1999; Atkinson and Sinclair, 2000; Froneman *et al.*, 2000), dominating total zooplankton biomass (Conover and Huntley, 1991). Siegel *et al.* (1992) found *Rhincalanus gigas* to be the dominant copepod in open waters, but found that its importance decreased in the pack ice. Hopkins and Torres (1988) concluded that, overall, for all copepods and nauplii combined, although numerical density was lower under ice, no significant difference in the upper 150 m was detectable between ice-covered and open waters. Ice cover thus seems to impact more on vertical distribution of copepods than on species abundance and diversity. However, it may be too simplistic to consider sea-ice cover in isolation as a factor affecting copepod community composition, since the underlying zooplankton community may be indicative of the prevailing oceanographic regime rather than just the presence or absence of sea ice (see Siegel *et al.*, 1992). Boysen-Ennen *et al.* (1991) compared zooplankton biomass from various ice-covered regions within the Weddell Sea and found some significant differences in biomass and species composition between regions that were distinct from differences in sea-ice cover.

Subsequent to AMERIEZ, Burghart *et al.* (1999) returned to the Weddell Sea to study the distribution of biomass-dominant copepods in spring relative to a retreating ice edge. They found that the abundance of *Metridia gerlachei*, *Calanus propinquus* and *Calanoides acutus* was greater in open waters than in ice-covered regions. There were, however, between-species differences, which were explained in terms of life histories (also see Quetin *et al.*, 1996). Adult female *Calanus propinquus* and *Calanoides acutus* contributed a greater fraction to the total population under ice, where they have been observed producing eggs (Fransz, 1988) even in low chlorophyll conditions. These species time emergence from diapause, and commencement of reproductive output, to coincide with the primary productivity bloom associated with ice-edge retreat (Huntley and Escritor, 1991). *C. acutus* undergoes ontogenetic migrations, overwintering at depths of >250 m. It begins its upward migration in September and gravid females are present in surface waters in October to exploit bloom conditions. *Calanus propinquus* does not undergo an ontogenetic migration (Atkinson and Peck, 1988) but does appear to have an annual life cycle, with reproduction peaking in coincidence with the spring and summer blooms. *M. gerlachei*, on the other hand, depends less upon the ice-edge bloom. This species has a diverse diet throughout the year (Pasternak,

1995) and may produce more than one generation per year (Schnack-Schiel and Hagen, 1995; Atkinson, 1998). Spring reproduction is uncoupled from primary production, with mature females being present before the blooms (Schnack-Schiel and Hagen, 1995), most probably fuelled by internal lipid reserves. There is some evidence to suggest that ice-edge blooms do not always develop (e.g. Boyd *et al.*, 1995; Jochem *et al.*, 1995) (see section 5). In these circumstances the most ice-dependent species, *Calanoides acutus*, may adopt a 2-year life cycle, postponing reproduction until after a second winter in diapause (Atkinson *et al.*, 1997). Despite life cycle differences, female *Calanus propinquus*, *M. gerlachei*, *Oithona similis* and *Calanoides acutus* have all been observed producing eggs under pack in the Weddell Sea (Fransz, 1988). Chlorophyll concentrations were immeasurably low at the time of these observations, and egg production rates of just 1–15 eggs per female per day were much lower than those recorded during peak bloom (10–$30\,\mu g\,chl$-$a\,l^{-1}$) conditions (typically 30–40, but up to 120 eggs per day) (Lopez *et al.*, 1993). Conover and Huntley (1991) reviewed life cycles of some copepod species from the Arctic and Antarctic and concluded that there were some strategies common to species around both poles adapted to the highly seasonal sea-ice environment. Individual growth did not appear to be as important as high fecundity.

In addition to exploiting the phytoplankton blooms at the sea-ice edge (Burghart *et al.*, 1999), sea-ice algae may be a food of considerable importance to some copepod species (Conover and Huntley, 1991). In a review of life cycles of Southern Ocean copepods, Atkinson (1998) concluded that the significance of sea ice either as a refuge or a food source still remains largely unknown. Indeed it seems that the situation reported in his review is not greatly advanced beyond that of Conover and Huntley (1991), who lamented the poor state of knowledge of copepods in ice-covered Antarctic seas. Atkinson (1998) commented that, compared with the Arctic, literature on copepods feeding beneath sea ice in the Southern Ocean was sparse. He went on to suggest that this was perhaps because efforts had concentrated on feeding by juvenile krill (Daly, 1990; Smetacek *et al.*, 1990), leaving copepods overlooked, or alternatively that the role of copepods as under-ice feeders in the Arctic had been usurped in the Antarctic by krill, a species that is both well adapted to feed under ice and an effective copepod predator (Atkinson and Snÿder, 1997). Clearly, given the potential contribution that copepods make to Southern Ocean productivity (Voronina, 1998), there is a pressing need to learn more about the biology of copepods beneath Antarctic sea ice. In this regard, the logistical and temporal constraints inherent in conventional ship-based sampling need to be overcome. Russian-American collaboration at the Ice Station Weddell 1 (ISW-1) has provided temporally extensive sampling opportunities deep in the pack ice. Voronina *et al.* (2001) collected zooplankton by net from depths as great as 1000 m from ISW-1 between March and May while the station drifted some 300 km south of the marginal ice zone. They found that, as in the open ocean (Voronina and Kolosova, 1999), copepods were dominant both in terms of mesozoo-plankton biomass (63%) and numerical abundance (62%). They also found that, as in AMERIEZ and EPOS studies, the depth distributions and relative contributions of different species to total community were significantly different to open-ocean locations. Voronina *et al.* (2001) similarly invoked life cycle differences and differences in food availability under sea ice compared with open water to explain their observations. Using concepts first aired by Ekman (1953), Voronina *et al.* (2001) went on to suggest that for *Calanus propinquus, Calanoides acutus* and *Rhincalanus gigas* the perennial sea-ice environment was a "sterile zone of expatriation" within which population losses exceeded production. They thought that for *Metridia gerlachei*, on the other hand, the under-sea ice zone was a "normal" part of its distribution range.

At the same time as Voronina *et al.* (2001) were conducting their net sampling, Menshenina and Melnikov (1995) collected zooplankton from ISW-1 by SCUBA from a layer extending from immediately beneath the sea ice to a water depth of 5 m. They followed the seasonal dynamics of the most abundant copepod species and found that development usually progressed more slowly in the pack ice than in inshore locations. The difference was ascribed to the "oasis effect".

4.3.2.2. *Micronekton* As a component of the AMERIEZ programme, Lancraft *et al.* (1991) compared micronekton and macrozooplankton communities sampled in sea-ice-covered and open sectors of the Weddell Sea. Sea-ice cover was found to affect community structure, although it remains possible that apparent sea-ice-related differences were confounded by aliasing of summer-to-winter seasonal differences: as is often the case, it is difficult from limited sampling to tease apart the relative effects of time, space and physical factors on community composition. Some species including mesopelagic fish were found closer to the surface under sea ice than in open water (also see Ainley *et al.*, 1986). The response of euphausiid species was not uniform; Antarctic krill, *Euphausia superba*, was, for example, more abundant under sea ice whereas *Euphausia frigida* (Hansen) and *Thysanoessa macrura* (G. O. Sars) were less abundant. *Euphausia crystallorophias* (Holt and Tattersall), the ice krill, is generally considered to increase in abundance at more southerly latitudes in neritic locations (e.g. Hosie *et al.*, 2000), although there is a zone of overlap between this species and *E. superba*. Abundances of the salp *Salpa thompsoni* (Foxton) and other gelatinous zooplankton were also reduced under sea ice compared with open water.

The importance of sea ice to Antarctic krill, and the relationship between sea-ice cover and the relative abundances of salps and krill, have received much recent attention (e.g. Siegel and Loeb, 1995; Quetin *et al.*, 1996; Loeb *et al.*, 1997; Nicol *et al.*, 2000; Brierley *et al.*, 2002) and we shall return to these later in this review (Section 4.3.3).

While some amphipods including *Paramoera walkeri* (Stebbing) (Lønne and Gullicksen, 1991), *Eusirus antarcticus* (Thomson) (Hopkins and Torres, 1988) and *Abyssorchomene rossi* (Walker) (Kaufman *et al.*, 1995) are common on the underside of sea-ice floes, Lancraft *et al.* (1991) and others (Ainley *et al.*, 1988; Daly and Macaulay, 1988) reported that pelagic species including *Cyllopus lucasii* (Bate) were less abundant under sea ice than in open water. As with copepods, Lancraft *et al.* (1991) conclude that overall in the micronekton and macrozooplankton there is no change in biomass with sea-ice condition. They argue that this is consistent with multi-year longevity of many of the species contributing to total biomass. The major difference that Lancraft *et al.* (1991) detected was between the winter and summer depth distributions between sea-ice-covered and open waters, with carnivores descending to greater depths under sea ice in winter whereas herbivores remained nearer the surface.

For maximum resolution of the differences between sea-ice-covered and open-water pelagic communities year-round studies need to be conducted in both ice-free and ice-covered waters in order to fully elucidate the role of sea ice, and sea ice alone. In an attempt to disengage from the temporal limitations that typically constrain ship-based sampling, Kaufman *et al.* (1995) conducted a feasibility study in which bottom-moored acoustic instruments were deployed to observe the water column in a seasonally sea-ice-covered and an open region of the Weddell Sea. Their sampling was augmented with baited traps and traditional netting activities. Acoustic observations revealed more targets in open water than under pack ice. Under ice, however, pronounced patterns of diel vertical migration were apparent, which were not seen in open water. Net sampling in the pack ice suggested that these patterns were caused by mesopelagic fish including *Pleuragramma antarcticum* (Boulenger) and *Electrona antarctica* (Günther). The fish might well have been making these migrations to feed on shallow-water zooplankton living in association with the ice.

Weddell Sea macrozooplankton and their relationships to sea ice were also studied extensively during the EPOS programme, for example by Siegel *et al.* (1992). They penetrated deep into the consolidated pack and sampled a wide range of zooplankton size classes with a variety of nets. Using multivariate analysis, they distinguished two distinct clusters of zooplankton, one extending southwards from the marginal ice zone into the consolidated pack, the other extending northwards from the marginal ice zone to open water. Similarly to the AMERIEZ studies, Siegel *et al.* (1992) could find no grounds to define the two clusters as distinct communities because there was a very large overlap in species composition between the two: sea-ice conditions were not a significant controller of community structure. Differences between numerical abundances and biomasses of species in the two clusters were evident, as were differences in the vertical distributions of species. Species richness and diversity were lowest under the closed pack ice. Only Antarctic krill showed an increase in abundance under ice.

4.3.3. *Large-scale krill/sea-ice interactions*

Antarctic krill (*Euphausia superba*) are central to the Southern Ocean food web (Miller and Hampton, 1989). They consume a large proportion of primary production (Voronina *et al.*, 1994), feed carnivorously on copepods (Atkinson and Snÿder, 1997; Atkinson *et al.*, 1999) and are a vital food source for many species of predator (e.g. Croxall *et al.*, 1999; Reid and Croxall, 2001). Krill are also fished commercially (Nicol and Endo, 1999). It has been apparent since the early decades of the last century, when the British *Discovery* investigations were underway in the Southern Ocean (see Hardy, 1967) and the link between the large-scale distribution of krill and the distribution of annual sea ice was first made, that the life cycle of krill is closely linked to sea ice (Marr, 1962; Mackintosh, 1972). Both juvenile and adult krill have subsequently been observed directly beneath sea ice (e.g. (Hamner *et al.*, 1983; Spirodonov *et al.*, 1985; Daly and Macaulay, 1988, 1991; Marschall, 1988; Stretch *et al.*, 1988; Bergström *et al.*, 1990; Quetin *et al.*, 1996; Brierley and Watkins, 2000; Hamner and Hamner, 2000; Brierley *et al.*, 2002) and there is now much information on the interactions between krill and the sea-ice environment. Very recent research on krill and sea ice has been motivated, at least in part, by the notion that sea-ice extent may eventually be reduced in the face of ongoing regional climatic warming (cf. King, 1994) and that, as a consequence, krill populations may be impacted (see section 6.2). Some of the key facts pertaining to krill and sea ice are summarized below.

Garrison *et al.* (1986) observed adult krill foraging under sea ice. From both laboratory experiments and *in situ* observations, others have also reported krill grazing microalgae directly from the under-surface of sea ice and foraging on algae released to the water column during sea-ice melt (Stretch *et al.*, 1986, 1988; Marschall, 1988; Huntley *et al.*, 1994). When krill encounter downwelling streams beneath melting sea ice that contain high concentrations of microalgae (chl-*a* concentrations of $\geq 65\,\mu\mathrm{g\,l^{-1}}$) (Stretch *et al.*, 1986) they display "feeding frenzy" behaviour typical of krill foraging in high food concentration patches (Hamner, 1984). This is often followed by periods of ice grazing, during which krill rake algal cells directly from the under sea-ice surface with their thoracic endopodites (Hamner et al., 1983). These observations altered the previously held view that sea-ice algae were of only limited value to krill (Holm-Hansen and Huntley, 1984; McConville et al., 1986). Despite the fact that overall primary production rates are low under sea ice (Kottmeier and Sullivan, 1987), levels of surface algae and algal biomass within the ice itself may be high (see section 4.1 of this review), and it is now established that sea-ice algal communities are a vital nutrient resource for krill (e.g. Hofmann and Lascara, 2000).

Adult krill require food concentrations of between 1 and $5\,\mu g\,chl\text{-}a\,l^{-1}$ to maintain sufficient metabolic activity for gonad development and reproduction (Ross and Quetin, 1986). Although these concentrations are uncommon in oceanic waters (Holm-Hansen and Huntley, 1984), they are often exceeded in the vicinity of sea-ice edges and within sea ice (Clarke and Ackley, 1984; Smith and Nelson, 1986). Adult krill may migrate towards the marginal ice zone in late winter (Huntley et al., 1994) and the blooms that develop may contribute significantly to the metabolic requirements of reproductive females. Adult krill are often found under sea ice in spring and summer, and spawning may coincide with the sea-ice edge phytoplankton bloom (Quetin et al., 1996): if phytoplankton concentrations are above average then females may produce multiple batches of eggs. A consequence of spawning at the sea-ice edge is that krill larvae arrive back at the sea surface, after their ontogenetic migration from the deep waters where they hatch, very distant from any obvious features. The location of high concentrations of larvae may mark the historic location of the sea-ice edge, which will have receded during the intervening developmental period, although the distribution will have been influenced by ocean currents. In the Weddell Sea the sea-ice edge may recede at $1.6\,km\,h^{-1}$ (see section 2.1). This is faster than krill can swim horizontally (sustained speed of about 0.35–$0.55\,km\,h^{-1}$) (Marr, 1962; Kils, 1981; Kanda et al., 1982; Siegel, 1988; Lascara et al., 1999), and observations that krill migrated north away from the sea-ice edge in spring (Sprong and Schalk, 1992) ought perhaps be recast to say that sea-ice receded south from above the krill.

Adult krill are opportunistic feeders and are able to operate carnivorously (Cripps and Atkinson, 2000). They are also able to survive extended periods of starvation and may shrink in order to do so (see review by Nicol, 2000). Juvenile krill, however, require a constant supply of food (Quetin et al., 1996; Daly, 1998) and are unable to survive starvation conditions for long (Hofmann and Lascara, 2000). Sea-ice algae appear to provide a vital resource for juvenile krill, particularly in winter, and aggregations have often been observed foraging on ice algae patches at this time (Kottmeier and Sullivan, 1987; Hamner et al., 1989; Daly, 1990). Quetin et al. (1996) have described sea ice as a "winter pasture" for krill, noting how, in the presence of sea ice and ice-associated food, larval growth rates are elevated and survival is probably increased (also see Quetin et al., 1994). Quetin et al. (1996) have emphasized the differences between the distribution of juvenile krill which, in winter, are tightly coupled to the sea ice, and adults, which are not (Lascara et al., 1999).

Krill recruitment to the west of the Antarctic Peninsula varies markedly from one year to the next. In an attempt to understand this variation Siegel and Loeb (1995) analysed multidisciplinary biological oceanographic data collected between 1977 and 1994 from the Elephant Island region. They could detect no significant relationships between recruitment and total krill stock density or spawning stock density – a common phenomenon in marine population dynamics (see Jennings et al., 2001). They did find that good krill year-class success was directly related to sea-ice conditions during the preceding winter season. The proportion of recruits (R_1 – the ratio of the numbers of krill in the first age class (1+) to the numbers of all age classes combined) in the Elephant Island area was correlated significantly with the time of local sea-ice retreat ($P < 0.05$), and with the duration of sea-ice cover ($P < 0.025$), suggesting that recruitment was increased during years when sea-ice cover was prolonged and sea-ice melt and subsequent retreat occurred later in the season. One-year-old krill were largely absent in years when sea-ice conditions had been poor the previous winter. Loeb, Siegel and colleagues (Loeb et al., 1997) went on to develop a conceptual model linking the environmental and biological processes that controlled krill recruitment (also see Siegel, 2000b). Their hypothesis suggests that long duration of heavy sea-ice cover during winter and late opening of the seasonal pack ice in spring favours earlier onset of the krill spawning season in the Antarctic Peninsula region, and high krill recruitment (see Figure 4 in Hewitt and Linen Low, 2000). Brierley et al. (1999b) subsequently demonstrated that krill abundance at Elephant Island could, in the short term at least, be predicted on the basis of cyclical variations in sea-ice extent.

Siegel and Loeb (1995) found that the abundance of the salp Salpa thompsoni was correlated negatively and significantly ($P = 0.0002$) with the combined duration and concentration of winter sea-ice cover. This suggested that high salp densities occurred after winters with reduced ice extent, years when krill recruitment was also low. Loeb et al. (1997) developed this idea further, suggesting that, through the mediation of sea-ice extent, either domination of krill or salps occurred off the Peninsula, with krill benefitting when ice extent had been elevated. Salps live for <1 year and, being phytoplankton filter feeders, are not able to use the sea-ice algal resource. In the absence of krill, however, salps are able to exploit the water column spring phytoplankton bloom that goes unconsumed by krill, and they undergo explosive population growth. Conversely, long periods of high concentration sea-ice cover delay the seasonal peak of phytoplankton production (El-Sayed, 1988) and dense winter sea-ice conditions provide good feeding conditions for krill, leading to advanced gonad development and early, possibly multiple, spawning. Under these latter conditions krill get the upper hand on salps. Quetin *et al.* (1996) question the proposal that extended sea-ice duration leads to an early start to the krill spawning season, because adult krill do not depend directly on sea-ice algae for winter survival. Instead, Quetin *et al.* (1996) placed a different emphasis on available data and suggested that more extensive winter sea-ice cover resulted in a greater spatial extent and temporal duration of sea-ice algae released from melting sea ice, and consequently to more extensive areas where ice-edge blooms could develop. This, they argued, provided more food in spring to fuel krill reproduction. Whatever the exact mechanism, there is a consensus of opinion that extended sea-ice coverage fosters increased krill recruitment. Loeb *et al.* (1997) also stated explicitly that a second winter of extensive sea ice further bolstered the krill population already elevated from 1 year of sea-ice conditions, both by promoting early and continued spawning and ensuring survival of larvae spawned the previous year. Early spawned larvae are able to feed and grow over a longer proportion of the summer than are late spawned larvae, and are more advanced and better able to survive food-limited winter conditions (Ross and Quetin, 1989; Quetin and Ross, 1991), leading to high recruitment of 1+ age group krill the following year. Abundance of small krill has independently been reported to be high after a season of heavy ice cover (Kawaguchi and Satake, 1994). Rapid early growth enables krill larvae to reach maturity a year earlier (in their third summer as opposed to their fourth) than larvae in food-limited conditions (Quetin *et al.*, 1996). Given this apparent dependence of krill upon sea ice, concerns have been voiced as to the possible implications for krill populations of long-term change in sea-ice extent. This topic is addressed in section 6 of this review.

Much of the evidence outlined above suggesting that sea ice is important for krill is either circumstantial or is from localized *in situ* SCUBA observations or from laboratory experiments. As Stretch *et al.* (1988) have pointed out, estimates of temporal and spatial distributions of krill beneath ice are required before the importance of ice algal communities to the reproductive biology and persistence of krill throughout the winter can be evaluated. Siegel *et al.* (1990) have also stressed how important it is to gain knowledge of krill distribution and abundance under sea ice. Because of the constraints imposed by sea-ice cover on conventional ship-based sampling techniques such as netting (Siegel *et al.*, 1990, 1992) and underway acoustic surveys (Daly and Macaulay, 1988, 1991; Lascara *et al.*, 1999), detailed information on the abundance and distribution of krill under sea ice has been constrained to interpolation from discrete spot samples. Krill distributions are notoriously patchy (Murray, 1996) and sparse point samples are unlikely to provide a robust estimation of regional mean abundance. Indeed, this patchiness may explain some of the apparent contradictions in the above studies, that have variously reported higher krill abundance under ice than in open water, and *vice versa*, often in very similar geographic locations. With the maturation of autonomous underwater vehicle (AUV) technology (Millard *et al.*, 1998; Fernandes and Brierley, 1999; Fernandes *et al.*, 2000) the means for direct observations beneath sea ice over the scale required for quantitative evaluation of krill adundance are now available. AUVs offer advantages over SCUBA divers and remotely operated vehicles (ROVs): they are not physiologically constrained as are divers, and are less hampered by safety considerations (maximum depth 30 m) (Siegel *et al.*, 1990; Ross *et al.*, 1996); neither are AUVs restricted by the umbilical cables that are necessary for ROV operation (Marschall, 1987, 1988; Bergström *et al.*, 1990). Brierley *et al.* (2002) deployed the *Autosub-2* AUV under sea ice in the Powell Basin, northern Weddell Sea, in January and February 2001 to survey for krill. They equipped *Autosub-2* with an EK500 scientific echosounder (Fernandes and Brierley, 1999) and conducted a line survey of seven pairs of parallel transects totalling >210 km under ice. Comparison of under ice AUV data and ship data along reciprocal transects in adjacent ice-free waters revealed krill density to be significantly higher under ice. Overall, three times more krill were detected under ice than in open water ($61.6 \, \mathrm{g \, m^{-2}}$ cf. $20.6 \, \mathrm{g \, m^{-2}}$), with the vast majority of krill biomass concentrated in a band between 1 and 13 km south of the sea-ice edge. These surveys, for the first time, provided continuous data on the distribution of krill under sea ice and highlight clearly the importance of the sea-ice edge to krill (Plate 1B). Acoustic data collected by ice-strengthened research vessels in ice had previously revealed that the behaviour of krill was modified by sea-ice cover (Lascara *et al.*, 1999; Brierley and Watkins, 2000), possibly owing to predation effects.

4.4. Predation

Numerous predators depend either directly or indirectly upon the sea-ice environment. In addition to some of the protozoan, zooplankton and micronekton species described earlier in section 4, these include fish, seabirds and marine mammals. Variations in regional distribution and abundance of predators impact upon processes at lower trophic levels in the sea-ice environment (e.g. Quetin *et al.* 1996; Alonzo and Mangel, 2001), and the behaviour and breeding success of many predators themselves may be influenced by prevailing sea-ice conditions (e.g. Ribic *et al.*, 1991; Wilson *et al.*, 2001).

Mesopelagic fish that usually inhabit deep water have been found shallower under pack ice than in open water (Kaufmann *et al.*, 1995). These fish may migrate from deeper water to feed on zooplankton living in association with the under-surface of sea ice. Ainley *et al.* (1986, 1988) have found such fish, including *Pleurogramma antarcticum* and *Electrona antarctica*, in the stomachs of seabirds foraging to shallow depths in open water amongst heavy pack. It appears, therefore, that the presence of sea-ice cover mediates a trophic coupling between two otherwise disjunct communities, enabling surface-foraging seabirds to exploit deep-water fish that in ice-free areas do not migrate to within the birds' foraging range.

Whilst sea-ice cover may provide a feeding opportunity for some deep-water mesopelagic fish, it forms a physical barrier to air-breathing predators that dive from above. This has consequences both for predator distributions and for behavioural responses of prey beneath sea ice. Brierley and Watkins (2000) examined acoustic records collected on a transect through open-water and sea-ice-covered regions of the Weddell and Bellingshausen Seas in an attempt to determine any impacts of sea-ice cover on the swarming behaviour of krill. Although their data were in the form of ageing paper traces, they were able to determine some major differences in swarm density (number of individual krill per unit volume) in sea-ice-covered and open-water regions (A. S. Brierley and J. L. Watkins, unpublished data). In the Weddell Sea, krill swarms were significantly less dense under sea ice than in open water (19 times fewer individual krill per unit volume of water). In terms of distance run, however, krill swarms were encountered six times more frequently under ice (6 swarms per 100 km) than they were in open water (1 swarm per 100 km). There was no detectable difference between swarm density in ice-covered and open regions of the Bellingshausen Sea. Lascara *et al.* (1999) have also reported differences in krill swarm characteristics in sea-ice-covered and open-water regions to the west of the Antarctic Peninsula. It is possible that these differences were a result of predator effects. In the following paragraphs these observations are used as a framework for discussion of krill predation in sea-ice-covered sections of the Southern Ocean.

One of the factors believed to drive formation of krill swarms is predator pressure (Miller and Hampton, 1989). It is possible, therefore, that further into the pack in the southern Weddell Sea the threat from predation under ice is less than in the open waters to the north, and as a consequence krill are less tightly aggregated there. The marginal ice zone is an important feeding area for krill predators including minke whales *Balaenoptera acutorostrata* (Lacépède) (Kasamatsu *et al.*, 1998; Murase *et al.*, 2001) and Antarctic fur seals *Arctocephalus gazella* (Peters) (Ribic *et al.*, 1991). These and other sea-ice-associated predators such as chinstrap penguins *Pygoscelis antarctica* (Forster, J. R.) (Fraser *et al.*, 1992), Adélie penguins *P. adeliae* (Hombron and Jacquinot) (Ainley *et al.*, 1998) and crabeater seals *Lobodon carcinophagus* (Hombron and Jacquinot) (Nordoy *et al.*, 1995) are air breathing and are unable to forage far beyond the ice edge in areas of unbroken sea-ice cover (see Figure 20 in Daly and Macaulay, 1991; Ainley *et al.*, 1998). Seal numbers generally increase in pack ice with the advance of summer (Bester *et al.*, 1995), possibly because with the break-up of the sea-ice canopy into smaller floes the environment becomes more favourable for their foraging activities. Indeed, loosely aggregated and banded sea ice may provide a staging post from which air-breathing predators can make their foraging dives, and a platform where they can rest between bouts. Predatory fish, of course, suffer no restrictions from a closed sea surface (Hopkins and Torres, 1988).

In the Weddell Sea, solid ice cover would have prevented foraging by some predators further south from the sea-ice edge. Although the relative impacts of various krill predators on overall krill mortality are generally unknown, the assumption that birds and mammals make a major contribution offers a possible explanation for the observations of differential swarm density within and without sea ice in the Weddell Sea. There was, however, no apparent difference in swarm density between sea-ice-covered and open regions of the Bellingshausen Sea (Brierley and Watkins, unpublished data), and therefore the assumption must be that if predation pressure influences swarm density, there were larger numbers of predators in the ice-covered region of the Bellingshausen Sea than the Weddell Sea. Predator numbers at sea, and in particular in ice-covered seas, are notoriously hard to establish (Green *et al.*, 1995). Seal numbers are usually assessed by line transect surveys from aircraft or ice-breakers, and from the few such studies that have been published there is no strong evidence to suggest that crabeater seal density is different in the Weddell or Bellingshausen Seas (Erickson and Hanson, 1990). Joiris (1991), for example, observed mean densities of 1.2 crabeater seals per km^2 in the Weddell Sea and Gerlatt and Siniff (1999) report 0.76 seals per km^2 in the Bellingshausen Sea. Estimates of whale abundance in the Southern Ocean are sporadic, but sighting surveys have indicated that some regions around Antarctica are noteworthy for higher density (Thiele *et al.*, 2000), and the eastern Weddell Sea is one such region (IWC, 1991): the Bellingshausen Sea is not highlighted as a region of high density. Whales are known to associate with the sea-ice edge (de la Mare, 1997), and the body fat condition of minke whales is related to prey (krill) availability there (Ichii *et al.*, 1998b). Knowledge of distributions and feeding behaviour of cetaceans within the pack ice is sparse, although minke whales have been sighted in the sea ice south of Australia in wintertime (Thiele and Gill, 1999), and they may exploit leads and polynyas to penetrate the pack and indeed to overwinter there. Penguin numbers are generally well quantified because birds in breeding colonies can be counted directly. There are considerably more emperor penguins (*Aptenodytes forsteri* Gray, G. R.) breeding around the Weddell Sea than the Bellingshausen Sea, but Adélie penguins which, as a species are far more numerous, breed in greater numbers on the shore of the Antarctic Peninsula to the east of the Bellingshausen Sea (Woehler, 1993). Krill is a major component of the diet of Adélie penguins feeding under ice (Watanuki *et al.*, 1994), and they overwinter on sea ice, foraging through leads (Fraser and Trivelpiece, 1996). Although penguin breeding colony size can be enumerated, how exactly this translates to predator pressure throughout a particular sea is hard to quantify because diets vary and some species (particularly emperors) have enormous foraging ranges (e.g. Kooyman *et al.*, 1996; Croxall, 1997). On balance, it is difficult to conclude whether predator pressure is higher in the Weddell Sea than the Bellingshausen Sea.

The above discussion pertains specifically to predation of adult krill in the pelagic phase. To the west of the Antarctic Peninsula krill larvae are encountered more regularly under sea ice in winter than are adults, and the distribution of adults is not so tightly coupled to the underside of ice (see section 4.3.3). The difference in association between juvenile and adult krill and sea ice may result from these different maturity stages having different optimal balance points between food acquisition and predator avoidance (Daly and Macaulay, 1991; Quetin *et al.*, 1996). Juveniles are less attractive to mammalian and avian predators because of their small size (Hamner *et al.*, 1989; also see Alonzo and Mangel, 2001), and can thus satisfy their continual requirement for food by grazing at the ice interface with little risk, enabling them to survive over winter when pelagic phytoplankton communities are depleted (Ikeda, 1985).

Some species of penguins, seals and cetaceans are themselves preyed upon by the top predators, leopard seals *Hydrurga leptonyx* (De Blainville) and killer whales *Orcinus orca* (Linnaeus), that forage in the sea-ice zone. Leopard seals actively predate Adélie penguins (Lowry *et al.*, 1988; Rogers and Bryden, 1995), and it has been estimated that in Prydz Bay as few as six seals could consume 2.7% of the penguin population in 120 days. Leopard seal predation may also constrain growth of Antarctic fur seal populations (Boveng *et al.*, 1998). The Weddell seal *Leptonychotes weddellii* (Lesson) is generally considered to consume krill and fish including *Pleurogramma antarcticum* (Plötz *et al.*, 2001), although there is a record of this species taking a chinstrap penguin (*Pygoscelis antarctica* Forster, J. R.) (see Todd, 1988): in an environment as potentially harsh as the sea-ice realm, opportunistic feeding clearly plays a role. Killer whales have been observed deep in the pack ice (Gill and Thiele, 1997) in association with minke whales that are a major prey item for them (Thiele and Gill, 1999). Seals and whales have themselves in turn been the subject of predation by man, and the following section deals with that exploitation.

4.5. Exploitation

Commercial fishing for krill and fin-fish (Everson, 1992), sealing and whaling (Laws, 1977; Brierley and Reid, 1999) have all taken place in the Southern Ocean. Of these, fishing for krill and commercial whaling occurred in direct geographical association with the sea-ice edge.

Krill fishing originally occurred year-round, moving northwards as the sea-ice edge advanced (Ichii, 1990; Everson and Goss, 1991), but catches are now much reduced compared with those before the break-up of the former Soviet Union. Present catches are around 100,000 tonnes per annum, compared with a peak catch of just over 500,000 tonnes in 1982 (Nicol and Endo, 1999), and krill fishing activities tend to be concentrated in continental shelf regions (Murphy *et al.*, 1997; Ichii *et al.*, 1998a).

Whereas krill fishing moved progressively northwards with the onset of autumn to grounds that were not engulfed by the advancing ice edge, whaling, which usually began in October, actively tracked the sea-ice edge southwards as it retreated through the summer (de la Mare, 1997). Floating factory ships were introduced in 1905 but were at first unable to haul their catches aboard for processing. However, whalers could process the catch on ice floes and this led to the discovery that blue (*Balaenoptera musculus* Linnaeus), fin (*Balaenoptera physalus* Linnaeus), humpback (*Megaptera novaengliae* Borowski) and minke whales were common at the sea-ice edge (Shimadzu and Katabami, 1984; Tønneson and Johnsen, 1984). These species of whale feed predominantly upon krill, which are abundant at the ice edge (Ichii, 1990; Brierley *et al.*, 2002), and tend to aggregate in areas where krill abundance is elevated (Reid *et al.*, 2000). Pelagic whaling subsequently concentrated at the sea-ice edge for much of the remainder of the commercial era (until the 1960s). With the exception of Japanese "scientific whaling" (Ichii *et al.*, 1998b) that takes around 400 minke whales annually, reported capture of whales by man in the Southern Ocean has ceased. In 1982 the International Whaling Commission (IWC) adopted a resolution calling for an indefinite moratorium on commercial whaling, which became effective in 1986. Furthermore, since 1994 much of the Southern Ocean has been designated as a whale sanctuary (MacKenzie, 1994), although at recent meetings of the IWC several nations have supported a proposal to overturn the complete ban on whaling within the sanctuary that is presently in place. Whale numbers in the Southern Ocean may be recovering post-exploitation, but the species composition may have changed (Kasamatsu, 2000). Despite its devastating impact, the whaling era did contribute much to our knowledge of biology at the sea-ice edge and has also left a legacy of data on sea-ice distribution that may, ironically, prove a key to understanding major environmental change in the Southern Ocean sea-ice realm (Murphy and King, 1997). There was a requirement to record the geographic positions of whale catches and, since catches were at the sea-ice edge, these records provide a unique circumpolar record of sea-ice edge position year-round from 1931 to 1987. Interpretation of this dataset by de la Mare (1997) is discussed in section 7.2 of this review with reference to evidence for major reductions in sea-ice extent, but it should be noted that his is not the only ecosystem interpretation that has been put on these data. Tynan (1998) used the same whale distribution data from the same month (January) to demonstrate the ecological importance of the oceanographic front at the Southern Boundary of the Antarctic Circumpolar Current: this was achieved using the argument that "the highest concentrations of . . . whales coincide with the Southern Boundary". De la Mare (1997) had used the observation that blue, fin and humpback whales "tended to con-centrate near the ice edge" to infer the position of the sea-ice edge, and described the Southern Boundary as being "well north of the ice edge". In contradiction, Tynan (1998) stated that by January the sea-ice edge had retreated southwards, away from the Southern Boundary, leaving "the highest densities of whales . . . in ice free areas". Clearly the whales could not have been in two places at once. The discrepancy between these two studies appears to lie in the fact that Tynan (1998) compared the mean ice-edge position for the period 1979–1987 (Orsi *et al.*, 1995) with pre-1970s whale distributions (her Figure 3). However, it has been shown that there may have been a large southward shift in the ice edge between these dates (de la Mare, 1997): therefore it is not surprising that Tynan (1998) concluded that whales were caught to the north of the ice edge. Interestingly, in an earlier publication Tynan (1997) had acknowledged that it was likely that "the proximity of the Southern Boundary to the ice edge may synergistically result in increased productivity and greater biomass of krill . . . for whales to feed upon". Sea-ice edges and ocean fronts are key environments in the polar oceans that have, in the Antarctic, usually been considered separately.

The interaction of sea ice and ocean fronts is likely to be of major importance to biological and physical oceanographic processes, and they should not be viewed in isolation. Bounding physically different water masses, ocean fronts will act as important transition regions, influencing sea-ice extent (Muench, 1990). Fronts may retain signatures of the winter sea-ice occurrence for some time after the ice retreat (Pollard *et al.*, 1995), and sea-ice influence may thus be extended temporally. Furthermore, both fronts and sea-ice edges are dynamic and highly mobile (Mackintosh, 1972; Moore *et al.*, 1999) and, because of the mesoscale physical variability in such areas, their positions will not always coincide exactly. Although lines on charts provide reassuringly solid indications of such features it is naïve to consider the features themselves as static. We do not believe that historic whale catch data can resolve the relative ecological importance of the Southern Boundary front and the sea-ice edge. During key periods of the year, such as the austral spring, these sea-ice/front systems will be highly coupled and their interactions are likely to be of major ecological significance. The above discrepancy in interpretation of whale catch position data highlights the need for integrated studies of ocean fronts and sea ice. Such studies have been notoriously difficult with conventional research vessels but, with the advent of new technologies such as *Autosub* (Millard *et al.*, 1998; Brierley *et al.*, 2002) may now be tractable.

5. EXPORT OF BIOGENIC MATERIAL FROM SEA ICE

The release of brine into surface waters during sea-ice consolidation, and of fresh water during ablation, have major effects on the mixed layer depth and irradiative fluxes (Eicken, 1992), and consequently a profound influence on the ecology of organisms living in sea-ice-associated waters. The release of the high concentrations of biological matter contained within the sea-ice matrix upon ice melt is another important event in the seasonal sea-ice cycle. The fate of this material has consequences for biogeochemical cycling (see Thomas and Dieckmann, 2002a, b), bentho-pelagic coupling and ultimately for sequestration of organic carbon in the sediments (Legendre *et al.*, 1992).

Biological material released from sea ice can be in the form of dissolved or particulate organic matter, or living cells or aggregations of cells. The concentrations of particulate and dissolved organic matter can be extremely high within the sea-ice matrix (see section 4), but upon release to surface waters these concentrations are soon diluted. Cell numerical densities are also diluted upon ice melt, but nevertheless there is a widely held view that released sea-ice organisms may act as "seed" populations for ice-edge plankton blooms, the development of which is facilitated by meltwater stabilization of the shallow mixed layer (Legendre *et al.*, 1992;

Palmisano and Garrison, 1993; Dunbar *et al.*, 1998; Leventer, 1998; Park *et al.*, 1999). Michel *et al.* (1996) have described ice algal blooms as special case diatom blooms, in which biomass accumulated over the growing season is released suddenly to the water column. Such seeding is by no means ubiquitous, but is dependent among other things on interactions of prevailing winds and vertical mixing (Savidge *et al.*, 1995; Bathmann *et al.*, 1997). Bianchi *et al.* (1992) observed no similarity between sea-ice algal assemblages and the algal assemblage in the adjacent stabilized surface water column. This cast doubt on the notion that the water column population had originated from melt-released stock. Ice algae have a tendency to form large aggregates, and Riebesell *et al.* (1991) concluded that sea-ice algae released from melting sea ice were subject to rapid sedimentation because of this. They also suggested that the selective melting of algal-rich sea-ice layers owing to enhanced absorption of solar radiation could cause the release of pulses of some sea-ice algae before the larger-scale sea-ice melting that induces stabilization of the surface layer. This, and high grazing pressure, they proposed, resulted in the negligible seeding effects they observed in their study area. Others have recorded sea-ice algae in the water column associated with melting ice (Bathmann *et al.*, 1997; Scott *et al.*, 2000; Pakhomov *et al.*, 2001), although it does not necessarily follow that these algae subsequently seeded blooms. The possible seeding effect of sea ice is not restricted just to diatoms. Kivi and Kuosa (1994) showed how heterotrophic flagellates, large bacteria and heterotrophic euglenophytes seed water underlying late winter sea-ice floes in the Weddell Sea. The complex exchange between sea ice and the sea-ice/water interface of the microbial network is illustrated by Krembs and Engel (2001).

Dead algal cells and their solid remains, phytodetritus, accumulate in sea ice. Gutt *et al.* (1998) have shown that large amounts of phytodetritus can be deposited at shelf locations in water depths up to 1300 m. Short pulses of organic material from overlying sea ice may result in adaptive feeding strategies for benthic feeders, which may receive occasional pulses of material that are particularly enriched in lipids (Pusceddu *et al.*, 1999). Seasonal and stochastic falls of material to the sea bed are important events in the deep sea (Billett *et al.*, 1983), and are exploited opportunistically (cf. Jones *et al.*, 1998). Studies of seasonal variation in the biochemical composition of sedimenting material are scarce, despite the fact that such information is pertinent to understanding benthic feeding processes, especially in shallow coastal regions (cf. Cripps and Clarke, 1998). Generally the rates of flux of biogenic material to depths are very low indeed (Fischer *et al.*, 1988) and flux is restricted to short bursts that are highly dependent on the degree of sea-ice cover (Wefer and Fischer, 1991). The fluxes measured around the Antarctic are highly variable. Dunbar *et al.* (1998), for example, working in the Ross Sea, recorded annual mid-depth fluxes of particulate organic matter that were 20–30 times the values recorded by Fischer *et al.* (1988) in the Weddell Sea. Many studies have shown that these sporadic events are associated with the break-up of sea ice, or with sea-ice retreat, and that common dominant components of sediment trap material are frustules of two diatoms *Fragilariopsis curta* and *F. cylindrus* encased in faecal pellets (Bathmann *et al.*, 1991; Leventer, 1998; Thomas *et al.*, 2001b). These may come directly from the sea ice itself (Thomas *et al.*, 2001b) or from bloom events associated with the melting ice (Cunningham and Leventer, 1998). Both of these diatom species are very common components of sea-ice assemblages, and dominate the circum-Antarctic ice-edge zone. Whether their origin is within the sea ice or with an associated bloom, the flux of biogenic material out of the euphotic zone is mediated either by the formation of aggregates as shown by Riebesell *et al.* (1991) or, more often, by packaging into faecal pellets. Mean settling velocities of between 60 and $200 \, \mathrm{m \, d^{-1}}$ are common (Dunbar *et al.*, 1998), although values up to $1500 \, \mathrm{m \, d^{-1}}$ have been reported (Leventer, 1998). Whereas diatom frustules encased in krill faecal pellets are often broken and damaged, there are numerous reports of faecal pellets from other grazers containing unbroken frustules, often of monospecific origin (Leventer, 1998; Thomas *et al.*, 2001b), reaching the sediments.

Krill grazing does account for much of the deposition of faecal pellets collected in sediment trap studies under ice (González, 1992a; González *et al.*, 1994) and may be a result of the krill feeding on the undersides of the ice (Bathmann *et al.*, 1991). Krill faecal pellets may be easily broken down though, and their efficiency as a major flux mediator to great depths is questionable, despite them having potentially high settling velocities (Cadée *et al.*, 1992, González 1992a). However, a sediment trap under a krill swarm recorded a flux of $660 \, \mathrm{mg \, C \, m^{-2} \, d^{-1}}$, which is the greatest flux recorded for faecal matter of herbivorous plankton (Cadée, 1992). Other forms of faecal pellets collected from sea-ice-associated waters tend to be smaller and more robust, being produced by smaller metazoans and protozoan grazers (Nöthig and von Bodungen, 1989; González, 1992b; Gowing and Garrison, 1992; Garrison and Gowing, 1993; Thomas *et al.*, 2001b).

Although many diatom species are destroyed within faecal pellets, or their frustules are broken up, many do become incorporated into the sediments. In particular, the dissolution-resistant *Fragilariopsis* species (Smetacek, 1999a, b) are members of a group of diatom species that can be used as palaeo-environmental indicators of past sea-ice cover (Zielinski and Gersonde, 1997; Leventer, 1998). Other indicators of sea-ice edge events include resting spores and vegetative cells of *Chaetoceros* species, as well as organisms such as the siliceous nanoplankton species *Penta-lalamina corona* (Booth and Marchant) (Zielinski, 1997) that has often been reported in sea ice and that has also been found in krill faecal material underlying sea ice (González, 1992a). Likewise the dinoflagellate cysts formed in sea ice and released seasonally may also be excellent markers in the sedimentary record for annual sea ice (Stoecker *et al.*, 1992, 1997, 1998).

Enrichment of ^{13}C in sediments is used as a proxy for past carbon dioxide concentrations in surface waters (Rau *et al.*, 1989, 1991a, b). However, because carbon dioxide is sometimes limiting in closed sea-ice systems (Gleitz *et al.*, 1995, 1996b), stable carbon isotopic values of diatoms there can become significantly enriched in ^{13}C (Bathmann *et al.*, 1991; Fischer, 1991; Dunbar and Leventer, 1992; Gibson *et al.*, 1999; McMinn *et al.*, 1999; Kennedy *et al.*, submitted). If this isotopic signature is preserved in the sediments it will provide a record of past sea-ice cover. It is not always as straightforward as this, however, as shown by the study of Thomas *et al.* (2001b) in which no ^{13}C enrichment was detected in material, including sea-ice diatoms, sedimenting from an ice sheet. Before the use of ^{13}C-enriched sediments as a proxy for sea-ice cover can be employed, more extensive seasonal studies must be conducted, and more rigorous attention paid to efficiency of collection and characterization of the biogeochemical processes taking place in sea ice and platelet layers.

It is not only particulate matter that is released from sea ice, but also the dissolved constituents. Localized sea-ice inputs may provide concentrated supplies of micronutrients such as iron (de Baar *et al.*, 1995; Löscher *et al.*, 1997). Dissolved organic matter released from sea ice may enhance bacterial and algal productivity, at least in the waters at the sea-ice/ water interface. Brandini and Baumann (1997) demonstrated increased growth rates in Antarctic diatom cultures after addition of DOM from sea ice, and Kähler *et al.* (1997) showed that bacterial activity in surface waters was significantly enhanced following enrichment with DOM released from melted sea ice. Giannelli *et al.* (2001) have shown that DOM behaves conservatively during sea-ice formation, that is it expelled in the same way as inorganic salts are expelled from the ice. Therefore DOM incorporated into the ice as well as that produced by the sea-ice assemblages will be released into surface waters upon ice melt. Considering the high concentrations of DOM in sea ice (Thomas *et al.*, 2001a; Carlson and Hansell, in press), it seems likely that in a stabilized sea-ice/water interface the inoculum of DOM will be very important in the structuring of the complex microbial interactions taking place there (see schematic of Krembs and Engel, 2001). Work in the Arctic has shown that transparent exopolymer particles (TEP) produced by sea-ice diatoms are concentrated in the melt water. Larger TEP particles settled out of the sea-ice/water interface, but smaller particles remained in suspension. Advection of neutrally-buoyant TEP may take place over large distances mediated by prevailing under-ice currents and sea-ice drift (Krembs and Engel, 2001), thus seeding more distant waters.

Gleitz *et al.* (1998) suggested that the dominant diatoms may have life histories geared to the annual sea-ice cycle: the high standing stocks of these species will introduce dense populations into ice-edge waters in spring and summer. This in turn may ensure an increased probability of a population being incorporated into new sea ice formed later in the season.

6. VARIABILITY IN THE SEA-ICE ECOSYSTEM

6.1. Responses to seasonal ozone depletion

Springtime ozone depletion around Antarctica is a well documented phenomenon (e.g. Farman *et al.*, 1985; Uchino *et al.*, 1999). Associated with this depletion is a seasonal increase in the amount of ultraviolet radiation (UV; UVA wavelength 320–400 nm, UVB 280–320 nm) reaching the ocean surface. The consequences of this increase have been the source of much speculation (Arrigo, 1994) and include the possibility of harmful effects on organisms living in surface waters of the world's oceans, in particular the Southern Ocean (Karentz, 1991; Smith *et al.*, 1992; Vincent and Roy, 1993; Kirst and Wiencke, 1995; Buma *et al.*, 2001). In terms of the ecology of sea ice and ice-associated waters, the ecosystem components that may suffer the largest influence of an increase in UV radiation are the blooms that occur in the stabilized waters of the marginal ice zone during spring and summer. The meltwater-stabilized shallow mixed depths may be 20 m or less, and UVB can penetrate to depths in excess of 50 m (Davidson and Marchant, 1994; Prézelin *et al.*, 1994; Buma *et al.*, 2001). The possible impacts of ozone depletion on sea-ice organisms have received little direct attention (see review by Prézelin *et al.*, 1998), and most of the speculation about the possible effects of increased UV radiation on these organisms has to be extrapolated from studies with open-water samples.

One of the biggest unknowns facing scientists interested in the impact of UV radiation on sea-ice ecology is knowledge of how much UV radiation actually penetrates sea ice to influence the biology within it. Perovich and Govoni (1991) and Perovich (1993) modelled UVA, UVB and biologically effective irradiance in different classes of sea ice including

Weddell Sea first-year ice, multi-year ice and ice from McMurdo Sound. Transmittance of all three radiation classes was greatest in the Weddell Sea first-year sea ice, with UVB transmittance >0.01%. In the other two types of sea ice transmittance was usually <0.005%. Snow cover just 0.1 m thick reduced penetration of UV radiation by almost two orders of magnitude. These results show that UV light levels under sea-ice cover will be low. However, since much of the most intense biological activity in sea ice takes place in surface or near-surface regions of the ice, predicting or even measuring transmission of UV radiation to the bottom of, and beneath sea-ice floes, may not actually provide much relevant information. As Prézelin *et al.* (1998) conclude, our present abilities to ascertain *in situ* UV radiation photoecology of sea-ice algae are inadequate, severely restricting our ability to make broader interpretations.

Trodahl and Buckley (1990) measured clear spring-time increases in UV radiation at the bottom of 1.7 m sea ice that was mostly snow free. They found that 1–2% of incident UVB, and >5% of UVA, was transmitted through the ice. Snow cover of just 18 mm reduced the transmission of all wavelengths by a factor of 0.3, although this was partly reversed by a reduced turbidity in the top of the sea ice in the presence of snow cover. Ryan and Beaglehole (1994) measured a 10% transmission of UVA at McMurdo during the spring maximum of ozone depletion. Perovich *et al.* (1998) showed decreases in UVB transmittance from 0.3% to 0.03% between April and June in Arctic sea ice: they attributed this to a bloom of sea-ice algae reducing the transmittance of UV radiation.

Other work in the Arctic has shown transmittance of between 2 and 13% of incident UVB radiation through snow, sea ice and algae within sea ice (Belzile *et al.*, 2000). The sea ice in that study ranged from 0.5 to 1.3 m thick, with snow cover between 0.1 and 0.9 m. Belzile *et al.* (2000) also showed that both dissolved and particulate organic matter in the sea ice contributed significantly to the attenuation of UV radiation in ice. Perhaps the most significant finding from these works is not the actual transmission of UV radiation through sea ice, but the effect that sea ice has on the ratio of UV radiation to photosynthetically active radiation (PAR). The biological impact of ozone depletion may be exacerbated because of the relative increase of UV radiation relative to PAR that is brought about by the high attenuation of PAR by sea-ice algae and DOM in the ice. UV radiation damage and subsequent repair depends on the ratio of UVB to longer wavelengths (Vincent and Roy, 1993), and shade-adapted sea-ice algae are very susceptible to UV damage (Prezelin *et al.*, 1998). As pointed out by Belzile *et al.* (2000), in sea ice the influence of vertical mixing on photoacclimation and repair processes that are important in plankton populations (Vincent and Roy, 1993) are removed, further increasing any harmful effects of transmitted radiation.

The prymnesiophyte *Phaeocystis antarctica* plays a large role in the seasonal dynamics of many parts of the Southern Ocean, in particular in coastal waters and waters of the marginal ice zone (Davidson and Marchant, 1994). However, it can also reach high standing stocks in sea-ice floes, especially surface ponds and surface gap layers (Garrison and Buck, 1991; Palmisano and Garrison, 1993; Garrison and Mathot, 1996). The colonial stage of *Phaeocystis* contains high concentrations of UVB-absorbing compounds which are highly efficient in protecting *Phaeocystis* colonies from UV damage. In contrast, flagellate stages of the *Phaeocystis* life cycle suffer severe mortality from UVA (Marchant *et al.*, 1991; Davidson and Marchant, 1994). These workers have also shown that despite possessing UV-absorbing compounds, *Phaeocystis* is more vulnerable to UV radiation than are diatoms. The conclusions of Riegger and Robinson (1997) are, in this regard, somewhat in conflict with these findings, since they report the presence of UV-protecting compounds in diatoms and *Phaeocystis antarctica*, and that these compounds were most effective in *Phaeocystis* where they were thought to be stored in the extracellular matrix of the colonies. Riegger and Robinson (1997) identified the UV-absorbing compounds in diatoms (including *Fragilariopsis cylindrus* that is commonly found in sea ice) as the mycosporine-like amino acids (MAAs) but were unable positively to identify these in their *Phaeocystis* cultures. They conclude that many phytoplankton species have the potential to respond to the increases in UV radiation resulting from seasonal changes in solar zenith angle by increasing cellular MAA levels, which has been shown to be the case in other Antarctic diatoms during short- and long-term photoacclimation to UV radiation (Hernando *et al.*, 2001). However, the ability to elicit a similar response to the elevated UVB radiation stemming from depleted ozone concentrations may be limited to *P. antarctica*.

One of the few generalizations that it seems possible to make with reference to susceptibility to UV damage is that smaller cells are more vulnerable (Karentz *et al.*, 1991; Buma *et al.*, 2001). This is because of the increased surface area to volume ratios, and the low effectiveness of UV-screening pigments in small cells. This said, however, Helbing *et al.* (1992) found that UV radiation was more inhibitory to microplankton than to nanoplankton.

In the most comprehensive work on effects of UV radiation affecting ice algae to date Prézelin *et al.* (1998) showed that sea ice provided incomplete protection from UV radiation damage, and that DNA damage occurred despite the elevated levels of MAAs within the algal assemblages. However, they did not find any evidence of the UV radiation-enhanced production of photoprotective carotenoids often found in microalgal cells. They also confirmed that UVB sensitivity was most pronounced at low light levels, and that most UVB damage was related to photosystem II damage, which accounted for most of the inhibition of carbon fixation at the cell level.

Response to increased UV radiation is the result of a complex suite of cellular mechanisms including protection, repair, cell size, growth rates and photacclimation (Karentz *et al.*, 1991; Davidson and Marchant 1994; Riegger and Robinson 1997; Prezelin *et al.*, 1998). Interspecific differences in response to this complex of factors will dictate any changes in phytoplankton or sea-ice algal composition. However, in the discussion about potential effects of increased UV radiation resulting from ozone depletion, the most likely scenario due to this environmental change is a shift in species composition or successional patterns (Karentz *et al.*, 1991; Vincent and Roy, 1993). Although there is likely to be little effect on sea-ice bottom assemblages, or on assemblages under snow cover (Ryan and Beaglehole, 1994), there is the potential for such changes to influence surface or near-surface assemblages.

Hefu and Kirst (1997) showed that increased UV radiation significantly reduced the production of DMSP by *Phaeocystis antarctica*, and that this was coupled to an increase in the conversion rate of DMSP to DMS and acrylic acid. Considering the ecological role that DMSP cleavage may have in the ice (see section 4.2), this is an interesting avenue for future investigations, especially as it is not known if UV damage to diatoms may also alter their production of DMSP and ultimately DMS and acrylic acid levels in the ice.

Bacterioplankton, including those in the marginal ice zone (Jeffrey and Mitchell, 2001), may suffer DNA damage from UV exposure (Jeffrey *et al.*, 1996) and although their activity may be impaired by exposure to normal sunlight (Herndl *et al.*, 1993) there can be very quick recovery from UV-induced stress (Kaiser and Herndl, 1997). We are unaware of any reported work on the effects of increased UV radiation on sea-ice bacteria. In temperate waters UVA and UVB radiation lead to only minor alterations in bacterioplankton species composition, since only approximately 10% of the species there are sensitive to UV radiation (Winter *et al.*, 2001). In Antarctic waters Helbing *et al.* (1995) and Buma *et al.* (2001) showed a link between significant UV radiation increase and reduced viability of natural bacterial assemblages. Davidson and van der Heijden (2000) also found significant inhibition of bacterial growth with increasing UV irradiance, although UV-induced damage was repaired. Because of effects on the whole community after strong UV treatment, there was an increased bacterial production, probably resulting from phytoplankton mortality, although virus inactivation, changes in bacterial species composition and acclimation to high UV radiation may also have contributed to the low UV-induced inhibition of bacterioplankton. Davidson and van der Heijden

(2000) also measured increased concentrations of bacterial grazers, ciliates, heterotrophic nanoflagellates and choanoflagellates following high UV radiation.

Coupled strongly to the effects of UV radiation on the bacterioplankton is the potential for photochemical reactions with DOM (Kieber et al., 1989; Mopper et al., 1991; Kieber and Mopper, 1996). Working with Antarctic diatom-derived DOM, Thomas and Lara (1995) indicated that this pool may be quite resistant to photodegradation, although it is clear that the photochemical effects are highly dependent on the nature of the DOM before exposure (Tranvik and Kokalj, 1998). Obernoster et al. (1999) have shown that freshly released labile DOM might be rendered more biologically refractory photochemically whereas, if the DOM is more biologically refractory, the bioavailability of the DOM can actually be enhanced upon exposure to solar radiation. Little work has been conducted on the effects of DOM and UV radiation in sea ice, although Belzile et al. (2000) showed a strong absorption of UV radiation in Arctic sea ice by DOM, and concluded that significant photochemical reactions could occur. In model predictions for surface waters, primary production is enhanced by the presence of chromophoric DOM where predicted increases in production due to the removal of damaging UV radiation offset its reduction resulting from the absorption of photosynthetically usable radiation by the DOM (Arrigo and Brown, 1996). A fascinating future development in sea-ice research will be to determine how UV-induced photochemistry and bacterioplankton damage or enhancement interact within the sea-ice matrix and the subsequent consequences for sympagic grazing organisms.

Within the sea-ice zone, UV-related damage aggravated by ozone depletion may potentially not just be limited to microorganisms, either in the sea-ice matrix or in the water column. Jarman et al. (1999) have reported that *Euphausia superba* DNA has the lowest percentage of guanine-cytosine base pairs (32%) known for any metazoan. A corollary of this is that the krill genome contains a large abundance of adjacent thymine residues, which are particularly susceptible to damage by UVB. Jarman et al. (1999) suggest that, as a consequence, krill may be more susceptible to damage than other Antarctic organisms. However, even krill larvae foraging on the underside of ice floes are unlikely to suffer much damage given the above-mentioned attenuation of UV by sea ice. In the pelagial, krill are most often found well below the sea surface during daylight and there too are unlikely to be exposed to high UV levels. Interestingly, it has been shown recently that vertical migration by zooplankton is influenced by UV radiation (Rhode et al., 2001), with deeper migrations occurring in the presence of UVB. It remains possible, therefore, that the depth and duration of krill diel vertical migrations may be influenced by the ozone-related increases in UV exposure, and this in turn may impact predators that dive for krill, including predators on the ice.

Because of the "keystone" status of krill in the Southern Ocean ecosytem, and its importance to fisheries, much effort has been devoted to explaining the observed variability in krill abundance. Naganobu *et al.* (1999) have proposed that one of the environmental factors influencing krill abundance is ozone depletion. They report a significant correlation between krill density on the Antarctic Peninsula region and total ozone as measured at Faraday/Vernadsky Station. Correlation, however, does not prove cause and effect. Links between krill abundance and sea-ice cover are explored further in the following section.

6.2. Responses to changes in the extent of sea ice

As described in sections 2.1 and 3.1, sea-ice extent around Antarctica oscillates not only within years with the passage of the seasons, but also between years with the passage of the Antarctic Circumpolar Wave (Murphy *et al.*, 1995; White and Peterson, 1996). There is also some evidence for a major reduction in sea-ice extent in the middle of the last century (de la Mare, 1997). There is no evidence of a systematic decline in sea-ice extent around Antarctica as a whole (Houghton *et al.*, 1996), but concern has been voiced in some quarters that sea-ice extent may change in the face of ongoing regional warming (King, 1994; Smith *et al.*, 1999) that may itself be a manifestation of global climate change. Evidence exists of an ongoing warming to the west of the Antarctic Peninsula, with an approximate 0.05°C per year rise in air temperature over the past 50 years (see Smith *et al.*, 1999; Comiso, 2000). A trend of decreasing fast-ice duration in the South Orkneys over the past 90 years has also been identified (Murphy *et al.*, 1995). Here we explore some possible ecosystem consequences of cyclical change and possible future reductions in Antarctic sea-ice extent. Because of the high profile role that krill play in the Southern Ocean ecosystem, and the perhaps disproportionate effort that has been expended in their study (see Conover and Huntley, 1991), this is inevitably biased towards krill.

White and Peterson (1996) demonstrated how, between 1979 and 1991, sea-ice extent around Antarctica oscillated as the Antarctic Circumpolar Wave (ACW) propagated around the continent. They suggested that the sea-ice field was effectively a mobile dipole (think of an oval record – the ice – rotating on a turntable – Antarctica). The approximately 7 years that it takes for the ACW dipole to complete one revolution of Antarctica manifests at any given location, for example the Antarctic Peninsula, as an increase (or a decrease) in ice extent every 3–4 years, or roughly twice a decade. In section 4.3.3 it was shown how krill appear to depend upon sea ice for food and as a spawning environment, and how krill recruitment and abundance at the Antarctic Peninsula are apparently elevated after seasons of extended sea-ice extent (Loeb *et al.*, 1997). Krill are a vital food resource for predators throughout the Southern Ocean, even at latitudes well to the north of the seasonal advance of sea ice, and ice-mediated changes in krill abundance have consequences for these predators. A 23-year time-series of predator performance data at South Georgia, for example, shows how breeding success of a suite of predators is linked directly to krill availability (Croxall *et al.*, 1988, 1999; Brierley *et al.*, 1997, 1999a; Reid and Croxall, 2001). Changes in krill abundance at South Georgia correspond directly with those at the Antarctic Peninsula (Brierley *et al.*, 1999b) and, despite the large distance across which krill must be advected to reach the island (Murphy *et al.*, 1998b), there does not appear to be a seasonal lag in propagation. Changes in sea-ice extent at the Antarctic Peninsula thus influence, via krill, predators very distant from the sea-ice edge (see Atkinson *et al.*, 2001 for a review of South Georgia's pelagic marine ecosystem). Knowledge of the response of krill and other components of the ecosystem to changes in sea-ice extent may convey some predictive capacity (see Brierley *et al.*, 1999b). It may enable ecosystem managers, for example, to restrict krill fishing effort in proximity to predator breeding colonies in advance of slumps in krill abundance resulting from reductions in sea-ice extent. Changing krill abundance could have implications for the krill fishery, and for fisheries on krill-dependent species (Everson *et al.*, 1999; Hewitt and Linen Low, 2000). Brierley *et al.* (1999b) predicted on the basis of sea-ice cyclicity that the 2000/2001 summer season in the south-west Atlantic would be characterized by low krill abundance: acoustic surveys at this time revealed this prediction to be correct.

Siegel and Loeb (1995) and Loeb *et al.* (1997) presented evidence that the frequency of cold winters with extensive sea-ice cover at the Antarctic Peninsula had decreased from 4 out of every 5 years in the middle of the last century to just 1 or 2 out of every 5 years since the 1970s (note that this is roughly the same frequency expected under the ACW scenario). They suggested that this reduced incidence of extensive sea-ice cover may have been the cause of the significantly lower krill population sizes they observed during the late 1980s and early 1990s. Year to year variation in sea-ice extent is unlikely to lead to a change in the krill standing stock size in the short term because the multi-year age structure of krill will buffer the population as a whole against the success or failure of a single year-class. Indeed, the longevity of krill itself, up to 7 years, may have evolved to provide maximum buffering capacity against unfavourable conditions brought periodically on the ACW. However, a sequence of ≥ 3 years when sea-ice extent is reduced may lead to a gradual decline in krill abundance. The apparent change in sea-ice conditions since the 1970s to which Loeb *et al.* (1997) refer could therefore potentially have impacted krill abundance. Acoustic survey estimates of krill abundance are lacking before the early 1980s, and it is difficult to comment directly on changes in krill abundance since that time. Although a major acoustic survey of krill was conducted throughout the south-west Atlantic in 1981 (Trathan *et al.*, 1995) (biomass 35.8 million tonnes, mean density $76.0 \, g \, m^{-2}$) and another was conducted in the same region in 2000 (Hewitt *et al.*, submitted) (biomass 44.3 million tonnes, mean density $21.4 \, g \, m^{-2}$), differences in survey techniques, timing and exact survey locations mean that the two data points cannot be interpreted together as evidence of an intervening reduction in krill abundance. More recently, Siegel *et al.* (1998) have presented data that show an upturn in the abundance of krill at the Antarctic Peninsula, and have concluded that variability in krill abundance is associated with "high interannual fluctuations and not with a persistent change in krill density". This conclusion is consistent with a time-series of acoustic observations of krill density at South Georgia (Brierley *et al.*, 1999b). It is possible that the ACW has ceased to exist as a defined and periodic feature in recent years: evidence for its continued existence in the sea-ice field at least is lacking (W. Connolly, British Antarctic Survey, personal communication). As Reid and Croxall (2001) note, given the substantial within and between year variation in marine ecosystems, detecting the consequences of progressive physical processes such as climate warming is difficult. However, Mangel and Nicol (2000) point out that biological variation is not noise and this, allied with Siegel (2000b) who concludes that it remains to be established whether observations of changes in krill abundance just represent high interannual variability or the beginning of a downward trend, makes it clear that much more needs to be done to verify the existence of, and to understand, changes in krill abundance (Siegel, 2000a). Despite these caveats, Reid and Croxall (2001) suggest that there might well have been a regime shift (see Steele, 1998) with relation to krill populations since the mid-1980s. Downturns in predator breeding success have coincided with changes in krill demography. They suggest that the "krill surplus" (Laws, 1977) brought about by slaughter of large numbers of whales that consumed krill may now be at an end: the original release of this krill surplus may have coincided with the reduction in sea-ice extent suggested by de la Mare (1997). This sea-ice reduction may have resulted in reduced krill production, and an irreversible change may thus have taken place with krill abundance now having settled at a new equilibrium point (see Murphy and King, 1997).

Away from South Georgia, cohorts of krill-eating leopard, crabeater and Weddell seals show strong approximate 5-year periodicity (Testa *et al.*,

1991) and it may be that this reflects periodic increases of krill abundance. Adélie penguins in the Ross Sea show a 5-year lag between annual population growth and sea-ice extent (Wilson *et al.*, 2001) but this may be more a feature of the time it takes for this species to reach maturity – 5 years – than the ACW periodicity. Increased sea-ice cover in winter reduced survival of subadults, which appear less able to cope with the difficult foraging conditions imposed by heavy sea-ice cover (Ainley *et al.*, 1998). Five years after a harsh winter, the reduction in numbers of birds recruiting to the summer breeding population results in reduced annual population growth. Adélie penguin demographics suggest a marked change in sea-ice extent over recent decades. Changes in krill abundance and sea-ice conditions have also been implicated in changes in chinstrap penguin populations at the Antarctic Peninsula (Fraser *et al.*, 1992). Whilst it is evident that there are many potential links between sea ice, krill and predators, and evidence of climatic warming (Comiso, 2000), caution should be exercised before extrapolating too widely. Barbraud and Weimeskirch (2001) report decreases in emperor penguins coinciding with a prolonged abnormally warm period and reductions in sea ice. This period coincided with the collapse of a glacier which afforded protection from wind to the site of the breeding colony under study. Although the collapse of the glacier may have been warming-related, the consequences of the collapse may well just have been restricted to penguins breeding in that specific colony and are not necessarily indicative of a widespread circum-Antarctic effect.

   The abrupt 25% decline in sea-ice extent between the 1950s and 1970s claimed by de la Mare (1997) could have had a major impact on krill populations and connected trophic processes. Indeed Barbraud and Weimeskirch (2001) suggest that their reported decline in emperor penguins is a manifestation of this change, although they suggest a mid-1970s date. There remain substantial doubts as to the existence of this decline in sea-ice extent. The sudden southwards jump of whale catches may simply reflect development of the ability of whalers to penetrate behind the tongue of ice that often protrudes eastward across the northern Weddell Sea. This tongue, and the open-water embayment to the south, may have existed earlier but may have remained undiscovered by whalers until the mid-twentieth century when improving vessel technology enabled them to hunt further afield. Even if the phenomenon of the dramatic mid-century sea-ice reduction is real, the impact may not have been as great as the 25% decline implies. Line transect acoustic surveys under ice by the *Autosub-2* autonomous underwater vehicle (Brierley *et al.*, 2002) have shown that it is the ice edge, rather than ice extent *per se*, that is important for krill. The 25% reduction in sea-ice area proposed by de la Mare (1997) would equate to only a 9% reduction in sea-ice edge extent. There is no direct census of krill density before the mid-twentieth century and, as outlined above, potential ecosystem changes are complicated by the fact that this period coincided with whaling, and removal of whales may have reduced substantially predation on krill (Laws, 1985).

The influence of sea ice upon krill is not restricted just to the immediate ice environs. Transport of krill from breeding grounds at the Antarctic Peninsula to outlying islands, such as South Georgia, may well be mediated by sea-ice cover (Murphy et al., 1998b). Modelling studies have shown that juvenile krill would not have the energetic reserves to make the journey entirely in a pelagic environment because of food limitations: an ice ceiling is required for part of the journey (Hofmann and Lascara, 2000). The possibility exists, therefore, that if ice extent has reduced by the amount that de la Mare (1997) suggests, whilst the impact on recruitment may not be as great as the 25% reduction suggests, the impact on supply of krill to northern latitudes could be significant.

Although it is apparent that a reduction in sea ice has the potential to precipitate many ecosystem changes, it is worth bearing in mind that at present there is no evidence for such a reduction Antarctic-wide. Sea-ice extent is increasing in some regions (Stammerjohn and Smith, 1997; Watkins and Simmonds, 2000; Vaughan et al., 2001), and the potential for complex feedback mechanisms between climate change and ice extent exists (Cai and Gordon, 1999). Furthermore, ecological implications of a reduction in sea-ice extent might not be all bad. Barbraud and Weimeskirch (2001), for example, note that paradoxically varying sea-ice extent has some opposing impacts on emperor penguins. Increased ice extent improves adult survival by increasing food availability, but is detrimental to hatching success because it increases the distance that adults have to walk between the breeding colony and feeding grounds. This and the perhaps surprising finding that sea-ice melt may serve to thicken ice shelves (Nicholls, 1997) illustrates the complexity of inter-actions in the Southern Ocean sea-ice realm. We already know that Southern Ocean sea ice affects thermohaline circulation, climate, eco-systems and fisheries, but we must accept that much more needs to be discovered: the sum of our knowledge probably represents just the tip of the aptly proverbial iceberg.

7. SUMMARY AND CONCLUSIONS

At its maximum extent sea ice can cover up to 13% of the Earth's surface. This area is similar to that covered by deserts and tundra, and makes the sea-ice environment one of the largest biomes on the planet.

Simply by weight of area alone, the sea-ice environment is clearly an important zone for study. The 19 million square kilometres of Antarctic sea ice that dominates the Southern Ocean in winter is the largest global expanse of sea ice, and its formation, consolidation and subsequent melt contribute significantly to global climate control and ocean circulation. These fundamental roles of Southern Ocean sea ice in Earth system physical function provide further major justification for its study. The sea-ice environment presents considerable challenges to life. Knowledge of how organisms are adapted to these extreme conditions (Thomas and Dieckmann, 2002a) is essential for a full understanding of life on Earth, and may provide useful clues to life elsewhere. As with physical processes, ecological interactions in Southern Ocean sea ice have relevance in a global context. It is imperative that sea-ice ecological studies continue if a full understanding of the continuum of environments on Earth is to be achieved (Chown and Gaston, 1998), and understanding of potential responses to environmental change is to be gained.

Southern Ocean sea-ice extent changes seasonally and interannually, and there is additional concern in some quarters that it may change in the longer term as a consequence of ongoing climate change. Assessments of long-term trends in the physical sea-ice realm are made difficult by the large inter-annual and cyclical anomalies in the length of the sea-ice season, sea-ice distribution and maximum sea-ice extent. From the two decades of satellite observations of sea ice now available, however, it is evident that overall sea-ice extent around Antarctica is increasing slightly, even though in some regions such as the Antarctic Peninsula significant decreases of sea-ice extent have been measured (Vaughan et al., 2001). Detection of these significant changes within the framework of large inter-annual variability illustrates well the worth of long-term observational data. Particular effort must be expended to ensure the continuing flow of these valuable data streams. Further research into the historic extent of sea ice, perhaps using ^{13}C, may extend the time-series of sea-ice extent backwards, but it seems unlikely at this stage that sufficiently detailed data will be forthcoming to confirm or refute the contention of a dramatic mid-twentieth century decline in sea-ice extent.

As with physical processes, extended time-series of observation are likely to be key to our understanding of biological processes in the Southern Ocean sea-ice realm. The longest series of continuous observations in the pack ice available so far are from the drifting ice stations, but these extend for no longer than just a few months. The drift of these stations may also result in a spatial aliasing of temporal variation. As in other marine environments, spatial patchiness, which can sometimes be extreme (Murphy, 1995), is an important component of sea-ice ecosystem function – yet few studies have been able to cover sufficiently wide geographic areas to encompass all relevant spatial scales. The importance of resolving temporal and spatial variability in marine ecosystems is becoming widely recognized, and will be particularly important in the highly dynamic Southern Ocean sea-ice system. In the pelagic environment, increasing use is being made of automated, free-drifting sensors (e.g. Argo Science Team, 1998) to this end. Although these types of sensors measure predominantly physical variables, technological advances are enabling ever more complex biological measurements to be automated. Autonomous underwater vehicle technology, for example, has now matured to the stage where these vehicles can be deployed under sea ice, and they have already delivered some substantial advances on biological sampling compared with SCUBA divers and ROVs. Further impetus for the development of these types of sensors may come from the desire to investigate more extreme under-ice environments such as the Antarctic Lake Vostok (Bell and Karl, 1998) and perhaps even the frozen seas of Jupiter's moon Europa (Kivelson et al., 2000). We know little of processes in the pack ice in winter, and automated sensors may improve data availability from this season. It is essential, for example, that efforts are made to gather direct observations of the interactions of krill and sea ice in winter because information from this time will be necessary for a full understanding of the importance of sea ice to this species.

It is not just at the large scale that deficiencies in available data exist. The microstructure of sea ice and physical/biological interactions at that scale remain relatively unknown. It is at this small scale that physical and biological interactions may be most significant for sea-ice ecology as a whole. Technological developments are required to produce probes for measuring small-scale physical and chemical parameters within ice, in particular within cold, hard ice. As with small-scale physical/chemical processes, we need to know more of small-scale and small-size biological processes in sea ice. The importance of marine viruses, for example, is being increasingly recognized (Fuhrman, 1999) but we know little of viral activity in sea ice.

In conclusion, although there have been many substantial advances in our understanding of the ecology of Southern Ocean sea ice over the past 15 years much remains to be discovered. More information is required on small and large spatial-scale processes, and longer time-series are required, particularly including winter. Scientific curiosity motivated "The worst journey in the world" in the depths of the Antarctic winter (Cherry-Garrard, 1922) and, although technological developments should help eliminate many of the perils of that journey, it may be necessary to embrace again that exploratory ethos if we are to gain full understanding of the ecology of Southern Ocean sea ice.

ACKNOWLEDGEMENTS

We thank S. Harangozo, J. Turner and T. Lachlan-Cope (British Antarctic Survey) and their partners in the EU-funded PELICON project (Project for Estimation of Long-term variability in Ice CONdition, see http://www-iup.physik.uni-bremen.de/iuppage/pelicon_e.html) for providing us with the data used to plot Figure 1 and Plate 1A. We also thank F. Cottier (Scottish Association for Marine Science) for the photographs of brine channel structure presented here as Figure 3. A.S.B. is grateful to J. L. Watkins and M. T. Wilkinson for assistance in the collection and analysis of data on krill swarms in sea ice, C. Phillips (British Antarctic Survey library) for her efforts in obtaining much of the literature cited here, and J. L. Brierley for assistance in assembling the reference list. A.S.B. was supported in part by grant no. GST022151 (Under Sea Ice and Pelagic Surveys) awarded by the UK's Natural Environment Research Council to A.S.B, P. G. Fernandes and M. A. Brandon under the *Autosub* Science Missions Thematic Programme, and he acknowledges the efforts of these collaborators together with F. Armstrong and the SOC *Autosub* team for work on krill under sea ice. D.T. would like to thank G. S. Dieckmann, V. Smetacek, C. Haas, S. Schiel, H. Eicken, R. Lara, G. Kattner, C. Krembs, M. Gleitz, S. Grossmann, T. Mock, A. Belem, K.-U. Evers, H. Kennedy, J.-L. Tison, V. Giannelli and S. Günther for their excellent collaboration. D.T.'s work was funded partially by the Alfred Wegener Institute Germany, the UK's Natural Environment Research Council, the British Council/DAAD (ARC Programme), the Hamburg Ship Model Basin (HSVA) through the European Commission Training and Mobility of Researchers Programme, the Nuffield Foundation and the Hanse Institute of Advanced Study. This review was improved by suggestions from an anonymous referee and A. J. Southward, for which we are grateful.

REFERENCES

Ackley, S. F. and Sullivan, C. W. (1994). Physical controls on the development and characteristics of Antarctic sea ice biological communities – a review and synthesis. *Deep Sea Research* **41**, 1583–1604.

Ainley, D. G., Fraser, W. R., Sullivan, C. W., Torres, J. J, Hopkins, T. L. and Smith, W. O. (1986). Antarctic mesopelagic micronekton: evidence from seabirds that pack ice affects community structure. *Science* **232**, 847–849.

Ainley, D. G., Fraser, W. R. and Daly, K. L. (1988). Effects of pack ice on the composition of micronektonic communities in the Weddell Sea. *In* "Antarctic Ocean and Resources Variability" (D. Sahrhage, ed.), pp. 140–146. Springer-Verlag, Berlin.

Ainley, D. G., Wilson, P. R., Barton, K. J., Ballard, G., Nur, N. and Karl, B. (1998). Diet and foraging effort of Adélie penguins in relation to pack-ice conditions in the southern Ross Sea. *Polar Biology* **20**, 311–319.

Aletsee, L. and Jahnke, J. (1992). Growth and productivity of the psychrophilic marine diatoms *Thalassiosira antarctica* Comber and *Nitzschia frigida* Grunow in batch cultures at temperatures below the freezing point of sea water. *Polar Biology* **11**, 643–647.

Alonzo, S. H. and Mangel, M. (2001). Survival strategies and growth of krill: avoiding predators in space and time. *Marine Ecology Progress Series* **209**, 203–217.

Archer, S. D., Leakey, R. J. G., Burkill, P., Sleigh, M. A. and Appleby, C. J. (1996). Microbial ecology of sea ice at a coastal Antarctic site: community composition, biomass and temporal change. *Marine Ecology Progress Series* **135**, 179–195.

Argo Science Team. (1998). On the design and implementation of ARGO. An initial plan for a Global Array of Profiling Floats. ICPO Report **21**, (Godae Report **5**), 32 pp. Bureau of Meteorology, Melbourne, Australia.

Arrigo, K. R. (1994). Impact of ozone depletion on phytoplankton growth in the Southern Ocean: large-scale spatial and temporal variability. *Marine Ecology Progress Series* **114,** 1–12.

Arrigo, K. R. and Brown, C. W. (1996). Impact of chromophoric dissolved organic matter on UV inhibition of primary productivity in the sea. *Marine Ecology Progress Series* **140**, 207–216.

Arrigo, K. R. and Lizotte, M. P. (1998). Antarctic sea ice: biological processes, interactions and variability. Preface. *Antarctic Research Series* **73**, xi.

Arrigo, K. R. and Sullivan, C. W. (1992). The influence of salinity and temperature covariation on the photophysiogical characteristics of Antarctic sea ice micro-algae. *Journal of Phycology* **28**, 746–756.

Arrigo, K. R., Dieckmann, G. S., Gosselin, M., Robinson, D. H., Fritsen, C. H. and Sullivan, C. W. (1995). High resolution study of the platelet ice ecosystem in McMurdo Sound, Antarctica: biomass, nutrient, and production profiles within a dense microalgal bloom. *Marine Ecology Progress Series* **127**, 255–268.

Arrigo, K. R., Worthern, D. L., Lizotte, M. P., Dixon, P. and Dieckmann, G. S. (1997). Primary production in Antarctic sea ice. *Science* **276**, 394–397.

Arrigo, K. R., Worthern, D. L., Dixon, P. and Lizotte, M. P. (1998a). Primary productivity of near surface communities within Antarctic pack ice. *Antarctic Research Series* **73**, 23–43.

Arrigo, K. R., Worthen, D., Schnell, A. and Lizotte, M. P. (1998b). Primary production in Southern Ocean waters. *Journal of Geophysical Research* **103**, 15587–15600.

Atkinson, A. (1998). Life cycle strategies of epipelagic copepods in the Southern Ocean. *Journal of Marine Systems* **15**, 289–311.

Atkinson, A. and Peck, J. M. (1988). A summer-winter comparison of zooplankton in the oceanic area around South Georgia. *Polar Biology* **8**, 463–473.

Atkinson, A. and Sinclair, D. (2000). Zonal distribution and seasonal vertical migration of copepod assemblages in the Scotia Sea. *Polar Biology* **23**, 46–58.

Atkinson, A. and Snÿder, R. (1997). Krill-copepod interactions at South Georgia, Antarctica, I. Omnivory by *Euphausia superba*. *Marine Ecology Progress Series* **160**, 63–76.

Atkinson, A., Schnack Schiel, S. B., Ward, P. and Marin, V. (1997). Regional differences in the life cycle of *Calanoides acutus* (Copepoda: Calanoida) within the Atlantic sector of the Southern Ocean. *Marine Ecology Progress Series* **150**, 99–111.

Atkinson, A., Ward, P., Hill, A., Brierley, A. S. and Cripps, G. C. (1999). Krill-copepod interactions at South Georgia, Antarctica, II. Possible control by *Euphausia superba* of copepod abundance. *Marine Ecology Progress Series* **176**, 63–79.

Atkinson, A., Whitehouse, M. J., Priddle, J., Cripps, G. C., Ward, P. and Brandon, M. A. (2001). South Georgia, Antarctica: a productive, cold water, pelagic ecosystem. *Marine Ecology Progress Series* **216**, 279–308.

Barbraud, C. and Weimeskirch, H. (2001). Emperor penguins and climate change. *Nature* **411**, 183–186.

Bathmann, U., Fischer, G., Muller, P. J. and Gerdes, D. (1991). Short-term variations in particulate matter sedimentation off Kapp Norvegia, Weddell Sea, Antarctica: relation to water mass advection, ice cover, plankton biomass and feeding activity. *Polar Biology* **11**, 185–195.

Bathmann, U., Scharek, R., Klass, C., Dubischer, C. D. and Smetacek, V. (1997). Spring development of phytoplankton biomass and composition in major water masses of the Atlantic sector of the Southern Ocean. *Deep-Sea Research* **44**, 51–67.

Bell, R. E. and Karl, D. M. (1998). Lake Vostok: a curiosity or a focus for interdisciplinary study? Final Report of the National Science Foundation Sponsored Workshop, 83 pp. National Science Foundation, Washington.

Belzile, C., Johannessen, S. C., Gosselin, M., Demers, S. and Miller, W. L. (2000). Ultraviolet attenuation by dissolved and particulate constituents of first-year ice during late spring in an Arctic polynya. *Limnology and Oceanography* **46**, 1265–1273.

Bergström, B. I., Hempel, G., Marschall, H.-P., North, A. W., Siegel, V. and Strömberg, J.-O. (1990). Spring distribution, size composition and behaviour of krill *Euphausia superba* in the western Weddell Sea. *Polar Record* **26**, 85–89.

Bester, M. N., Erickson, A. W. and Ferguson, J. W. H. (1995). Seasonal change in the distribution and density of seals in the pack ice off Princess Martha Coast, Antarctica. *Antarctic Science* **7**, 357–364.

Bianchi, F., Bolrin, A., Cioce, F., Dieckmann, G. S., Kuosa, H., Larsson, A.-M., Nöthig, E.-M., Sehlstedt, P.-I., Socal, G. and Syvertsen, E. E. (1992). Phyto-plankton distribution in relation to sea ice, hydrography and nutrients in the northwestern Weddell Sea in early spring 1988 during EPOS. *Polar Biology* **12**, 225–235.

Billett, D. S. M., Lampitt, R. S., Rice, A. L. and Mantoura, R. F. C. (1983). Seasonal sedimentation of phytoplankton to the deep-sea benthos. *Nature* **302**, 520–522.

Blome, D. and Riemann, F. (1999). Antarctic sea ice nematodes, with description of *Geomonhystera glaciei* sp. nov. (Monhysteridae). *Mitteilung des Hamburg-ischen Zoologischen Museum Instituts* **96**, 15–20.

Boveng, P. L., Hiruki, L. M., Schwartz, M. K. and Bengtson J. L. (1998). Population growth of Antarctic fur seals: limitation by a top predator, the leopard seal? *Ecology* **79**, 2863–2877.

Boyd P. W. and Laws, C. S. (2001). The Southern Ocean Iron Release Experiment (SOIREE) – introduction and summary. *Deep-Sea Research II* **48**, 2425–2438.

Boyd, P. W., Robinson, C., Savidge, G. and Williams, P. J. le B. (1995). Water column and sea-ice primary production in the Bellingshausen Sea during the Austral spring of 1992. *Deep-Sea Research II* **42**, 1177–1200.

Boyd, P. W., Watson, A. J., Law, C. S., Abraham, E. R., Trull, T., Murdoch, R., Bakker, D. C. E., Bowie, A. R., Buesseler, K. O., Chang, H., Charette, M., Croot, P., Downing, K., Frew, R., Gall, M., Hadfield, M., Hall, J., Harvey, M., Jameson, G., LaRoche, J., Liddicoat, M., Ling, R., Maldonado, M. T., McKay, R. M., Nodder, S., Pickmere, S., Pridmore, R., Rintoul, S., Safi, K., Sutton, P., Strzepek, R., Tanneberger, K., Turner, S., Waite, A. and Zeldis, J. (2000). A mesoscale phytoplankton bloom in the polar Southern Ocean stimulated by iron fertilization. *Nature* **407**, 695–702.

Boysen-Ennen, E., Hagen, W., Hubold, G. and Piatkowski, U. (1991). Zooplankton biomass in the ice-covered Weddell Sea, Antarctica. *Marine Biology* **111**, 227–235.

Brandini, F. P. and Baumann, M. E. M. (1997). The potential role of melted "brown ice" as sources of chelators and ammonia to the surface waters of the Weddell Sea, Antarctica. *Proceedings of the NIPR Symposium on Polar Biology* **10**, 1–13.

Brierley, A. S. and Reid, K. (1999). Kingdom of the krill. *New Scientist* **2182**, 36–41.

Brierley, A. S. and Watkins, J. L. (2000). Effects of sea ice cover on the swarming behaviour of Antarctic krill, *Euphausia superba*. *Canadian Journal of Fisheries and Aquatic Sciences* **57** (Suppl. 3), 24–30.

Brierley, A. S., Watkins, J. L. and Murray, A. W. A. (1997). Interannual variability in krill abundance at South Georgia. *Marine Ecology Progress Series* **150**, 87–98.

Brierley, A. S., Watkins, J. L., Goss, C., Wilkinson, M. T. and Everson, I. (1999a). Acoustic estimates of krill density at South Georgia, 1981 to 1998. *Commission for the Conservation of Antarctic Marine Living Resources Science* **6**, 47–57.

Brierley, A. S., Demer, D. A., Watkins, J. L. and Hewitt, R. P. (1999b). Concordance of interannual fluctuations in acoustically estimated densities of Antarctic krill around South Georgia and Elephant Island: biological evidence of same-year teleconnections across the Scotia Sea. *Marine Biology* **134**, 675–681.

Brierley, A. S., Fernandes, P. G., Brandon, M. A., Armstrong, F., Millard, N. W., McPhail, S. D., Stevenson, P., Pebody, M., Perrett, J., Squires, M., Bone, D. G. and Griffiths, G. (2002). Antarctic krill under sea ice: elevated abundance in a narrow band just south of the ice edge. *Science* (in press).

Buck, K. R., Bolt, P. A. and Garrison, D. L. (1990). Phagotrophy and faecal pellet production by an athecate dinoflagellate in Antarctic sea ice. *Marine Ecology Progress Series* **60**, 75–84.

Budillon, G. and Spezie, G. (2000). Thermohaline structure and variability in the Terra Nova Bay polynya, Ross Sea. *Antarctic Science* **12**, 493–508.

Buma, A. G. J., de Boer, M. K. and Boelen, P. (2001). Depth distributions of DNA damage in Antarctic marine phyto- and bacterioplankton exposed to summertime UV radiation. *Journal of Phycology* **37**, 200–208.

Burghart, S. E., Hopkins, T. L., Vargo, G. A. and Torres, J. J. (1999). Effects of rapidly receding ice edge on the abundance, age structure and feeding of three dominant calanoid copepods in the Weddell Sea, Antarctica. *Polar Biology* **22**, 279–288.

Cadée, G. C. (1992). Organic carbon in the upper layer and its sedimentation during the ice-retreat period in the Scotia-Weddell Sea, 1988. *Polar Biology* **12**, 253–259.

Cadée, G. C., González, H. and Schnack-Schiel, S. B. (1992). Krill diet affects faecal string settling. *Polar Biology* **12**, 75–80.

Cai, W. J. and Gordon, H. B. (1999). Southern high-latitude ocean climate drift in a coupled model. *Journal of Climate* **12**, 132–146.

Carey, A. G. (1985) Marine ice fauna: Arctic. *In* "Sea Ice Biota" (R. A. Horner, ed.), pp. 173–190. CRC Press, Florida.

Carlson, C. A. and Hansell, D. A. (in press). The contribution of DOM to the biogeochemistry of the Ross Sea. *Antarctic Research Series*.

Cavalieri, D. J., Gloersen, P. and Campbell, W. J. (1984). Determination of sea ice parameters with Nimbus-7 SMMR. *Journal of Geophysical Research – Atmospheres* **89**, 5355–5369.

Cherry-Garrard, A. (1922). "The Worst Journey in the World". Constable, London, 585 pp.

Chown, S. L. and Gaston, K. J. (1998). Including the Antarctic: insights for ecologists everywhere. *New Zealand Natural Sciences* **23** (Suppl.), 32.

Clarke, D. B. and Ackley, S. F. (1984). Sea ice structure and biological activity in the Antarctic marginal ice zone. *Journal of Geophysical Research* **89**, 2087–2095.

Cole, D. M. and Shapiro, L. H. (1998). Observations of brine drainage networks and microstructure of first-year sea ice. *Journal of Geophysical Research* **103**, 739–750.

Comiso, J. C. (1991) Satellite remote sensing of the Polar Oceans. *Journal of Marine Systems* **2**, 395–434.

Comiso, J. C. (2000). Variability and trends in Antarctic surface temperatures from *in situ* and satellite infrared measurements. *Journal of Climate* **13**, 1674–1696.

Comiso, J. C. and Gordon, A. L. (1998). Interannual variability in summer sea ice minimum, coastal polynyas and bottom water formation. *Antarctic Research Series* **74**, 293–315.

Comiso, J. C., Grenfell, T. C., Lange, M., Lohanick, A. W., Moore, R. K. and Wadhams, P. (1992). Microwave remote sensing of the Southern Ocean ice cover. *Geophysical Monograph Series* **68**, 243–259.

Conover, R. J. and Huntley, M. (1991) Copepods in ice covered seas – distribution, adaptations to seasonally limited food, metabolism, growth patterns and life cycle strategies in polar seas. *Journal of Marine Systems* **2**, 1–41.

Copley, J. (2000). The great ice mystery. *Nature* **408**, 634–636.

Corliss, J. O. and Sneider, R. A. (1986). A preliminary description of several new ciliates from the Antarctica, including *Cohnilembus grassei* n. sp. *Protistologica* **22**, 39–46.

Cottier, F. (1999). Brine channel distribution in young sea ice. PhD thesis, Scott Polar Research Institute, University of Cambridge.

Cottier, F., Eicken, H. and Wadhams, P. (1999). Linkages between salinity and brine channel distribution in young sea ice. *Journal of Geophysical Research* **104**, 15859–15871.

Cripps, G. C. and Atkinson, A. (2000). Fatty acid composition as an indicator of carnivory in Antarctic krill. *Canadian Journal of Fisheries and Aquatic Sciences* **57** (Suppl. 3), 31–37.

Cripps, G. C. and Clarke, A. (1998). Seasonal variation in the biochemical composition of particulate material collected by sediment traps at Signy Island, Antarctica. *Polar Biology* **20**, 414–423.

Croxall, J. P. (1997). Emperor ecology in the Antarctic winter. *Trends in Ecology and Evolution* **12**, 333–334.

Croxall, J. P., McCann, T. S., Prince, P. A. and Rothery, P. (1988). Reproductive performance of seabirds and seals at South Georgia and Signy Island, South Orkney Islands, 1976–1987: implications for Southern Ocean monitoring studies.

In "Antarctic Ocean and Resources Variability" (D. Sahrhage, ed.), pp. 261–285. Springer-Verlag, Berlin.

Croxall, J. P., Reid, K. and Prince, P. A. (1999). Diet, provisioning and productivity responses of marine predators to differences in availability of Antarctic krill. *Marine Ecology Progress Series* **177**, 115–131.

Cunningham, W. L. and Leventer, A. (1998). Diatom assemblages in surface sediments of the Ross Sea: relationship to present oceanographic conditions. *Antarctic Science* **10**, 134–146.

Dahms, H.-U., Bergmans, M. and Schminke, H. K. (1990). Distribution and adaptations of sea ice inhabiting Harpacticoida (Crustacea, Copepoda) of the Weddell Sea (Antarctica). *PSZNI: Marine Ecology* **11**, 207–226.

Daly, K. L. (1990). Overwintering development, growth, and feeding of larval *Euphausia superba* in the Antarctic marginal ice zone. *Limnology and Oceanography* **35**, 1564–1576.

Daly, K. L. (1998). Physioecology of juvenile Antarctic krill (*Euphausia superba*) during spring in ice-covered seas. *Antarctic Research Series* **73**, 183–198.

Daly, K. L. and Macaulay, M. C. (1988). Abundance and distribution of krill in the ice edge zone of the Weddell Sea, austral spring 1983. *Deep-Sea Research I* **35**, 21–41.

Daly, K. L. and Macaulay, M. C. (1991). Influence of physical and biological mesoscale dynamics on the seasonal distribution and behavior of *Euphausia superba* in the Antarctic marginal ice zone. *Marine Ecology Progress Series* **79**, 37–66.

Davidson, A. T. and Marchant, H. J. (1994). The impact of ultraviolet radiation on *Phaeocystis* and selected species of Antarctic marine diatoms. *Antarctic Research Series* **62**, 187–205.

Davidson, A. T. and van der Heijden, A. (2000). Exposure of natural Antarctic marine microbial assemblages to ambient UV radiation: effects on bacterioplankton. *Aquatic Microbial Ecology* **21**, 257–264.

de Baar, H. J. W., de Jong, J. T. M., Bakker, D. C. E., Löscher, B. M., Veth, C., Bathmann, U. and Smetacek, V. (1995). Importance of iron for plankton blooms and carbon dioxide drawdown in the Southern Ocean. *Nature* **373**, 412–415.

Decho, A. W. (1990). Microbial exopolymer secretions in ocean environments – their roles in food webs and marine processes. *Oceanography and Marine Biology, An Annual Review* **28**, 73–153.

Decho, A. W. (2000). Microbial biofilms in intertidal systems: an overview. *Continental Shelf Research* **20**, 1257–1273.

de la Mare, W. K. (1997). Abrupt mid-twentieth-century decline in Antarctic sea-ice extent from whaling records. *Nature* **389**, 57–59.

DeLong, E. F. (1998). Archaeal means and extremes. *Science* **280**, 542–543.

Delille, D. (1992). Marine bacterioplankton at the Weddell Sea ice edge, distribution of psychrophilic and psychrotrophic populations. *Polar Biology* **12**, 205–210.

Delille, D. and Rosiers, C. (1996). Seasonal changes of Antarctic marine bacterioplankton and sea ice bacterial assemblages. *Polar Biology* **16**, 27–34.

Dieckmann, G. S., Rohardt, G., Hellmer, H. and Kipfstuhl, J. (1986). The occurrence of ice platelets at 250 m depth near the Filchner Ice Shelf and its significance for sea ice biology. *Deep Sea Research* **33**, 141–148.

Dieckmann, G. S., Lange, M. A., Ackley, S. F. and Jennings Jr., J. C. (1991a). The nutrient status in sea ice of the Weddell Sea during winter: effects of sea ice texture and algae. *Polar Biology* **11**, 449–456.

Dieckmann, G. S., Spindler, M., Lange, M., Ackley, S. F. and Eicken, H. (1991b). Antarctic sea ice: as habitat for the foraminferan *Neogloboquadrina pachyderma*. *Journal of Foraminiferal Research* **21**, 182–189.

Dieckmann, G. S., Arrigo, K. and Sullivan, C. W. (1992). A high resolution sampler for nutrient and chlorophyll a profiles of the sea ice platelet layer and underlying water column below fast ice in Polar oceans – preliminary results. *Marine Ecology Progress Series* **80**, 291–300.

Dieckmann, G. S., Eicken, H., Haas, C., Garrison, D. L., Gleitz, M., Lange, M., Nöthig, E.-M., Spindler, M., Sullivan, C. W., Thomas, D. N. and Weissenberger, J. (1998). A compilation of data on sea ice algal standing crop from the Bellingshausen, Amundsen and Weddell Seas from 1983 to 1994. *Antarctic Research Series* **73**, 85–92.

DiTullio, G., Garrison, D. L. and Mathot, S. (1998). Dimethylsulphoniopropionate in sea ice algae from the Ross Sea polynya. *Antarctic Research Series* **73**, 139–146.

Drinkwater, M. R. (1998). Active microwave remote sensing observations of Weddell Sea ice. *Antarctic Research Series* **74**, 187–212.

Duarte, C., Gasol, J. M. and Vaque, D. (1997). Role of experimental approaches in marine microbial ecology. *Aquatic Microbial Ecology* **13**, 101–111.

Dunbar, R. B. and Leventer, A. (1992). Seasonal variation in carbon isotopic composition of Antarctic sea ice and open water plankton communities. *Antarctic Journal of the United States* **27**, 79–81.

Dunbar, R. B., Leventer, A. R. and Mucciarone, D. A. (1998). Water column sediment fluxes in the Ross Sea, Antarctic: atmospheric and sea ice forcing. *Journal of Geophysical Research* **103**, 30741–30759.

Eicken, H. (1992). The role of sea ice in structuring Antarctic ecosystems. *Polar Biology* **12**, 3–13.

Eicken, H. (1998). Deriving modes and rates of ice growth in the Weddell Sea from microstructural, salinity and stable isotope data. *Antarctic Research Series* **74**, 89–122.

Eicken, H., Ackley, S. F., Richter-Menge, J. A. and Lange, M. A. (1991a). Is the strength of sea ice related to its chlorophyll content? *Polar Biology* **11**, 347–350.

Eicken, H., Lange, M. A. and Dieckmann, G. S. (1991b). Spatial variability of sea ice properties in the Northwestern Weddell Sea. *Journal of Geophysical Research* **96**, 10603–10615.

Eicken, H., Lange, M. A., Hubbertan, H.-W. and Wadhams, P. (1994). Character-istics and distribution patterns of snow and meteoric ice in the Weddell Sea and their contribution to the mass balance of sea ice. *Annales Geophysicae* **12**, 80–93.

Eicken, H., Weissenberger, J., Cottier, F., Evers K.-U., Krembs, C., Kuosa, H., Hall, R., Jochmann, P., Leonard, G. H., Lindemann, F., Reisemann, M., Shen, H., Smedsrud, L. H., Ackley, S. F., Bussmann, I., Freitag, J., Gradinger, R., Ikävalko, J., Schuster, W., Valero Delgado, F. and Wadhams, P. (1998). Ice tank studies of physical and biological sea-ice processes. *In* "Ice In Surface Waters" (H. T. Shen, ed.). Proceedings of the 14th International Symposium on Ice, Potsdam, New York, pp. 363–370. A. A. Balkema, Rotterdam.

Eicken, H., Bock, C., Wittig, R., Miller, H. and Poertner, H.-O. (2000). Magnetic resonance imaging of sea-ice pore fluids: methods and thermal evolution of pore microstructure. *Cold Regions Science and Technology*, **31**, 207–225.

Ekman, S. (1953). "Zoogeography of the Sea". Sidgwick and Jackson, London.

El-Sayed, S. Z. (1988). Seasonal and interannual variabilities in Antarctic phyto-plankton with reference to krill distribution. *In* "Antarctic Ocean and Resources Variability" (D. Sahrhage, ed.), pp. 101–119. Springer-Verlag, Berlin.

Erickson, A. W. and Hanson, M. B. (1990). Continental estimates and population trends of Antarctic ice seals. *In* "Antarctic Ecosystems. Ecological Change and Conservation" (K. R. Kerry and G. Hempel, eds), pp. 253–264. Springer-Verlag, Berlin.

Everson, I. (1992). Managing Southern Ocean krill and fish stocks in a changing environment. *Philosophical Transactions of the Royal Society of London Series B* **338**, 311–317.

Everson, I. and Goss, C. (1991). Krill fishing activity in the southwest Atlantic. *Antarctic Science* **3**, 351–358.

Everson, I., Parkes, G., Kock, K.-H. and Boyd, I. L. (1999). Variation in standing stock of the mackerel icefish *Champsocephalus gunnari* at South Georgia. *Journal of Applied Ecology* **36**, 591–603.

Fahl, K. and Kattner, G. (1993). Lipid content and fatty acid composition of algal communities in sea ice and water from the Weddell Sea (Antarctica). *Polar Biology* **13**, 405–409.

Farman, J. C., Gardiner, B. G. and Shanklin, J. D. (1985). Large losses of total ozone in Antarctica reveal seasonal clox/nox interaction. *Nature* **315**, 207–210.

Fernandes, P. G. and Brierley, A. S. (1999). Using an Autonomous Underwater Vehicle as a platform for mesoscale acoustic sampling in marine environments. International Council for the Exploration of the Sea, Council Meeting paper CM 1999/M:01. (Mimeo).

Fernandes, P. G., Brierley, A. S., Simmonds, E. J., Millard, N. W., McPhail, S. D., Armstrong, F., Stevenson, P. and Squires, M. (2000). Fish do not avoid survey vessels. *Nature* **404**, 35–36.

Fischer, G. (1991). Stable carbon isotope ratios of plankton carbon and sinking organic matter from the Atlantic sector of the Southern Ocean. *Marine Chemistry* **35**, 581–596.

Fischer, G., Fütterer, D., Gersonde, R., Honjo, S., Ostermann, D. and Wefer, G. (1988). Seasonal variability of particle flux in the Weddell Sea and its relation to ice cover. *Nature* **335**, 426–428.

Fransz, H. G. (1988). Vernal abundance, structure and development of epipelagic copepod populations of the eastern Weddell Sea (Antarctica). *Polar Biology* **9**, 107–114.

Fraser, W. R. and Trivelpiece, W. Z. (1996). Factors controlling the distribution of seabirds: winter-summer heterogeneity in the distribution of Adélie penguin populations. *Antarctic Research Series* **70**, 257–272.

Fraser, W. R., Trivelpiece, W. Z., Ainley, D. G. and Trivelpiece, S. G. (1992). Increases in Antarctic penguin populations: reduced competition with whales or a loss of sea ice due to environmental warming? *Polar Biology* **11**, 525–531.

Fritsen, C. H. and Sullivan, C. W. (1997). Distributions and dynamics of microbial communities in the pack ice of the Western Weddell Sea, Antarctica. *In* "Antarctic Communities: Species, Structure and Survival" (B. Battglia, J. Valencia and D. W. H. Walton, eds), pp. 101–106. Cambridge University Press, Cambridge.

Fritsen, C. H., Coale, S. L., Neenan, D. R., Gibson, A. H. and Garrison, D. L. (2001). Biomass, production and microhabitat characteristics near the freeboard of ice floes in the Ross Sea, Antarctica, during the austral summer. *Annals of Glaciology* **33**, 280–286.

Fritsen, C. H., Lytle, V. I., Ackeley, S. F. and Sullivan, C. W. (1994). Autumn bloom of Antarctic pack-ice algae. *Science* **266**, 782–784.

Fritsen, C. H., Ackeley, S. F., Kremer, J. N. and Sullivan, C. W. (1998). Flood-freeze cycles and microalgal dynamics in Antarctic pack ice. *Antarctic Research Series* **73**, 1–21.

Froneman, P. W., Pakhomov, E. A., Perissinotto, R. and McQuaid, C. D. (2000). Zooplankton structure and grazing in the Atlantic sector of the Southern Ocean in late austral summer 1993 – Part 2. Biochemical zonation. *Deep-Sea Research I* **47**, 1687–1702.

Fuhrman, J. A. (1999). Marine viruses and their biogeochemical and ecological effects. *Nature* **399**, 541–548.

Fukuchi, M., Tanimura, A. and Hoshiai, T. (1979). "NIPR-I", a new plankton sampler under sea ice. *Bulletin of the Plankton Society of Japan* **26**, 104–109.

Garrison, D. L. (1991). Antarctic sea ice biota. *American Zoologist* **31**, 17–33.

Garrison, D. L. and Buck, K. R. (1986). Organism losses during ice melting: a serious bias in sea ice community studies. *Polar Biology* **6**, 237–239.

Garrison, D. L. and Buck, K. R. (1989). The biota of Antarctic pack ice in the Weddell Sea and Antarctic peninsula regions. *Polar Biology* **10**, 211–219.

Garrison, D. L. and Buck, K. R. (1991). Surface-layer sea ice assemblages in Antarctic pack ice during the austral spring: environmental conditions, primary production and community structure. *Marine Ecology Progress Series* **75**, 161–172.

Garrison, D. L. and Close, A. R. (1993). Winter ecology of the sea ice biota in Weddell Sea pack ice. *Marine Ecology Progress Series* **96**, 17–31.

Garrison, D. L. and Gowing, M. M. (1993). Protozooplankton. *In* "Antarctic Microbiology" (E. I. Friedman, ed.), pp. 123–165. Wiley-Liss, New York.

Garrison, D. L. and Mathot, S. (1996). Pelagic and sea ice microbial communities. *Antarctic Research Series* **70**, 155–172.

Garrison, D. L., Ackley, S. F. and Buck, K. R., (1983). A physical mechanism for establishing algal populations in frazil ice. *Nature* **306**, 363–365.

Garrison, D. L., Sullivan, C. W. and Ackley, S. F. (1986). Sea ice microbial communities in Antarctica. *BioScience* **36**, 243–250.

Garrison, D. L., Close, A. R. and Reimnitz, E. (1989). Algae concentrated by frazil ice: evidence from laboratory experiments and field measurements. *Antarctic Science* **1**, 313–316.

Gerlatt, T. S. and Siniff, D. B. (1999). Line transect survey of crabeater seals in the Amundsen-Bellingshausen Seas, 1994. *Wildlife Society Bulletin* **27**, 330–336.

Giannelli, V., Thomas, D. N., Haas, C., Kattner, G., Kennedy, H. A. and Dieckmann, G. S. (2001). Behaviour of dissolved organic matter and inorganic nutrients during experimental sea ice formation. *Annals of Glaciology* **33**, 317–321.

Gibson, J. A. E., Trull, T., Nichols, P. D., Summons, R. E. and McMinn, A. (1999). Sedimentation of C-13 rich organic matter from Antarctic sea ice algae: a potential indicator of past sea ice extent. *Geology* **27**, 331–334.

Giesenhagen, H. C., Detma, A. E., de Wall, J., Weber, A., Gradinger, R. and Jochem, F. J. (1999). How are Antarctic planktic microbial food webs and algal blooms affected by melting of sea ice? Microcosm simulations. *Aquatic Microbial Ecology* **20**, 183–201.

Gill, P. C. and Thiele, D. (1997). A winter sighting of killer whales (*Orcinas orca*) in Antarctic sea ice. *Polar Biology* **17**, 401–404.

Gleitz, M. and Kirst, G. O. (1991). Photosynthesis-irradiance relationship and carbon metabolism of different ice algal assemblages collected from Weddell Sea pack ice during austral spring (EPOS 1). *Polar Biology* **11**, 385–392.

Gleitz, M. and Thomas, D. N. (1992). Physiological responses of a small Antarctic diatom (*Chaetoceros* sp.) to simulated environmental constraints associated with sea-ice formation. *Marine Ecology Progress Series* **88**, 271–278.

Gleitz, M. and Thomas, D. N. (1993). Variation in phytoplankton standing stock, chemical composition and physiology during sea ice formation in the southeastern Weddell Sea, Antarctica. *Journal of Experimental Marine Biology and Ecology* **173**, 211–230.

Gleitz, M., Rutgers vd Loeff, M., Thomas, D. N., Dieckmann, G. S. and Millero, F. J. (1995). Comparison of summer and winter inorganic carbon, oxygen and nutrient concentrations in Antarctic sea ice brine. *Marine Chemistry* **51**, 81–91.

Gleitz, M., Grossmann, S., Scharek, R. and Smetacek, V. (1996a). Ecology of diatom and bacterial assemblages in water associated with melting summer sea ice in the Weddell Sea, Antarctica. *Antarctic Science* **8**, 135–146.

Gleitz, M., Kukert, H., Riebesell, U. and Dieckmann, G. S. (1996b). Carbon acquisition and growth of Antarctic sea ice diatoms in closed bottle incubations. *Marine Ecology Progress Series* **135**, 169–177.

Gleitz, M., Bartsch, A., Dieckmann, G. S. and Eicken, H. (1998). Composition and succession of sea ice diatom assemblages in the eastern and southern Weddell Sea, Antarctica. *Antarctic Research Series* **73**, 107–120.

Gloersen, P. and Mernicky, A. (1998). Oscillatory behaviour in Antarctic sea ice conditions. *Antarctic Research Series* **74**, 161–171.

Gloersen, P., Campbell, W. J., Cavalieri, D. J., Comiso, J. C., Parkinson, C. L. and Zwally, H. H. (1992). "Arctic and Antarctic Sea Ice, 1978–1987: Satellite Passive-microwave Observations and Analysis". National Aeronautics and Space Administration SP-511, Washington DC.

González, H. E. (1992a). The distribution and abundance of krill faecal material and oval pellets in the Scotia and Weddell Seas (Antarctica) and their role in particle flux. *Polar Biology* **12**, 81–91.

González, H. E. (1992b). Distribution and abundance of minipellets around the Antarctic Peninsula. Implications for protistan feeding behaviour. *Marine Ecology Progress Series* **90**, 223–236.

González, H. E., Kurbjeweit, F. and Bathmann, U. (1994). Occurrence of cyclopoid copepods and faecal material in the Halley Bay region, Antarctica, during January–February 1991. *Polar Biology* **14**, 331–342.

Gordon, A. and Comiso, J. C. (1988). Polynyas in the Southern Ocean. *Scientific American* **256**, 89–96.

Gowing, M. M. and Garrison, D. L. (1992). Abundance and feeding ecology of larger protozooplankton in the ice edge zone of the Weddell and Scotia Seas during austral winter. *Deep-Sea Research* **39**, 893–919.

Gradinger, R. (1999). Integrated abundance and biomass of sympagic meiofauna in Arctic and Antarctic pack ice. *Polar Biology* **22**, 169–177.

Gradinger, R. and Schnack-Schiel, S. B. (1998). Potential effect of ice formation on Antarctic pelagic copepods: salinity induced mortality of *Calanus propinquus* and *Metridia gerlachei* in comparison to sympagic acoel turbellarians. *Polar Biology* **20**, 139–142.

Green, K., Burton, H. R., Wong, V., McFarlane, R. A., Flaherty, A. A., Pahl, B. C. and Haigh, S. A. (1995). Difficulties in assessing population status of ice seals. *Wildlife Research* **22**, 193–199.

Grossmann, S. (1994). Bacterial activity in sea ice and open waters of the Weddell Sea, Antarctica: a microautoradiographic study. *Microbial Ecology* **28**, 1–18.

Grossmann, S. and Dieckmann, G. S. (1994). Bacterial standing stock, activity, and carbon production during formation and growth of sea ice in the Weddell Sea, Antarctica. *Applied Environmental Microbiology* **60**, 2746–2753.

Grossmann, S. and Gleitz, M. (1993). Microbial responses to experimental sea-ice formation: implications for the establishment of Antarctic sea-ice communities. *Journal of Experimental Marine Biology and Ecology* **173**, 273–289.

Grossmann, S., Lochte, K. and Scharek, R. (1996). Algal and bacterial processes in platelet ice during late austral summer. *Polar Biology* **16**, 623–633.

Guglielmo, L., Carrada, G. C., Catalano, G., Dell'Anno, A., Fabiano, F., Lazzara, L., Mangoni, O., Pusceddu, A. and Saggiomo, V. (2000). Structural and functional properties of sympagic communities in the annual sea ice at Terra Nova Bay (Ross Sea, Antarctica). *Polar Biology* **23**, 137–146.

Günther, S. and Dieckmann, G. S. (1999). Seasonal development of algal biomass in snow-covered fast ice and the underlying platelet layer in the Weddell Sea, Antarctica. *Antarctic Science* **11**, 305–315.

Günther, S. and Dieckmann, G. S. (2001). Vertical zonation and community transition of sea-ice diatoms in fast ice and platelet layer, Weddell Sea, Antarctica. *Annals of Glaciology* **33**, 287–296.

Günther, S., George, K. H. and Gleitz, M. (1999a). High sympagic metazoan abundance in platelet layers at Drescher Inlet, Weddell Sea, Antarctica. *Polar Biology* **22**, 82–89.

Günther, S., Gleitz, M. and Dieckmann, G. S. (1999b). Biogeochemistry of Antarctic sea ice: a case study on platelet ice at Drescher Inlet, Weddell Sea. *Marine Ecology Progress Series* **177**, 1–13.

Gutt, J. (1995). The occurrence of sub-ice algal aggregations off northeast Greenland. *Polar Biology* **15**, 247–252.

Gutt, J. (2001). On the direct impact of ice on marine benthic communities, a review. *Polar Biology* **24**, 553–564.

Gutt, J., Starmans, A. and Dieckmann, G. (1998). Phytodetritus deposited on the Antarctic shelf and upper slope: its relevance for the benthic system. *Journal of Marine Systems* **17**, 435–444.

Haas, C. (1997). Sea-ice thickness measurements using seismic and electro-magnetic-inductive techniques. *Berichte zur Polarforschung* **223**, 1–159.

Haas, C. (1998). Evaluation of ship-based electromagnetic-inductive thickness measurements of summer sea-ice in the Bellingshausen and Amundsen Seas. *Cold Regions Science and Technology* **27**, 1–16.

Haas, C. (2001). The seasonal cycle of ERS scatterometer signatures over perennial Antarctic sea ice and associated surface properties and processes. *Annals of Glaciology* **33**, 69–73.

Haas, C., Gerland, S., Eicken, H., and Miller, H. (1997). Comparison of sea ice thickness measurements under summer and winter conditions in the Arctic using a small electromagnetic induction device. *Geophysics* **62**, 749–757.

Haas, C., Cottier, F., Smedsrud, L. H., Thomas, D. N., Buschmann, U., Dethleff, D., Gerland, S., Giannelli, V., Hoelemann, J., Tison, J.-L. and Wadhams, P. (1999a). Multidisciplinary ice tank study shedding new light on sea ice growth processes. *EOS, Transactions of the American Geophysical Union* **80**, 507–513.

Haas, C., Liu, Q. and Martin, T. (1999b). Retrieval of Antarctic sea-ice pressure ridge frequencies from ERS SAR imagery by means of *in-situ* laser profiling and usage of a neural network. *International Journal of Remote Sensing* **20**, 3111–3123.

Haas, C., Thomas, D. N. and Bareiss, J. (2001). Surface properties and processes of perennial Antarctic sea ice in summer. *Journal of Glaciology* **47**, 613–625.

Hamner, W. M. (1984) Aspects of schooling of *Euphausia superba*. *Journal of Crustacean Biology* **4**, 67–74.

Hamner, W. M. and Hamner, P. P. (2000). Behaviour of Antarctic krill (*Euphausia superba*): schooling, foraging and antipredatory behaviour. *Canadian Journal of Fisheries and Aquatic Sciences* **57**, 192–202.

Hamner, W. M., Hamner, P. P., Strand, S. W. and Gilmer, R. W. (1983). Behavior of Antarctic krill, *Euphausia superba*: chemoreception, feeding, schooling, and molting. *Science* **220**, 433–435.

Hamner, W. M., Hamner P. P., Obst, B. S. and Carleton, J. H. (1989). Field observations on the ontogeny of schooling of *Euphausia superba* furciliae and its relationship to ice in Antarctic waters. *Limnology and Oceanography* **34**, 451–456.

Hardy, A. (1967). "Great Waters". Collins, London.

Hefu, Y. and Kirst, G. O. (1997). Effect of UV-radiation on DMSP content and DMS formation of *Phaeocystis antarctica*. *Polar Biology* **18**, 402–409.

Helbing, E. W., Villafañe, V. E., Ferrario, M. and Holm-Hansen, O. (1992). Impact of natural ultraviolet radiation on rates of photosynthesis and on specific marine phytoplankton species. *Marine Ecology Progress Series* **80**, 89–100.

Helbing, E. W., Marguet, E. R, Villafañe, V. E. and Holm-Hansen, O. (1995). Bacterioplankton viability in Antarctic waters as affected by solar ultraviolet radiation. *Marine Ecology Progress Series* **126**, 293–298.

Helmke, E. and Weyland, H. (1995). Bacteria in sea ice and underlying water of the eastern Weddell Sea in midwinter. *Marine Ecology Progress Series* **117**, 269–287.

Herborg, L.-M., Thomas, D. N., Kennedy, H., Haas, C. and Dieckmann, G. S. (2001). Dissolved carbohydrates in Antarctic sea ice. *Antarctic Science* **13**, 119–125.

Herman, A. W., Knox, D. F., Conrad, J. and Mitchell, M. R. (1993). Instruments for measuring subice algal profiles and productivity *in situ*. *Canadian Journal of Fisheries and Aquatic Science* **50**, 359–369.

Hernando, M., Carreto, J. I., Carignan, M. O., Ferreyra, G. A. and Gross, C. (2002). Effects of solar radiation on growth and mycosporine-like amino acids content in *Thalassiosira* sp., an Antarctic diatom. *Polar Biology* **25**, 12–20.

Herndl, G. J., Müller-Niklas, G. and Frick, J. (1993). Major role of ultraviolet-B in controlling bacterioplankton growth in the surface layer of the ocean. *Nature* **361**, 717–719.

Hewitt, R. P. and Linen Low, E. H. (2000). The fishery on Antarctic krill: defining an ecosystem approach to management. *Reviews in Fisheries Science* **83**, 235–298.

Hewitt, R. P., Watkins, J. L., Naganobu, M., Sushin, V., Brierley, A. S., Demer, D. A., Kasatkina, S., Takao, Y., Goss, C., Malyshko, A., Brandon, M. A., Kawaguchi, S., Siegel, V., Trathan, P., Emery, J., Everson, I. and Miller, D. (Submitted). Biomass and dispersion of Antarctic krill across the Scotia Sea, as estimated from multiship acoustic and net survey conducted in January and February 2000. *Deep Sea Research*.

Hofmann, E. E. and Lascara, C. M. (2000). Modeling the growth dynamics of Antarctic krill *Euphausia superba*. *Marine Ecology Progress Series* **194**, 219–231.

Holland, D. M. (2001). Explaining the Weddell polynya – a large ocean eddy shed at Maud Rise. *Science* **292**, 1697–1700.

Holm-Hansen, O. and Huntley, M. (1984). Feeding requirements of krill in relation to food sources. *Journal of Crustacean Biology* **4** (Special Issue 1), 156–173.

Hooker, J. D. (1847). "The Botany of the Antarctic Voyage of H.M. Discovery Ships Erebus and Terror in the Years 1838–1843. Part 1. Flora Antarctica". Reeve Brothers, London.

Hopkins, T. L. and Torres J. J. (1988). The zooplankton community in the vicinity of the ice edge, western Weddell Sea, March 1986. *Polar Biology* **9**, 79–87.

Horner, R. A. (1985a). Ecology of sea ice algae. *In* "Sea Ice Biota" (R. A. Horner, ed.), pp. 83–104. CRC Press, Florida.

Horner, R. A. (editor) (1985b). "Sea Ice Biota". CRC Press, Florida.

Horner, R. A. (1985c). Taxonomy of sea ice algae. *In* "Sea Ice Biota" (R. A. Horner ed.). pp. 105–130. CRC Press, Florida.

Horner, R. A. (1990). Techniques for sampling sea ice. *In* "Polar Marine Diatoms" (L. K. Medlin and J. Priddle, eds), pp. 19–23. British Antarctic Survey, Cambridge.

Horner, R., Ackley, S. F., Dieckmann, G. S., Gullikson, B., Hoshai, T., Legendre, L., Melnikov, I. A., Reeburgh, W. S., Spindler, M. and Sullivan, C. W. (1992). Ecology of sea ice biota. 1. Habitat, terminology and methodology. *Polar Biology* **12**, 417–427.

Hoshiai, T., Tanimura, A. and Kudoh, S. (1996). The significance of autumnal sea ice biota in the ecosystem of ice-covered polar seas. *Proceedings of the NIPR Symposium on Polar Biology* **9**, 27–34.

Hosie, G. W., Schultz, M. B., Kitchener, J. A., Cochran, T. G. and Richards, K. (2000). Macrozooplankton community structure off East Antarctica (80–150 degrees E) during the Austral summer of 1995/1996. *Deep-Sea Research II* **47**, 2437–2463.

Houghton, J. T. and others (1996). Climate change 1995. In "The Science of Climate Change. Contribution of Working Group 1 to the Second Assessment Report of the Intergovernmental Panel on Climate Change" (J. T. Houghton, L. G. Meira Filho, B. A. Callander, N. Harris, A. Kattenberg and K. Maskell, eds), Cambridge University Press, Cambridge.

Huntley, M. and Escritor, F. (1991). Dynamics of *Calanoides acutus* (Copepoda: Calanoida) in Antarctic coastal waters. *Deep-Sea Research* **38**, 1145–1167.

Huntley, M. E., Nordhausen, W. and Lopez, M. D. G. (1994). Elemental composition, metabolic activity and growth of Antarctic krill *Euphausia superba* during winter. *Marine Ecology Progress Series* **107**, 23–40.

Ichii, T. (1990). Distribution of Antarctic krill concentrations exploited by Japanese krill trawlers and minke whales. *Proceedings of the NIPR Symposium on Polar Biology* **3**, 36–56.

Ichii, T., Katayama, K., Obitsu, N., Ishii, H. and Naganobu, M. (1998a). Occurrence of Antarctic krill (*Euphausia superba*) concentrations in the vicinity of the South Shetland Islands: relationship to environmental parameters. *Deep-Sea Research I* **45**, 1235–1262.

Ichii, T., Shinohara, N., Fujise, Y., Nishiwaki, S. and Matsuoka, K. (1998b). Interannual changes in body fat condition index of minke whales in the Antarctic. *Marine Ecology Progress Series* **175**, 1–12.

Ikeda, T. (1985). Life history of Antarctic krill *Euphausia superba*: a new look from an experimental approach. *Bulletin of Marine Science* **37**, 599–608.

International Whaling Commission (1991). "Report of the Sub-committee on Southern Hemisphere Minke Whales". Annex E. pp. 113–131. International Whaling Commission, Cambridge.

Janssen, H. H. and Gradinger, R. (1999). Turbellaria (Archoophora: Acoela) from Antarctic sea ice endofauna – examination of their micromorphology. *Polar Biology* **21**, 410–416.

Jarman, S., Elliott, N., Nicol, S., McMinn, A. and Newman, S. (1999). The base composition of the krill genome and its potential susceptibility to damage by UV-B. *Antarctic Science* **11**, 23–26.

Jefferies, M. O. (editor) (1998). "Antarctic Sea Ice. Physical Processes, Interactions and Variability". *Antarctic Research Series* **74**. American Geophysical Union, Washington, DC.

Jeffrey, W. H. and Mitchell, D. L. (2001). Measurement of UVB-induced DNA damage in marine planktic communities. *Methods in Microbiology* **30**, 469–485.

Jeffrey, W. H., Pledger, R. J., Aas, P., Hager, S., Coffin, R. B., VonHaven, R. and Mitchell, D. L. (1996). Diel and depth profiles of DNA photodamage in bacterioplankton exposed to ambient solar ultraviolet radiation. *Marine Ecology Progress Series* **137**, 283–291.

Jennings, S., Kaiser, M. J. and Reynolds, J. D. (2001). "Marine Fisheries Ecology". Blackwell Science, Oxford.

Jochem, F. J., Mathot, S. and Queguiner, B. (1995). Size-fractionated primary production in the open Southern Ocean in austral spring. *Polar Biology* **15**, 381–392.

Joiris, C. R. (1991). Spring distribution and ecological role of seabirds and marine mammals in the Weddell Sea, Antarctica. *Polar Biology* **11**, 415–424.

Jones, E. G., Collins, M. A., Bagley, P. M., Addison, S. and Priede, I. G. (1998). The fate of cetacean carcasses in the deep sea: observations on consumption rates and succession of scavenging species in the abyssal north-east Atlantic Ocean. *Proceedings of the Royal Society of London Series B* **265**, 1119–1127.

Junge, K., Krembs, C., Deming, J., Stierle, A. and Eicken, H. (2001). A microscopic approach to investigate bacteria under in situ conditions in sea ice samples. *Annals of Glaciology* **33**, 304–310.

Jürgens, K. and Güde, H. (1994).The potential importance of grazing-resistant bacteria in planktic systems. *Marine Ecology Progress Series* **112**, 169–188.

Kähler, P., Bjørnsen, P. K., Lochte, K. and Antia, A. (1997). Dissolved organic matter and its utilisation by bacteria during spring in the Southern Ocean. *Deep-Sea Research* **44**, 341–353.

Kaiser, E. and Herndl, G. J. (1997). Rapid recovery of marine bacterioplankton activity after inhibition by UV radiation in coastal waters. *Applied and Environmental Microbiology* **63**, 4026–4031.

Kanda, K., Takagi, K. and Seki, Y. (1982). Movement of the larger swarms of Antarctic krill *Euphausia superba* off Enderby Land during 1976–1977 season. *Journal of Tokyo University of Fisheries* **68**, 25–42.

Karentz, D. (1991). Ecological considerations of Antarctic ozone depletion. *Antarctic Science* **3**, 3–11.

Karentz, D., Cleaver, J. E. and Mitchell, D. L. (1991). Cell survival characteristics and molecular responses of Antarctic phytoplankton to Ultraviolet-B radiation. *Journal of Phycology* **27**, 326–341.

Kasamatsu, F. (2000). Species diversity of the whale community in the Antarctic. *Marine Ecology Progress Series* **200**, 297–301.

Kasamatsu, F., Ensor, P. and Joyce, G. G. (1998). Clustering and aggregation of minke whales in the Antarctic feeding grounds. *Marine Ecology Progress Series* **168**, 1–11.

Kaufmann, R. S., Smith, K. L., Jr, Baldwin, R. J., Glatts, R. C., Robison, B. H. and Reisenbichler, K. R. (1995). Effects of seasonal pack ice on the distribution of macrozooplankton and micronekton in the northwestern Weddell Sea. *Marine Biology* **124**, 387–397.

Kawaguchi, S. and Satake, M. (1994). Relationship between recruitment of the Antarctic krill and the degree of ice cover near the South Shetland Islands. *Fisheries Science* **60**, 123–124.

Kennedy, H., Thomas, D. N., Kattner, G., Haas, C. and Dieckmann, G. S. (submitted). Particulate organic carbon in Antarctic summer sea ice: concentration and stable carbon isotopic composition. *Marine Ecology Progress Series*.

Kieber, D. J. and Mopper, K. (1996). Photochemistry of Antarctic waters during the 1994 austral summer. *Antarctic Journal of the United States* **30**, 150–151.

Kieber, D. J., McDaniel, J. and Mopper, K. (1989). Photochemical source of biological substrates in sea water: implications for carbon cycling. *Nature* **341**, 637–639.

Kils, U. (1981). "Swimming Behaviour, Swimming Performance and Energy Balance of Antarctic Krill *Euphausia superba*". Biomass Scientific Report Series **3**. SCAR and SCOR, Cambridge, UK.

King, J. C. (1994). Recent climate variability in the vicinity of the Antarctic Peninsula. *International Journal of Climatology* **14**, 357–369.

Kirst, G. O. and Wiencke, C. (1995). Ecophysiology of polar algae. *Journal of Phycology* **31**, 181–199.

Kirst, G. O., Thiel, C., Wolff, H., Nothangel, J., Wanzek, M. and Ulmke, R. (1991). Dimethylsulfonioproprinate (DMSP) in ice-algae and its possible biological role. *Marine Chemistry* **35**, 381–388.

Kivelson, M. G., Khurana, K. K., Russell, C. T., Volwerk, M., Walker, R. J. and Zimmer, C. (2000). Galileo magnetometer measurements: a stronger case for a subsurface ocean at Europa. *Science* **289**, 1340–1343.

Kivi, K. and Kuosa, H. (1994). Late winter microbial communities in the western Weddell Sea (Antarctica). *Polar Biology* **14**, 389–399.

Klass, C. (1997). Microprotozooplankton distribution and their potential grazing impact in the Antarctic Circumpolar Current. *Deep-Sea Research* **44**, 375–394.

Knox, G. A., Waghorn, E. J. and Ensor, P. H. (1996). Summer plankton beneath the McMurdo ice shelf at White Island, McMurdo Sound, Antarctica. *Polar Biology* **16**, 87–94.

Kooyman, G. L., Kooyman, T. G., Horning, M. and Kooyman, C. A. (1996). Penguin dispersal after fledging. *Nature* **383**, 397.

Kottmeier, S. T. and Sullivan, C. W. (1987) Late winter primary production and bacterial production in sea ice and seawater west of the Antarctic Peninsula. *Marine Ecology Progress Series* **36**, 287–298.

Kottmeier, S. T., and Sullivan, C. W. (1990). Bacterial biomass and production in pack ice of Antarctic marginal ice edge zones. *Deep-Sea Research* **37**, 1311–1330.

Kovacs, A., Valleau, N. C. and Holladay, J. C. (1987). Airborne electromagnetic sounding of sea ice thickness and subice bathymetry. *Cold Regions Science and Technology* **14**, 289–311.

Kovacs, A., Holladay, J. S. and Bergeron, C. J., Jr. (1995). The footprint/ altitude ratio for helicopter electromagnetic sounding of sea ice thickness: comparison of theoretical and field estimates. *Geophysics* **60**, 374–380.

Krembs, C. and Engel, A. (2001). Abundance and variability of microorganisms and transparent exopolymer particles across the ice water interface of melting first-year sea ice in the Laptev Sea (Arctic). *Marine Biology* **138**, 173–185.

Krembs, C., Gradinger, R. and Spindler, M. (2000). Implications of brine channel geometry and surface area for the interaction of sympagic organisms in Arctic sea ice. *Journal of Experimental Marine Biology and Ecology* **243**, 55–80.

Krembs, C., Mock, T. and Gradinger, R. (2001). A mesocosm study of physical-biological interactions in artificial sea ice: effects of brine channel surface evolution and brine movement on algal biomass. *Polar Biology* **24**, 356–364.

Krembs, C., Tuschling, K. and v. Juterzenka, K. (2002). The topography of the ice water interface – its influence on the colonisation of sea ice by algae. *Polar Biology* **25**, 106–117.

Krembs, C., Junge, K., Eicken, H. and Deming, J. (in press). High concentrations of exopolymeric substances in Arctic winter sea ice: implications for the polar ocean carbon cycle and cryoprotection of diatoms. *Deep-Sea Research II* (in press).

Kristiansen, S., Syvertsen, E. E. and Farbrot, T. (1992). Nitrogen uptake in the Weddell Sea during late winter and spring. *Polar Biology* **12**, 245–251.

Kristiansen, S., Farbrot, T., Kuosa, H., Myklestad, S. and Quillfeldt, C. H. (1998). Nitrogen uptake in the infiltration community, an ice algal community in Antarctic pack-ice. *Polar Biology* **19**, 307–315.

Kühl, M., Glud, R. N., Borum, J., Roberts, R. and Rysgaard, S. (2001). Photo-synthetic performance of surface-associated algae below sea ice as measured with a pulse-amplitude-modulated (PAM) fluorometer and O_2 microsensors. *Marine Ecology Progress Series* **223**, 1–14.

Kuosa, H., Norrmann, B., Kivi, K. and Brandini, F. (1992). Effects of Antarctic sea ice biota on seeding as studied in aquarium experiments. *Polar Biology* **12**, 333–339.

Kurbjeweit, F., Gradinger, R. and Weissenberger, J. (1993). The life cycle of *Stephos longipes* – an example for cryopelagic coupling in the Weddell Sea (Antarctica). *Marine Ecology Progress Series* **98**, 255–262.

Lancraft, T. M., Torres, J. J. and Hopkins, T. L. (1989). Micronekton and macro-zooplankton in open waters near Antarctic ice edge zones (AMERIEZ 1983 and 1986). *Polar Biology* **9**, 225–233.

Lancraft, T. M., Hopkins, T. L., Torres, J. J. and Donnelly, J. (1991). Oceanic micronekton/macrozooplankton community structure and feeding in ice covered Antarctic waters during winter (AMERIEZ 1988). *Polar Biology* **11**, 157–167.

Lange, M. A., Ackley, S. F., Wadhams, P., Dieckmann, G. S. and Eicken, H. (1989). Development of sea ice in the Weddell Sea, Antarctica. *Annals of Glaciology* **12**, 92–96.

Lascara, C. M., Hofmann, E. E., Ross, R. M. and Quetin, L. B. (1999). Seasonal variability in the distribution of Antarctic krill, *Euphausia superba*, west of the Antarctic Peninsula. *Deep-Sea Research I* **46**, 951–984.

Laws, R. M. (1977). Seals and whales of the Southern Ocean. *Philosophical Transactions of the Royal Society Series B* **279**, 81–96.

Laws, R. M. (1985). The ecology of the Southern Ocean. *American Scientist* **73**, 26–40.

Legendre, L., Ackley, S. F., Dieckmann, G. S., Gulliksen, B., Horner, R., Hoshai, T., Melnikov, I. A., Reeburgh, W. S., Spindler, M. and Sullivan, C. W. (1992). Ecology of sea ice biota – 2. Global significance. *Polar Biology*, **12**, 429–444.

Leppäranta, M. (editor) (1998). "The Physics of Ice-Covered Seas", Vols 1 and 2. University of Helsinki Printing House, Helsinki.

Leventer, A. (1998). The fate of Antarctic "Sea ice diatoms" and their use as paleoenvironmental indicators. *Antarctic Research Series* **73**, 121–137.

Liu, H. and Buskey, E. J. (2000). Hypersalinity enhances the production of extra-cellular polymeric substance (EPS) in the Texas brown tide alga, *Aureoumbra lagunensis* (Pelagophyceae). *Journal of Phycology* **36**, 71–77.

Lizotte, M. P. (2001). The contributions of sea ice algae to Antarctic marine primary production. *American Zoologist* **41**, 57–73.

Lizotte, M. P. and Arrigo, K. R. (1998) "Antarctic Sea Ice. Biological Processes, Interactions and Variability". Antarctic Research Series **73**. American Geophysical Union, Washington, DC.

Lizotte, M. P. and Sullivan, C. W. (1992). Biochemical composition and photosynthesis distribution in sea ice microalgae of McMurdo Sound, Antarctica: evidence for nutrient stress during the spring bloom. *Antarctic Science* **4**, 23–30.

Lizotte, M. P., Robinson, D. H. and Sullivan, C. W. (1998). Algal pigment signatures in Antarctic sea ice. *In* "Antarctic Sea Ice. Biological Processes, Interactions and Variability". Antarctic Research Series **73**. (M. Lizotte and K. Arrigo, eds), pp. 93–105. American Geophysical Union, Washington DC.

Loeb, V., Siegel, V., Holm-Hansen, O., Hewitt, R., Fraser, W., Trivelpiece, W. and Trivelpiece, S. (1997). Effects of sea-ice extent and krill or salp dominance on the Antarctic food web. *Nature* **387**, 897–900.

Lønne, O. J. and Gulliksen, B. (1991). On the distribution of sympagic macro-fauna in the seasonally ice-covered Barents Sea. *Polar Biology* **11**, 457–469.

Lorenzen, S. (1986). *Odontobius* (Nematoda, Monhysteridae) from the baleen plates of whales and its relationship to *Gammarinema* living on crustaceans. *Zoologica Scripta* **15**, 101–106.

Lopez, M. D. G., Huntley, M. E. and Lovette, J. T. (1993). *Calanoides acutus* in Gerlache Strait, Antarctica. 1. Distribution of late copepodite stages and reproduction during spring. *Marine Ecology Progress Series* **100**, 153–165.

Löscher, B. M., de Baar, H. J. W., de Jong, J. T. M. and Dehairs, F. (1997). The distribution of Fe in the Antarctic circumpolar current. *Deep-Sea Research* **44**, 143–187.

Lowry, L. F., Testa, J. W. and Calvert, W. (1988). Notes on winter feeding of crab-eater and leopard seals near the Antarctic Peninsula. *Polar Biology* **8**, 475–478.

Macaulay, M. C. (1994). Applications of hydroacoustics in marine ecological studies: a perspective on the present status and future directions. *Proceedings of the NIPR Symposium on Polar Biology* **7**, 118–132.

McConville, M. J., Ikeda, T., Bacic, A. and Clarke, A. E. (1986). Digestive carbohydrases from the hepatopancreas of two Antarctic euphausiid species (*Euphausia superba* and *E. crystallorophias*). *Marine Biology* **90**, 371–378.

MacKenzie D. (1994). Whales win southern sanctuary. *New Scientist* **142**, 7.

Mackintosh, N. A. (1972). Life cycle of Antarctic krill in relation to ice and water conditions. *Discovery Reports* **36**, 1–94.

McMinn, A. and Ashworth, C. (1998). The use of oxygen microelectrodes to determine the net production by an Antarctic sea ice algal community. *Antarctic Science* **10**, 39–44.

McMinn, A., Skerratt, J., Trull, T., Ashworth, C. and Lizotte, M. (1999). Nutrient stress gradient in the bottom 5 cm of fast ice, McMurdo Sound, Antarctica. *Polar Biology* **21**, 220–227.

McMinn, A., Ashworth, C. and Ryan, K. G. (2000). *In situ* net primary productivity of an Antarctic fast ice bottom algal community. *Aquatic Microbial Ecology* **21**, 177–185.

Malin, G. and Kirst, G. O. (1997). Algal production of dimethyl sulfide and its atmospheric role. *Journal of Phycology* **33**, 889–896.

Mangel, M. and Nicol, S. (2000). Krill and the unity of biology. *Canadian Journal of Fisheries and Aquatic Sciences* **57**, 1–5.

Maranger, R., Bird, D. F. and Juniper, S. K. (1994). Viral and bacterial dynamics in Arctic sea ice during the spring algal bloom near Resolute, NWT, Canada. *Marine Ecology Progress Series* **111**, 121–127.

Marchant, H., Davidson, A. and Kelly, G. J. (1991). UV-B protecting compounds in the marine alga *Phaeocystis pouchetii* from Antarctica. *Marine Biology* **109**, 391–395.

Marchant, H., Davidson, A., Wright, S. and Glazebrook, J. (2000). The distribution and abundance of viruses in the Southern Ocean during spring. *Antarctic Science* **12**, 414–417.

Markus, T., Kottmeier, C. and Fahrbach, E. (1998). Ice formation in coastal polynyas in the Weddell Sea and their impact on oceanic salinity. *Antarctic Research Series* **74**, 273–292.

Marr, J. W. S. (1962). The natural history and geography of the Antarctic krill (*Euphausia superba* Dana). *Discovery Reports* **32**, 33–464.

Marschall, H. P. (1987). Underwater observations with a remotely operated vehicle. *Berichte zur Polarforschung* **39**, 141–145.

Marschall, H. P. (1988). The overwintering strategy of Antarctic krill under the pack-ice of the Weddell Sea. *Polar Biology* **9**, 129–135.

Massom, R. A., Eicken, H., Haas, C., Jeffries, M. O., Drinkwater, M. R., Sturm, M., Worby, A. P., Wu, X. R., Lytle, V. I., Ushio, S., Morris, K., Reid, P. A., Warren, S. G. and Allison, I. (2001). Snow on Antarctic Sea ice. *Reviews of Geophysics* **39**, 413–444.

Maykut, G. A. (1985) The ice environment. *In* "Sea Ice Biota" (R. A. Horner, ed.), pp. 21–82. CRC Press, Florida.

Melnikov, I. (1995). An *in-situ* experimental study of young sea-ice formation on an Antarctic lead. *Journal of Geophysical Research* **100**, 4673–4680.

Melnikov, I. A. (1998). Winter production of sea ice algae in the western Weddell Sea. *Journal of Marine Systems* **17**, 195–205.

Melnikov, I. A. and Spiridonov, V. A. (1996). Antarctic krill under perennial sea ice in the western Weddell Sea. *Antarctic Science* **8**, 323–329.

Menshenina, L. L. and Melnikov, I. A. (1995). Under-ice zooplankton of the western Weddell Sea. *Proceedings of the NIPR Symposium on Polar Biology* **8**, 126–138.

Meyer-Reil, L.-A. (1994). Microbial life in sedimentary biofilms – the challenge to microbial ecologists. *Marine Ecology Progress Series* **112**, 303–311.

Michel, C., Legendre, L., Ingram, R. G., Gosselin, M. and Levasseur, M. (1996). Carbon budget of sea-ice algae in spring: Evidence of a significant transfer to zooplankton grazers. *Journal of Geophysical Research* **101**, 18345–18360.

Millard, N. W., Griffiths, G., Finegan, G., McPhail, S. D., Meldrum, D. T., Pebody, M., Perrett, J. R., Stevenson, P. and Webb, A. T. (1998). Versatile autonomous submersibles – the realising and testing of a practical vehicle. *Journal of the Society for Underwater Technology* **23**, 7–17.

Miller, D. G. M. and Hampton, I. (1989). "Biology and Ecology of the Antarctic Krill (*Euphausia superba* Dana): A Review". Biomass Science Series **9**. SCAR and SCOR, Cambridge, UK.

Miralto, A., Barone, G., Romano, G., Poulet, S. A., Buttino, I., Mazzarella, G., Laabir, M., Cabrini, M. and Glacobbe, M. G. (1999). The insidious effect of diatoms on copepod reproduction. *Nature* **402**, 173–176.

Mitchell, C. and Beardall, J. (1996). Inorganic carbon uptake by an Antarctic sea-ice diatom, *Nitzschia frigida*. *Polar Biology* **16**, 95–99.

Mock, T. (2002). *In situ* primary production in young Antarctic sea ice. *Hydrobiologia* (in press).

Mock, T. and Gradinger, R. (1999). Determination of Arctic ice algal production with a new *in situ* incubation technique. *Marine Ecology Progress Series* **177**, 15–26.

Mock, T. and Gradinger, R. (2000). Changes in photosynthetic carbon allocation in algal assemblages of Arctic sea ice with decreasing nutrient concentrations and irradiance. *Marine Ecology Progress Series*, **202**, 1–11.

Mock, T. and Kroon, B. M. A. (Personal Communication).

Monfort, P., Demers, P. and Levasseur, M. (2000). Bacterial dynamics in first year sea ice and underlying seawater of Saroma-ko lagoon (Sea of Okhotsk, Japan) and Resolute Passage (High Canadian Arctic): inhibitory effects of ice algae on bacterial dynamics. *Canadian Journal of Microbiology* **46**, 623–632.

Montresor, M., Procaccini, G. and Stoecker, D. K. (1999). *Polarella glacialis*, gen. nov., sp. nov. (Dinophyceae): Suessiaceae are still alive. *Journal of Phycology* **35**, 186–197.

Moore, J. K., Abbot, M. R. and Richman, J. G. (1999). Location and dynamics of the Antarctic Polar Front from satellite sea surface temperature data. *Journal of Geophysical Research* **104**, 3059–3073.

Mopper, K., Zhou, X., Kieber, R. J., Kieber, D. J., Sikorski, R. J. and Jones, R. D. (1991). Photochemical degradation of dissolved organic carbon and its impact on the oceanic carbon cycle. *Nature* **353**, 60–62.

Morris, K., Jefferies, M. O. and Shusun, L. (1998). Seasonal characteristics and seasonal variability of ERS-1 SAR backscatter in the Bellingshausen Sea. *Antarctic Research Series* **74**, 213–242.

Muench, R. D. (1990). Mesoscale phenomena in the Polar Oceans. *In* "Polar Oceanography" (W. O. Smith, ed.), pp. 223–285. Academic Press, San Diego.

Murase, H., Matsuoka, K., Ichii, T. and Nishiwaki, S. (2002). Relationship between the distribution of euphausiids and baleen whales in the Antarctic (35°E–145°W). *Polar Biology* (in press).

Murphy, E. and King, J. (1997). Icy message from the Antarctic. *Nature* **389**, 20–21.

Murphy, E. J. (1995). Spatial structure of the Southern Ocean ecosystem: predator-prey linkages in Southern Ocean food webs. *Journal of Animal Ecology* **64**, 333–347.

Murphy, E. J., Clarke, A., Symon, C. and Priddle, J. (1995). Temporal variation in Antarctic sea-ice: analysis of a long term fast-ice record from the South Orkney Islands. *Deep-Sea Research I* **42**, 1045–1062.

Murphy, E. J., Trathan, P. N., Everson, I., Parkes, G. and Daunt, F. (1997). Krill fishing in the Scotia Sea in relation to bathymetry, including the detailed distribution around South Georgia. *Commission for the Conservation of Antarctic Marine Living Resources Science* **4**, 1–17.

Murphy, E. J., Boyd, P. W., Leakey, R. J. G., Atkinson, A., Edwards, E. S., Robinson, C., Priddle, J., Bury, S. J., Robins, D. B., Burkill, P. H., Savidge, G., Owens, N. J. P. and Turner, D. (1998a). Carbon flux in ice-ocean-plankton systems of the Bellingshausen Sea during a period of ice retreat. *Journal of Marine Systems* **17**, 207–227.

Murphy, E. J., Watkins, J. L., Reid, K., Trathan, P. N., Everson, I., Croxall, J. P., Priddle, J., Brandon, M. A., Brierley, A. S. and Hofmann, E. (1998b). Inter-annual variability of the South Georgia marine ecosystem: biological and physical sources of variation in the abundance of krill. *Fisheries Oceanography* **7**, 381–390.

Murray, A. W. A. (1996). Comparison of geostatistical and random sample survey analyses of Antarctic krill acoustic data. *ICES Journal of Marine Science* **53**, 415–421.

Naganobu, M., Kutsuwada, K., Sasai Y., Taguchi, S. and Siegel, V. (1999). Relationship between Antarctic krill (*Euphausia superba*) variability and westerly fluctuations and ozone depletion in the Antarctic Peninsula area. *Journal of Geophysical Research – Oceans* **104**, 20651–20665.

Nedwell, D. B. (1999). Effect of low temperature on microbial growth: lowered affinity for substrates limits growth at low temperature. *FEMS Microbiology Ecology* **30**, 101–111.

Nevitt, G. A., Veit, R. R. and Kareiva, P. (1995). Dimethyl sulfide as a foraging cue for antarctic procellariiform seabirds. *Nature* **376**, 680–682.

Nicholls, K. W. (1997). Predicted reduction in basal melt rates of an Antarctic ice shelf in a warmer climate. *Nature* **388**, 460–462.

Nichols, D. S., Nichols, P. D. and McMeekin, T. A. (1995). Ecology and physiology of psychrophyllic bacteria from Antarctic saline lakes and sea ice. *Science Progress* **78**, 311–347.

Nichols, D. S., Bowman, J., Sanderson, K., Mancuso-Nichols, C., McMeekin, T. A. and Nichols, P. D. (1999a). Developments with Antarctic microorganisms: culture collections, bioactivity screening, taxonomy, PUFA production and cold-adapted enzymes. *Current Opinion in Biotechnology* **10**, 240–246.

Nichols, D. S., Greenhill, A. R., Shadbolt, C. T., Ross, T. and McMeekin, T. A. (1999b). Physicochemical parameters for growth of the sea ice bacteria *Glaciecola punicea* ACAM 611T and *Gelidibacter* sp. Strain IC158. *Applied and Environmental Microbiology* **65**, 3757–3760.

Nichols, D. S., Olley, J., Garda, H., Brenner, R. R. and McMeekin, T. A. (2000). Effect of temperature and salinity stress on growth and lipid composition of *Shewanella gelidimarina*. *Applied and Environmental Microbiology* **66**, 2422–2429.

Nichols, P. D., Palmisano, A. C., Rayner, M. S., Smith, G. A. and White, D. C. (1989). Changes in the lipid composition of Antarctic sea ice diatom communities during a spring bloom: an indication of community physiological status. *Antarctic Science* **1**, 133–140.

Nicol, S. (2000). Understanding krill growth and ageing: the contribution of experimental studies. *Canadian Journal of Fisheries and Aquatic Sciences* **57**, 168–177.

Nicol, S. and Allinson, I. (1999). The frozen skin of the Southern Ocean. *American Scientist* **85**, 426–439.

Nicol, S. and de la Mare, W. K. (1993). Ecosystem management and the Antarctic krill. *American Scientist* **81**, 36–47.

Nicol, S. and Endo, Y. (1999). Krill fisheries: development, management and ecosystem implications. *Aquatic Living Resources* **12**, 105–120.

Nicol, S, Pauly, T., Bindhoff, N. L., Wright, S., Thiele, D., Hosie, G. W., Strutton, P. G. and Woehler, E. (2000). Ocean circulation off east Antarctica affects ecosystem structure and sea-ice extent. *Nature* **406**, 504–507.

Niebauer, H. J. and Alexander, V. (1985) Oceanographic frontal structure and biological production at an ice edge. *Continental Shelf Research* **4**, 367–388.

Noordkamp, D. J. B., Schotten, M., Gieskes, W. W. C., Forney, L. J., Gottschal, J. C. and von Rijssel, M. (1998). High acrylate concentrations in the mucus of *Phaeocystis globosa* colonies. *Aquatic Microbial Ecology* **16**, 45–52.

Nordoy, E. S., Folkow, L. and Blix, A. S. (1995). Distribution and diving behavior of crab-eater seals (*Lobodon carcinophagus*) off Queen-Maud-Land. *Polar Biology* **15**, 261–268.

Nöthig, E.-M. and von Bodungen, B. (1989). Occurrence and vertical flux of faecal pellets of probably protozoan origin in the southeastern Weddell Sea (Antarctica). *Marine Ecology Progress Series* **56**, 281–289.

Obernosterer, I., Reitner, B. and Herndl, G. J. (1999). Contrasting effects of solar radiation on dissolved organic matter and its bioavailability to marine bacterio-plankton. *Limnology and Oceanography* **44**, 1645–1654.

O'Brien, D. P. (1987) Direct observations of the behaviour of *Euphausia superba* and *Euphausia crystallorophias* (Crustacea: Euphausiacea) under pack ice during the Antarctic spring of 1985. *Journal of Crustacean Biology* **7**, 437–448.

Orsi, A. H., Whitworth, T. and Nowlin, W. D. (1995). On the meridional extent and fronts of the Antarctic Circumpolar Current. *Deep-Sea Research I* **42**, 641–673.

Pakhomov, E. A., Ratkova, T. N., Froneman, P. W. and Wassmann, P. (2001). Phytoplankton dynamics at the ice-edge zone of the Lazarev Sea (Southern Ocean) during austral summer 1994/1995 drogue study. *Polar Biology* **24**, 422–431.

Palmisano, A. C. and Garrison, D. L. (1993). Microorganisms in Antarctic sea ice. *In* "Antarctic Microbiology" (E. I. Friedman, ed.), pp. 167–218. Wiley-Liss, New York.

Palmisano, A. C. and Sullivan, C. W. (1985a). Growth, metabolism, and dark survival in sea ice microalgae. *In* "Sea Ice Biota" (R. A. Horner, ed.), pp. 131–146. CRC Press, Florida.

Palmisano, A. C. and Sullivan, C. W. (1985b). Physiological response of microalgae in the ice platelet layer to low light conditions. *In* "Antarctic Nutrient Cycles and Food Webs" (W. R. Siegfried, P. R. Condy and R. M. Laws, eds.), pp. 84–88. Springer, Berlin.

Park, M. G., Yang, S. R., Kang, S. H., Chung, K. H. and Shim, J. H. (1999). Phytoplankton biomass and primary production in the marginal ice zone of the northwestern Weddell Sea during austral summer. *Polar Biology* **21**, 251–261.

Parkinson, C. L. (1998). Length of the sea ice season in the Southern Ocean. *Antarctic Research Series* **74**, 173–186.

Parkinson, C. L. and Gloersen, P. (1993). Global sea ice coverage. *In* "Atlas of Satellite Observations Related to Global Change" (R. J. Gurney, J. L. Foster and C. L. Parkinson, eds), pp. 371–383. Cambridge University Press, Cambridge.

Passow, U. and Alldredge, A. L. (1999). Do transparent exopolymer particles (TEP) inhibit grazing by the euphausiid *Euphausia pacifica*? *Journal of Plankton Research* **21**, 2203–2217.

Pasternak, A. F. (1995). Gut contents and diel feeding rhythm in dominant copepods in the ice-covered Weddell Sea, March 1992. *Polar Biology* **15**, 583–586.

Penrose, J. D., Conde, M. and Pauly, T. J. (1994). Acoustic detection of ice crystals in Antarctic waters. *Journal of Geophysical Research* **99**, 12573–12580.

Perovich, D. K. (1993). A theoretical model of ultraviolet light transmission through Antarctic sea ice. *Journal of Geophysical Research* **98**, 22579–22587.

Perovich, D. K. and Govoni, J. W. (1991). Absorption coefficients of ice from 250 to 400 nm. *Geophysical Research Letters* **18**, 1233–1235.

Perovich, D. K., Roesler, C. S. and Pegu, W. S. (1998). Variability in Arctic sea ice properties. *Journal of Geophysical Research* **103**, 1193–1208.

Peters, E. and Thomas, D. N. (1996a). Prolonged nitrate exhaustion and diatom mortality – a comparison of polar and temperate *Thalassiosira* species. *Journal of Plankton Research* **18**, 953–968.

Peters, E. and Thomas, D. N. (1996b). Prolonged darkness and diatom mortality 1: Marine Antarctic species. *Journal of Experimental Marine Biology and Ecology* **207**, 25–41.

Petz, W., Song, W. and Wilbert, N. (1995). Taxonomy and ecology of the ciliate fauna (Protozoa, Ciliophora) in the endopagial and pelagial of the Weddell Sea, Antarctica. *Stapfia* **40**, 1–223.

Plötz, J., Bornemann, H., Knust, R., Schröder, A. and Bester, M. (2002). Foraging behaviour of Weddell seals, and its ecological implications. *Polar Biology* (in press).

Pollard, R. T., Read, J. F., Allen, J. T., Griffiths, G. and Morrison, A. I. (1995). On the physical structure of a front in the Bellingshausen Sea. *Deep-Sea Research II* **42**, 955–982.

Pomeroy, L. R. and Wiebe, W. J. (2001). Temperature and substrates as interactive limiting factors for marine heterotrophic bacteria. *Aquatic Microbial Ecology* **23**, 187–204.

Prézelin, B., Boucher, N. P. and Smith, R. C. (1994). Marine primary production under the influence of the Antarctic ozone hole: ICECOLORS '90. *Antarctic Research Series* **62**, 159–186.

Prézelin, B., Moline, M. A. and Matlick, H. A. (1998). ICECOLORS '93: Spectral UV radiation effects on Antarctic frazil ice algae. *Antarctic Research Series* **73**, 45–83.

Priddle, J., Smetacek, V. and Bathmann, U. (1992). Antarctic marine primary production, biogeochemical carbon cycles and climatic change. *Philosophical Transactions of the Royal Society Series B* **338**, 289–297.

Priddle, J., Leakey, R. J. G., Archer, S. D. and Murphy, E. J. (1996). Eukaryotic microbiota in the surface waters and sea ice of the Southern Ocean: aspects of physiology, ecology and biodiversity in a "two-phase" ecosystem. *Biodiversity and Conservation* **5**, 1473–1504.

Priscu, J. C., Priscu, L. R., Palmisano, A. C. and Sullivan, C. W. (1990). Estimation of neutral lipid levels in Antarctic sea ice microalgae by nile red fluorescence. *Antarctic Science* **2**, 149–155.

Pusceddu, A., Cattaneo-Vietti, R., Albertelli, G. and Fabiano, M. (1999). Origin, biochemical composition and vertical flux of particulate organic matter under the pack ice in Terra Nova Bay (Ross Sea, Antarctica) during late summer 1995. *Polar Biology* **22**, 124–132.

Quetin, L. B. and Ross, R. M. (1991). Behavioural and physiological characteristics of the Antarctic krill, *Euphausia superba*. *American Zoologist* **31**, 49–63.

Quetin, L. B., Ross, R. M. and Clarke, A. (1994). Krill energetics: seasonal and environmental aspects of the physiology of *Euphausia superba*. *In* "Southern Ocean Ecology: the BIOMASS Perspective" (S. Z. El-Sayed, ed.), pp. 165–184. Cambridge University Press, Cambridge.

Quetin, L. B., Ross, R. M., Frazer, T. K. and Haberman, K. L. (1996). Factors affecting distribution and abundance of zooplankton, with an emphasis on Antarctic krill, *Euphausia superba*. *Antarctic Research Series* **70**, 357–371.

Rau, G. H., Takahashi, T. and Des Mareis, D. J. (1989). Latitudinal variations in plankton $\delta^{13}C$: implications for CO_2 and productivity in past oceans. *Nature* **341**, 516–518.

Rau, G. H., Froelich, P. N., Takahashi, T. and Des Marais, D. J. (1991a). Does sedimentary organic $\delta^{13}C$ record variations in Quaternary ocean (CO_2 (aq))? *Paleoceanography* **6**, 335–347.

Rau, G. H., Sullivan, C. W. and Gordon, L. (1991b). $\delta^{13}C$ and $\delta^{15}N$ variations in Weddell Sea particulate organic matter. *Marine Chemistry* **35**, 355–369.

Raymond, J. A. (2000). Distribution and partial characterisation of ice-active molecules associated with sea ice diatoms. *Polar Biology* **23**, 721–729.

Raymond, J. A., Sullivan, C. W. and DeVries, A. L. (1994). Release of an ice-active substance by Antarctic sea ice diatoms. *Polar Biology* **14**, 71–75.

Reay, D. S., Nedwell, D. B., Priddle, J. and Ellis-Evans, J. C. (1999). Temperature dependence of inorganic nitrogen uptake: Reduced affinity for nitrate at sub-optimal temperatures in both algae and bacteria. *Applied and Environmental Microbiology* **65**, 2577–2584.

Reid, K. and Croxall, J. P. (2001). Environmental response of upper trophic-level predators reveals a system change in an Antarctic marine ecosystem. *Proceedings of the Royal Society of London Series B* **268**, 377–384.

Reid, K., Brierley, A. S. and Nevitt, G. A. (2000). An initial examination of the relationships between the distribution of whales and Antarctic krill *Euphausia superba* at South Georgia. *Journal of Cetacean Research and Management* **2**, 143–149.

Reimnitz, E., Kempema, E. W. and Barnes, P. W. (1987). Anchor ice, seabed freezing, and sediment dynamics in shallow Arctic seas. *Journal of Geophysical Research* **92**, 14671–14678.

Rhode, S. C., Pawlowski, M. and Tollrian, R. (2001). The impact of ultraviolet radiation on the vertical distribution of zooplankton of the genus *Daphnia*. *Nature* **412**, 69–72.

Ribic, C. A., Ainley, D. G. and Fraser, W. R. (1991). Habitat selection by marine mammals in the marginal ice-zone. *Antarctic Science* **3**, 181–186.

Riebesell, U., Schloss, I. and Smetacek, V. (1991). Aggregation of algae released from melting sea ice: implications for seeding and sedimentation. *Polar Biology* **11**, 239–248.

Riegger, L. and Robinson, D. (1997). Photoinduction of UV-absorbing compounds in Antarctic diatoms and *Phaeocystis antarctica*. *Marine Ecology Progress Series* **160**, 13–25.

Riemann, F. and Schaumann, K. (1993). Thraustochytrid protists in Antarctic fast ice? *Antarctic Science* **5**, 279–280.

Riemann, F. and Sime-Ngando, T. (1997). Note on sea ice nematodes (Monhysteroidea) from Resolute Passage, Canadian High Arctic. *Polar Biology* **18**, 70–75.

Rivkin, R. B. and Putt, M. (1987). Heterotrophy and photoheterotrophy in Antarctic microalgae: light-dependent incorporation of amino acids and glucose. *Journal of Phycology* **23**, 442–452.

Robinson, C., Hill, H. J., Archer, S., Leakey, R. J. G, Boyd, P. W. and Bury, S. J. (1995a). Scientific diving under sea ice in the Southern Ocean. *Underwater Technology* **21**, 21–27.

Robinson, D. H., Arrigo, K. R., Iturriaga, R. and Sullivan, C. W. (1995b). Adaptation to low irradiance and restricted spectral distribution by Antarctic microalgae from under-ice habitats. *Journal of Phycology* **31**, 508–520.

Rogers, T. and Bryden, M. M. (1995). Predation of Adélie penguins (*Pygoscelis adeliae*) by leopard seals (*Hydrurga leptonyx*) in Prydz Bay, Antarctica. *Canadian Journal of Zoology* **73**, 1001–1004.

Ross, R. M. and Quetin, L. B. (1986). How productive are Antarctic krill? *Bioscience* **36**, 264–269.

Ross, R. M. and Quetin, L. B. (1989). Energetic cost to develop to the first feeding stage of *Euphausia superba* Dana and the effect of delays in food availability. *Journal of Experimental Marine Biology and Ecology* **133**, 103–127.

Ross, R. M., Quetin, L. B. and Lascara, C. M. (1996). Distribution of Antarctic krill and dominant zooplankton west of the Antarctic Peninsula. *In* "Foundations for Ecological Research West of the Antarctic Peninsula" (R. M. Ross, L. B. Quetin and E. E. Hofmann, eds), pp. 199–217. American Geophysical Union, Washington, DC.

Ryan, K. G. and Beaglehole, D. (1994). Ultraviolet radiation and bottom-ice algae: laboratory and field studies from McMurdo Sound, Antarctica. *Antarctic Research Series* **62**, 229–242.

Rysgaard, S., Kühl, M., Glud, R. N. and Würgler Hansen, J. (2001). Biomass, production and horizontal patchiness of sea ice algae in a high-Arctic fjord (Young Sound, NE Greenland). *Marine Ecology Progress Series* **223**, 15–26.

Savidge, G., Harbour, D., Gilpin, L. C. and Boyd, P. W. (1995). Phytoplankton distribution and production in the Bellingshausen Sea, Austral spring 1992. *Deep-Sea Research* **42**, 1201–1224.

Schnack-Schiel, S. B. and Hagen, W. (1995). Life-cycle strategies of *Calanoides acutus*, *Calanus propinquus* and *Metridia gerlachei* (Copepoda: Calanoida) in the eastern Weddell Sea, Antarctica. *ICES Journal of Marine Science* **52**, 541–548

Schnack-Schiel, S. B., Thomas, D. N., Dieckmann, G. S., Eicken, H., Gradinger, R., Spindler, M., Weissenburger, J., Mizdalski, E. and Beyer, K. (1995). Life cycle strategy of the Antarctic calanoid copepod *Stephos longipes*. *Progress in Oceanography* **36**, 45–75.

Schnack-Schiel, S. B., Thomas, D. N., Haas, C., Mizdalski, E. and Dahms, H.-U. (1998). Copepods in Antarctic sea ice. *Antarctic Research Series* **73**, 173–182.

Schnack-Schiel, S. B., Dieckmann, G. S., Gradinger, R., Melnikov, I. A., Spindler, M. and Thomas, D. N. (2001a). Meiofauna in sea ice of the Weddell Sea (Antarctica). *Polar Biology* **24**, 724–728.

Schnack-Schiel, S. B., Thomas, D. N., Haas, C., Dieckmann, G. S. and Alheit, R. (2001b). The occurrence of the copepods *Stephos longipes* (Calanoida) and *Drescheriella glacialis* (Harpacticoida) in summer sea ice in the Weddell Sea. *Antarctic Science* **13**, 150–157.

Scott, F. J., Davidson, A. T. and Marchant, H. J. (2000). Seasonal variation in plankton, submicrometer particles and size-fractionated dissolved organic carbon in Antarctic coastal waters. *Polar Biology* **23**, 635–643.

Scott, F. J., Davidson, A. T. and Marchant, H. J. (2001). Grazing by the Antarctic sea-ice ciliate *Pseudocohnilembus*. *Polar Biology* **24**, 127–131.

Sheridan, P. P. and Brenchley, J. E. (2000). Characterisation of a salt-tolerant family 42 B-Galactosidase from a psychrophillic Antarctic *Planococcus* isolate. *Applied and Environmental Microbiology* **66**, 2438–2444.

Shimadzu, Y. and Katabami, Y. (1984). A note on the information on the pack ice edge obtained by Japanese catcher boats in the Antarctic. *Reports of the International Whaling Commission* **34**, 361–363.

Siegel, V. (1988) A concept of seasonal variation of krill (*Euphausia superba*) distribution and abundance west of the Antarctic peninsula. *In* "Antarctic Ocean and Resources Variability" (D. Sahrhage, ed.), pp. 219–230. Springer-Verlag, Berlin.

Siegel, V. (2000a). Krill (Euphausiacea) demography and variability in abundance and distribution. *Canadian Journal of Fisheries and Aquatic Sciences* **57**, 151–167.

Siegel, V. (2000b). Krill (Euphausiacea) life history and aspects of population dynamics. *Canadian Journal of Fisheries and Aquatic Sciences* **57**, 130–150.

Siegel, V. and Loeb, V. (1995). Recruitment of Antarctic krill *Euphausia superba* and possible causes for its variability. *Marine Ecology Progress Series* **123**, 45–56.

Siegel, V., Bergström, B., Stromberg, J.-O. and Schalk, P. H. (1990). Distribution, size frequencies and maturity stages of krill, *Euphausia superba*, in relation to sea-ice in the northern Weddell Sea. *Polar Biology* **10**, 549–557.

Siegel, V., Skibowski, A. and Harm, U. (1992). Community structure of the epipelagic zooplankton community under the sea-ice of the northern Weddell Sea. *Polar Biology* **12**, 15–24.

Siegel, V., Loeb, V. and Groger, J. (1998). Krill (*Euphausia superba*) density, proportional and absolute recruitment and biomass in the Elephant Island region (Antarctic Peninsula) during the period 1977 to 1997. *Polar Biology* **19**, 393–398.

Slezak, D. M., Puskaric, S. and Herndl, G. J. (1994). Potential role of acrylic acid in bacterioplankton communities in the sea. *Marine Ecology Progress Series* **105**, 191–197.

Smetacek, V. (1999a). Bacteria and silica cycling. *Nature* **397**, 475–476.

Smetacek, V. (1999b). Diatoms and the ocean carbon cycle. *Protist* **150**, 25–32.

Smetacek, V. (2001). A watery arms race. *Nature* **411**, 475.

Smetacek, V., Scharek, R. and Nöthig E.-M. (1990). Seasonal and regional variation in the pelagial and its relationship to the life history cycle of krill. *In* "Antarctic Ecosystems. Ecological Change and Conservation" (K. R. Kerry and G. Hempel, eds), pp. 103–114. Springer-Verlag, Berlin.

Smetacek, V., Scharek, R., Gordon, L. I., Eicken, H., Fahrbach, E., Rohardt, G. and Moore, S. (1992). Early spring phytoplankton blooms in ice platelet layers of the southern Weddell Sea, Antarctica. *Deep-Sea Research* **39**, 153–168.

Smith, D. C. (2001). Expansion of the marine Archaea. *Science* **293**, 56–57.

Smith, R. C., Prezelin, B. B., Baker, K. S., Bidigare, R. R., Boucher, N. P., Coley, T., Karentz, D., MacIntyre, S., Matlick, H. A., Menzies, D., Ondrusek, M., Wan, Z. and Waters, K. J. (1992). Ozone depletion: ultraviolet radiation and phytoplankton biology in Antarctic waters. *Science* **255**, 952–959.

Smith, R. C., Baker, K. S. and Stammerjohn, S.E. (1998). Exploring sea ice indexes for polar ecosystem studies. *Bioscience* **48**, 83–93.

Smith, R. C., Ainley, D. G., Baker, K., Domack, E., Emslie, S., Fraser, B., Kennete, J., Leventer, A., Mosley-Thompson, E., Stammerjohn, S. and Vernet, M. (1999). Marine ecosystem sensitivity to climate change. *Bioscience* **49**, 393–404.

Smith, R. E. H. and Herman, A. W. (1991). Productivity of sea ice algae: *In situ* vs. incubator methods. *Journal of Marine Systems* **2**, 97–110.

Smith, W. O. J. and Nelson, D. M. (1986). The importance of ice edge phytoplankton production in the Southern Ocean. *Bioscience* **36**, 251–257.

Song, W. and Wilbert, N. (2000). Ciliates from Antarctic sea ice. *Polar Biology* **23**, 212–222.

Spindler, M. (1990). A comparison of Arctic and Antarctic sea ice and the effects of different properties on sea ice biota. *In* "Geological History of the Polar Oceans: Arctic versus Antarctic" (U. Bleil and J. Thiede, eds), pp. 173–186. Kluwer Academic Publishers, Dordrecht, The Netherlands.

Spindler, M. and Dieckmann, G. S. (1986). Distribution and abundance of the planktic foraminfer *Neogloboquadrina pachyderma* in sea ice of the Weddell Sea (Antarctica). *Polar Biology* **5**, 185–191.

Spirodonov, V. A., Gruzov, E. N. and Pushkin, A. F. (1985). Investigations of schools of Antarctic *Euphausia superba* (Crustacea, Euphausiacea) under the ice. [Issledovaniia stai antarkticheskoi *Euphausia superba* (Crustacea, Euphausiacea) pod l'dom.] *Zoologicheski Zhurnal* **64**, 1655–1660.

Sprong, I. and Schalk, P. H. (1992). Acoustic observations on krill spring-summer migration and patchiness in the Weddell Sea. *Polar Biology* **12**, 261–268.

Squire, V. A., Dugan, J. P., Wadhams, P., Rottier, P. J. and Liu, A. K. (1995). Of ocean waves and sea-ice. *Annual Reviews of Fluid Mechanics* **27**, 115–168.

Staley, J. T. and Gosink, J. J. (1999). Poles apart: biodiversity and biogeography of sea ice bacteria. *Annual Review of Microbiology* **53**, 189–215.

Stammerjohn, S. E. and Smith, R. C. (1997). Opposing Southern Ocean climate patterns as revealed by trends in regional sea ice coverage. *Climatic Change* **37**, 617–639.

Stanwell-Smith, D., Peck, L. S., Clarke, A., Murray, A. W. A. and Todd, C. D. (1999). The distribution, abundance and seasonality of pelagic marine inverte-brate larvae in the maritime Antarctic. *Philosophical Transactions of the Royal Society of London Series B* **354**, 471–484.

Steele, J. H. (1998). Regime shifts in marine ecosystems. *Ecological Applications* **8** (Suppl.), 33–36.

Stevens, J. E. (1995a). Life on a melting continent. *Discover* (August), 71–75.

Stevens, J. E. (1995b) The Antarctic pack-ice ecosystem. *BioScience* **45**, 128–132.

Stevens, J. E. (1996). Exploring Antarctic ice. *National Geographic Magazine* **189**, 36–53.

Stoecker, D. K., Buck, K. R. and Putt, M. (1992). Changes in the sea ice brine community during the spring-summer transition, McMurdo Sound, Antarctica. I. Photosynthetic protists. *Marine Ecology Progress Series* **84**, 265–278.

Stoecker, D. K., Buck, K. R. and Putt, M. (1993). Changes in the sea ice brine community during the spring-summer transition, McMurdo Sound, Antarctica. II. Phagotrophic protists. *Marine Ecology Progress Series* **95**, 103–113.

Stoecker, D. K., Gifford, D. J. and Putt, M. (1994). Preservation of marine planktic ciliates: losses and cell shrinkage during fixation. *Marine Ecology Progress Series* **110**, 293–299.

Stoecker, D. K., Gustafson, D. E., Merrell, J. R., Black, M. M. D. and Baier, C. T. (1997). Excystment and growth of chryophytes and dinoflagellates at low temperatures and high salinities in Antarctic sea-ice. *Journal of Phycology* **33**, 585–595.

Stoecker, D. K., Gustafson, D. E., Black, M. M. D. and Baier, C. T. (1998). Population dynamics of microalgae in the upper land-fast sea ice at a snow free location. *Journal of Phycology* **34**, 60–69.

Stoecker, D. K., Gustafson, D. E., Baier, C. T. and Black, M. M. D. (2000). Primary production in the upper sea ice. *Aquatic Microbial Ecology* **21**, 275–287.

Stonehouse, B. (1989). "Polar Ecology". Blackie, Glasgow.

Stretch, J. J., Hamner, P. P., Hamner, W. M., Michel, C., Cook, J. and Sullivan, C. W. (1986). Ice algae foraging behavior of the antarctic krill, *Euphausia superba* Dana. *American Zoologist* **26**, A80.

Stretch, J. J., Hamner, P. P., Hamner, W. M., Michel, C., Cook, J. and Sullivan, C. W. (1988). Foraging behaviour of antarctic krill *Euphausia superba* in sea ice microalgae. *Marine Ecology Progress Series* **44**, 131–139.

Sullivan, C. W. (1985). Sea ice bacteria; reciprocal interactions of the organisms and their environment. *In* "Sea Ice Biota" (R. A. Horner, ed.), pp. 159–171. CRC Press, Florida.

Sullivan, C. W. and Palmisano, A. C. (1984). Sea ice microbial communities: distribution, abundance, and diversity of ice bacteria in McMurdo Sound, Antarctica in 1980. *Applied and Environmental Microbiology* **47**, 788–795.

Swadling, K. M. and Gibson, J. A. E. (2000). Grazing rates of a calanoid copepod (*Paralabidocera antarctica*) in a continental Antarctic lake. *Polar Biology* **23**, 301–308.

Swadling, K. M., Gibson, J. A. E., Ritz, D. A. and Nichols, P. D. (1997). Horizontal patchiness in sympagic organisms of the Antarctic fast ice. *Antarctic Science* **9**, 399–406.

Swadling, K. M., McPhee, A. D. and McMinn, A. (2000a). Spatial distribution of copepods in fast ice of eastern Antarctica. *Polar Bioscience* **13**, 55–65.

Swadling, K. M., Nichols, P. D., Gibson, J. A. E. and Ritz, D. A. (2000b). Role of lipid in the life cycles of ice-dependent and ice-independent populations of the copepod *Paralabidocera antarctica*. *Marine Ecology Progress Series* **208**, 171–182.

Syvertsen. E. E. and Kristiansen, S. (1993). Ice algae during EPOS leg 1: assemblages, biomass, origin and nutrients. *Polar Biology* **13**, 61–65.

Tang, K. W. and Dam, H. G. (2001). Phytoplankton inhibition of copepod egg hatching: test of an exudate hypothesis. *Marine Ecology Progress Series* **209**, 197–202.

Tanimura, A., Fukuchi, M. and Ohtsuka, H. (1984a) Occurrence and age composition of *Paralabidocera antarctica* (Calanoida, Copepoda) under the fast ice near Syowa Station, Antarctica. *Memoirs of the National Institute of Polar Research (Japan)* Special Issue **32**, 81–86.

Tanimura, A., Minoda, T., Fukuchi, M., Hoshiai, T. and Ohtsuka, H. (1984b) Swarm of *Paralabidocera antarctica* (Calanoida, Copepoda) under sea ice near Syowa Station, Antarctica. *Antarctic Record* **82**, 12–18.

Tanimura, A., Hoshiai, T. and Fukuchi, M. (1996). The life cycle strategy of the ice-associated copepod, *Paralabidocera antarctica* (Calanoida, Copepoda), at Syowa Station, Antarctica. *Antarctic Science* **8**, 257–266.

Tchesunov, A. V. and Riemann, F. (1995). Arctic sea ice nematodes (Monhysteroidea), with descriptions of *Cryonema crassum* Gen. N., Sp. N. and *C. tenue* Sp. N. *Nematologica* **41**, 35–50.

Testa, J. W., Oehlert, G., Ainley, D. G., Bengston, J. L., Siniff, D. B., Laws, R. M. and Rounsevell, D. (1991). Temporal variability in Antarctic marine ecosystems – periodic fluctuations in the phocid seals. *Canadian Journal of Fisheries and Aquatic Sciences* **48**, 631–639.

Thiele, D. and Gill, P. C. (1999). Cetacean observations during a winter voyage into Antarctic sea ice south of Australia. *Antarctic Science* **11**, 48–53.

Thiele, D., Chester, E. T. and Gill, P. C. (2000). Cetacean distribution off Eastern Antarctica (80–150 degrees E) during the Austral summer of 1995/1996. *Deep-Sea Research II* **47**, 2543–2572.

Thomas, D. N. (1996). Nature's ice-show. *BBC Wildlife Magazine* **14**, 30–36.

Thomas, D. N. and Dieckmann, G. S. (1994). Life in a frozen lattice. *New Scientist* **142**, 33–37.

Thomas, D. N. and Dieckmann, G. S. (2002a). Antarctic sea ice – a habitat for extremophiles. *Science* **295**, 641–644.

Thomas, D. N. and Dieckmann, G. S. (2002b). Biogeochemistry of Antarctic sea ice. *Oceanography and Marine Biology, An Annual Review* (in press).

Thomas, D. N. and Gleitz, M. (1993). Allocation of photoassimilated carbon into major algal metabolite fractions: variation between two diatom species isolated from the Weddell Sea (Antarctica). *Polar Biology* **13**, 281–286.

Thomas, D. N. and Lara, R. J. (1995). Photodegradation of algal derived dissolved organic carbon. *Marine Ecology Progress Series* **116**, 309–310.

Thomas, D. N., Baumann, M. E. M. and Gleitz, M. (1992). Efficiency of carbon assimilation and photoacclimation in a small unicellular *Chaetoceros* species from the Weddell Sea (Antarctica): influence of temperature and irradiance. *Journal of Experimental Marine Biology and Ecology* **157**, 195–209.

Thomas, D. N., Lara, R. J., Haas, C., Schnack-Schiel, S. B., Nöthig, E.-M., Dieckmann, G. S., Kattner, G. and Mizdalski, E. (1998). Biological soup within decaying summer sea ice in the Amundsen Sea, Antarctica. *Antarctic Research Series* **73**, 161–171.

Thomas, D. N., Engbrodt, R., Giannelli, V., Kattner, G., Kennedy, H., Haas, C. and Dieckmann, G. S. (2001a). Dissolved organic matter in Antarctic sea ice. *Annals of Glaciology* **33**, 297–303.

Thomas, D. N., Kennedy, H., Kattner, G., Gerdes, D., Gough, C. and Dieckmann, G. S. (2001b). Biogeochemistry of platelet ice: influence on particle flux under land fast sea ice during summer at Drescher Inlet, Weddell Sea, Antarctica. *Polar Biology* **24**, 486–496.

Todd, F. S. (1988). Weddell seal preys on chinstrap penguin. *Condor* **90**, 249–250.

Tønnenson, J. N. and Johnsen, A. O. (1984) "History of Modern Whaling". Hurst, London.

Tranvik, L. and Kokalj, S. (1998). Decreased biodegradability of algal DOC due to interactive effects of UV radiation and humic matter. *Aquatic Microbial Ecology* **14**, 301–307.

Trathan, P. N., Everson, I., Miller, D. G. M., Watkins, J. L. and Murphy, E. J. (1995). Krill biomass in the Atlantic. *Nature* **373**, 201–202.

Trenerry, L. J., McMinn, A. and Ryan, K. G. (2002). *In situ* oxygen microelectrode measurements of bottom-ice algal production in McMurdo Sound, Antarctica. *Polar Biology* **25**, 72–80.

Trodahl, H. J. and Buckley, R. G. (1990). Enhanced ultraviolet transmission of Antarctic sea ice during the austral spring. *Geophysical Research Letters* **17**, 2177–2179.

Turner, D. R. and Owens, N. J. P. (1995). A biogeochemical study in the Bellingshausen Sea: overview of the STERNA 1992 expedition. *Deep-Sea Research II*, **42**, 907–932.

Turner, S. M., Nightingale, P. D., Broadgate, W. and Liss, P. W. (1995). The distribution of dimethyl sulphide and dimethylsulphoniopropionate in Antarctic waters and sea ice. *Deep-Sea Research II* **42**, 1059–1080.

Tynan, C. T. (1997). Cetacean distributions and oceanographic features near the Kerguelen Plateau. *Geophysical Research Letters* **24**, 2793–2796.

Tynan, C. T. (1998). Ecological importance of the Southern Boundary of the Antarctic Circumpolar Current. *Nature* **392**, 708–710.

Uchino, O., Bojkov, R. D., Balis, D. S., Akagi, K., Hayashi, M. and Kajihara, R. (1999). Essential characteristics of the Antarctic-spring ozone decline: update to 1998. *Geophysical Research Letters* **26**, 1377–1380.

Vaughan, D. G., Marshall, G. J., Connolley, W. M., King, J. C. and Mulvaney, R. (2001). Devil in the detail. *Science* **293**, 1777–1779.

Vincent, W. F. and Roy, S. (1993). Solar ultraviolet-B radiation and aquatic primary production: damage, protection and recovery. *Environment Review* **1**, 1–12.

Voronina, N. M. (1998). Comparative abundance and distribution of major filter-feeders in the Antarctic pelagic zone. *Journal of Marine Systems* **17**, 375–390.

Voronina, N. M. (1999). The vertical distribution and seasonal migrations of populations of common copepods in the Pacific sector of the Antarctic and their dependence on the environmental factors. *Okeanologiya* **39**, 395–405.

Voronina, N. M. and Kolosova, E. G. (1999). Composition, distribution and changes in time of abundance and structure of under fast ice zooplankton in the eastern Weddell Sea in March–May 1992, compared with the open ocean zooplankton. *Okeanologiya* **39**, 80–86.

Voronina, N. M., Levin, L. A., Zadorina, L. A. and Sazhin, A. F. (1994). Planktic community in the Pacific sector of the Antarctic in February–March 1992. *Okeanologiya* **33**, 778–784.

Voronina, N. M., Kolosova, E. G. and Melnikov, I. A. (2001). Zooplankton life under the perennial Antarctic sea ice. *Polar Biology* **24**, 401–407.

Wadhams, P. (1995) Antarctic sea-ice extent and thickness. *Philosophical Transactions of the Royal Society Series A* **352**, 301–319.

Wadhams, P. (2000). "Ice in the Ocean". Gordon and Breach Science. Reading.

Wadhams, P., McLaren, A. S. and Weintraub, R. (1985). Ice thickness distribution in Davis Strait in February from submarine sonar profiles. *Journal of Geophysical Research – Oceans* **90**, 1069–1077.

Wadhams, P., Lange, M. and Ackley, S. F. (1987). The ice thickness distribution across the Atlantic sector of the Antarctic Ocean in midwinter. *Journal of Geophysical Research* **92**, 14535-14552.

Wakatsuchi, M. and Saito, T. (1985). On brine drainage channels of young sea ice. *Annals of Glaciology* **6**, 200–202.

Ward, P., Shreeve, R. S., Cripps, G. C. and Trathan, P. N. (1996). Mesoscale distribution and population dynamics of *Rhincalanus gigas* and *Calanus simillimus* in the Antarctic Polar Open Ocean and Polar Frontal Zone during summer. *Marine Ecology Progress Series* **140,** 21–32.

Watanuki, Y., Mori, Y. and Naito, Y. (1994). *Euphausia superba* dominates in the diet of Adélie penguins feeding under fast sea-ice in the shelf areas of Enderby Land in summer. *Polar Biology* **14**, 429–432.

Watkins, A. B. and Simmonds, I. (2000). Current trends in Antarctic sea ice: the 1990s impact on short climatology. *Journal of Climate* **13**, 4441–4451.

Weeks, W. F. and Ackley, S. F. (1982). The growth, structure and properties of sea ice. *Cold Regions Research and Engineering Laboratory Monograph* **82-1**, 1–130.

Wefer, G. and Fischer, G. (1991). Annual primary production and export flux in the Southern Ocean from sediment trap data. *Marine Chemistry* **35**, 597–613.

Weissenberger, J. (1992). The environmental conditions in the brine channels of Antarctic sea ice. *Berichte zur Polarforschung* **111**, 159.

Weissenberger, J. (1998). Arctic sea ice biota: design and evaluation of a mesocosm experiment. *Polar Biology* **19**, 151–159.

Weissenberger, J. and Grossmann, S. (1998). Experimental formation of sea ice: importance of water circulation and wave action for incorporation of phytoplankton and bacteria. *Polar Biology* **20**, 178–188.

Weissenberger, J., Dieckmann, G. S., Gradinger, R. and Spindler, M. (1992). Sea ice: a cast technique to examine and analyse brine pockets and channel structure. *Limnology and Oceanography* **37**, 179–183.

White, W. B. and Peterson, R. G. (1996). An Antarctic circumpolar wave in surface pressure, wind, temperature and sea-ice extent. *Nature* **380**, 699–702.

Wilson, P. R., Ainley, D. G., Nur, N., Jacobs, S. S., Barton, K. J., Ballard, G. and Comiso, J. C. (2001). Adélie penguin population change in the Pacific sector of Antarctica: relation to sea-ice extent and the Antarctic Circumpolar Current. *Marine Ecology Progress Series* **213**, 301–309.

Winter, C., Moeseneder, M. M. and Herndl, G. J. (2001). Impact of UV radiation on bacterioplankton community composition. *Applied and Environmental Microbiology* **67**, 665–672.

Woehler, E. J. (1993). "The Distribution and Abundance of Antarctic and Subantarctic Penguins". Scientific Committee on Antarctic Research, Cambridge.

Wolfe, G. V. (2000). The chemical defense ecology of marine unicellular plankton: constraints, mechanisms and impacts. *Biological Bulletin* **198**, 225–244.

Wolfe, G. V., Steinke, M. and Kirst, G. O. (1997). Grazing-activated chemical defence in a unicellular marine alga. *Nature* **387**, 894–897.

Yuang, X. J. and Martinson, D. G. (2000). Antarctic sea ice extent variability and its global connectivity. *Journal of Climate* **13**, 1697–1717.

Zaslavskia, L. A., Lippmeier, J. C., Shih, C., Ehrhardt, D., Grossman, A. R. and Apt, K. E. (2001). Trophic conversion of an obligate phototrophic organism through metabolic engineering. *Science* **292**, 2073–2075.

Zeebe, R. E., Eicken, H., Robinson, D. H., Wolf-Gladrow, D. and Dieckmann, G. S. (1996). Modeling the heating and melting of sea ice through light absorption by microalgae. *Journal of Geophysical Research* **101**, 1163–1181.

Zielinski, U. (1997). Parmales species (siliceous marine nanoplankton) in surface sediments of the Weddell Sea, Southern Ocean: indicators for sea ice environment? *Marine Micropaleontology* **32**, 387–395.

Zielinski, U. and Gersonde, R. (1997). Diatom distribution in Southern Ocean surface sediments (Atlantic sector): implications for paleoenvironmental reconstructions. *Palaeogeography, Palaeoclimatology, Palaeoecology* **129**, 213–250.

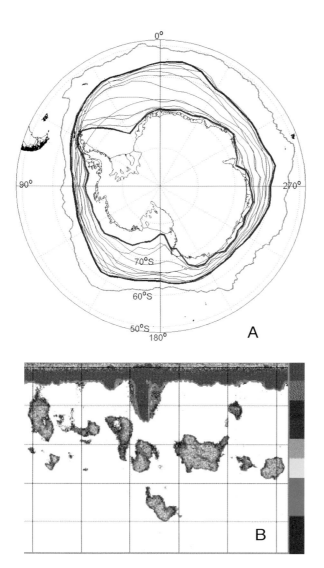

Plate 1 (A) Mean monthly latitude of the sea-ice edge (15% sea-ice concentration) around Antarctica as determined from satellite observations during the period from January 1979 to December 1993. The red line shows the mean position of the February sea-ice minimum and the blue line the mean position of the September maximum. Data sources as for Figure 1 The green line indicates the position of the Polar Front (data from Moore *et al.*, 1999). (B) Echogram of krill swarms under sea ice and an iceberg as detected by *Autosub-2*. Vertical lines are at 185 m intervals along transect, horizontal lines are at 25 m depth intervals. The green line marks the sea-ice/water interface. Colour scale shows echo intensity in 3 dB steps from -38 dB (brown) and -68 dB (light blue). Data from Brierley *et al.* (2002).

Plate 2 Sampling sea ice. (A) A sea-ice corer in use (photograph courtesy of C. Haas). (B) *Autosub*-2 equipped with 38 and 120 kHz echosounder transducers (forward) and an upward-looking acoustic doppler current profiler (ADCP) (aft). The vehicle is 6.8 m long and 0.9 m in diameter (photograph courtesy of M. A. Brandon.)

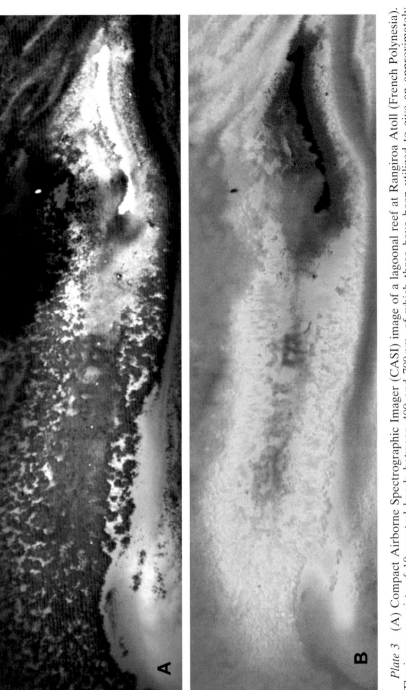

Plate 3 (A) Compact Airborne Spectrographic Imager (CASI) image of a lagoonal reef at Rangiroa Atoll (French Polynesia). The image consists of 10 spectral bands between 400 and 700 nm, of which three have been utilized to give an approximately "true" colour image. Pixel resolution is ∼ 1 m. (B) The same image transformed to show the full variation expressed by all 10 bands. This reveals spatial patterns in the spectral information that are not apparent in the upper image; it also shows the attenuation in the signals caused by depth of water.

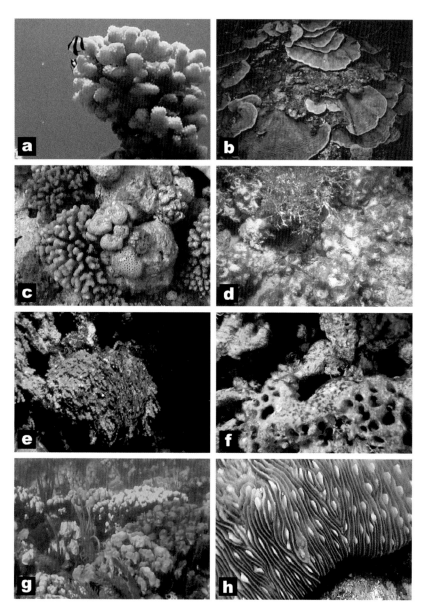

Plate 4 Examples of diverse morphology and pigmentation in both corals and algae. (a) A shallow-water branching coral in the genus *Pocillopora*, (b) deep-water colonies of the coral genus *Agaricia* with a plate-like morphology, (c) red and brown colour morphs of the coral genus *Pocillopora* (left of image), (d) two genera of Phaeophyta showing *Dictyota* (branching, centre of image) and *Lobophora* (fleshy, surrounding *Dictyota* in image), (e) calcified chlorophytes in the genus *Halimeda*, (f) crustose coralline rhodophytes in the genus *Porolithon*, (g) mound-shaped colonies of *Montastraea* sp. showing various intensities of coral bleaching and (h) fluorescence in the coral genus *Fungia*.

Biological and Remote Sensing Perspectives of Pigmentation in Coral Reef Organisms

John D. Hedley and Peter J. Mumby

*Tropical Coastal Management Studies, Department of Marine Sciences
and Coastal Management, Ridley Building, University of Newcastle,
NE1 7RU, UK
FAX: +44 (0)191 2227891, e-mail: j.d.hedley@ncl.ac.uk, p.j.mumby@ncl.ac.uk*

Coral reef communities face unprecedented pressures on local, regional and global scales as a consequence of climate change and anthropogenic disturbance. Optical remote sensing, from satellites or aircraft, is possibly the only means of measuring the effects of such stresses at appropriately large spatial scales (many thousands of square kilometres). To map key variables such as coral community structure, percentages of living coral or percentages of dead coral, a remote sensing instrument must be able to distinguish the reflectance spectra (i.e. "spectral signature", reflected light as a function of wavelength) of each category. For biotic classes, reflectance

ADVANCES IN MARINE BIOLOGY VOL. 43
ISBN 0-12-026143-X

is a complex function of pigmentation, structure and morphology. Studies of coral "colour" fall into two disparate but potentially complementary types. Firstly, biological studies tend to investigate the structure and significance of pigmentation in reef organisms. These studies often lack details that would be useful from a remote sensing perspective such as intraspecific variation in pigment concentration or the contribution of fluorescence to reflectance. Secondly, remote sensing studies take empirical measurements of spectra and seek wavelengths that discriminate benthic categories. Benthic categories used in remote sensing sometimes consist of species groupings that are biologically or spectrally inappropriate (e.g. merging of algal phyla with distinct pigments). Here, we attempt to bridge the gap between biological and remote sensing perspectives of pigmentation in reef taxa. The aim is to assess the extent to which spectral discrimination can be given a biological foundation, to reduce the ad hoc *nature of discriminatory criteria, and to understand the fundamental (biological) limitations in the spectral separability of biotic classes.*

Sources of pigmentation in reef biota are reviewed together with remote sensing studies where spectral discrimination has been effectively demonstrated between benthic categories. The basis of reflectance is considered as the sum of pigmented components, such as zooxanthellae, host tissues and skeletons of corals. Problems in the empirical in situ *measurement of reflectance are identified, such as the differing types of reflectance which can be measured, the interaction of the light field with morphology, and depth-dependent variability of measured reflectance due to fluorescence. The latter is estimated in some cases to introduce an error of up to 20% when depth differs by 8 m.*

*Spectral features useful in discriminating reef benthos are identified and related to pigmentation. The slope in the reflectance spectra between 650 and 690 nm is dependent on chlorophyll-*a *concentration and can be used to discriminate bare sand with no algal component from chlorophyll-*a *containing benthos (algae, corals). The slope in reflectance at various locations between 500 and 560 nm can be useful in discriminating bleached and unbleached corals, possibly due to reduced peridinin concentration. Rhodophyta may be discernible by the presence of a dip in reflectance at 570 nm, due to a phycoerythrin absorption peak. However, the utility of some discriminatory criteria in deeper waters is mitigated by the relatively poor transmission of light through water at longer wavelengths (especially >600 nm).*

Contrary to suggested categorizations of fluorescent pigments in coral host tissues, it is shown that these pigments form an almost continuous distribution with respect to their excitation and emission peaks. Remote sensing by induced fluorescence is a promising approach, but further details about the variation and distribution of these pigments are required.

It is hoped that this review will promote cross-disciplinary collaboration between pigment biologists and the reef remote sensing community. Where possible, the discriminative criteria adopted in remote sensing should be related to biological phenomena, thus lending an intuitive, process-orientated basis for interpreting spectral data. Similarly, remote sensing may provide a novel scaling perspective to biological studies of pigmentation in reef organisms.

1. INTRODUCTION

Significant components of the world's coral reefs are under threat from environmental and anthropogenic stress (Wilkinson, 1999). Remote sensing has the potential to offer cost-effective monitoring on large scales (Mumby *et al.*, 1997b; Green *et al.*, 2000), but its application to reefs presents several challenges that are not found in terrestrial environments (Holden and LeDrew, 1998a). In addition to the atmosphere, the intervening depth of water obfuscates the spectral signal (Lubin *et al.*, 2001). Also, reefs are typically heterogeneous at scales smaller than all but the highest resolution sensors can resolve, so the relationship between the measured spectral signal and benthos is not straightforward. Tackling either of these problems requires that the basic benthic components be inherently spectrally separable, and for this separability to be robustly carried through the water, atmosphere and sensor so that analytical techniques may resolve it in the data.

With this in mind, maximizing the effectiveness of remote sensing on reefs requires the determination of the following factors:

(1) The basic classes for which discrimination is ecologically meaningful or useful.
(2) The extent to which these classes are fundamentally spectrally separable (for biotic classes this involves an appreciation of between- and within-species variance in spectra, for example).
(3) The extent to which this separability is compromised by attenuation due to water depth and the atmosphere.
(4) Given the above information, the best way to extract this separability with the sensor (i.e. which are the best sensor and wavelength bands to choose).
(5) The most effective way to analyse the resultant data to maximize the discrimination of the chosen benthic classes.

Recent studies have demonstrated the ability to discriminate in airborne hyperspectral data (Plate 3), coral, algae and sand (Hochberg and

Atkinson, 2000). *In situ* spectral measurements have implied the possibility of discriminating bleached from non-bleached corals (Holden and LeDrew, 1998b), and the separation of live from dead coral in hyper-spectral data has been demonstrated by Mumby *et al.* (2001). However, up to now the approach has been somewhat *ad hoc*. The constant improve-ments in results give little idea where the absolute limits of reef remote sensing lie, and whether the most significant current limitations are in the sensor spectral resolution, spatial resolution, analysis techniques or funda-mental spectral separability of the components of interest. Although some of the limitations of current satellite sensors have been evaluated for reef remote sensing (Andréfouët *et al.*, 2001) hyperspectral airborne instruments such as CASI are still producing new results (Mumby *et al.*, 2001). Both satellite and airborne sensor technology will inevitably improve and become more cost-effective in time. Therefore, the funda-mental limitations in spectrally distinguishing relevant substrata are of greater importance in the long term than current technological limitations.

Any attempt to deal with the specific problems of remote sensing on reefs must be based on a sound understanding of the separability and variability of the reflectance spectra of reef components. A consistent methodology for measuring reference spectra of reef biota, and an under-standing of the causes and extent of spectral variation between and within species are currently lacking. As will be shown, various workers have used different techniques and equipment, and differing definitions of reflectance to derive spectra, which make cross-study comparisons difficult. Additionally, although pigments have been used to guide the selection of wavelength bands in reef remote sensing (de Vries, 1994), very little work has been done to relate spectral signals of reef benthic classes to functional distribution of their pigments. This situation can be contrasted with studies on phytoplankton, where analysis of the pigment composition of the various groups is well developed (Jeffery *et al.*, 1997a) and has led to some success in remote sensing (particularly with induced fluorescence techniques) (Yentsch and Yentsch, 1979). The contribution of fluorescent pigments to apparent reflectance in corals is a factor that may be significant, but has so far only been partially addressed (Mazel, 1997a; Fux and Mazel, 1999; Fuchs, 2001).

This review considers spectral properties of benthos from the point of view of their causes (pigmentation, structure), consequential variation and impact on separability with respect to remote sensing. It is hoped that the information provided will both be a useful first step when considering the resolution of the five factors above, and an indication of where further work is needed.

There are three sections. Firstly, after a brief discussion of the problem of choosing suitable benthic categories (factor 1, above), what is known of the pigments of reef macroalgae and corals is summarized (relates to factor 2). Secondly, the nature of "reflectance" and the practice of measuring it on benthic material is critically reviewed; several potential problems will be highlighted (this has implications for determining factor 2). Thirdly, reef surveys by remote sensing and *in situ* demonstrations are collated to show the spectral features and methods by which discrimination has been successfully achieved (encompasses factors 3, 4 and 5 above), and these findings will be related back to the fundamental causes of this spectral variation described in the first section.

2. BENTHIC CATEGORIES

A first step in mapping coral reefs is to decide how to subdivide benthic categories. Classes that have little biological or geomorphological meaning are of limited use to practitioners (Mumby and Harborne, 1999). For example, although algae are a fundamental category of reef organisms, to use this general class is to ignore the very different ecological functions (and spectra) of coralline red algae (which are a preferred substratum for settlement of some coral larvae) (Morse *et al.*, 1988); turf algae (which are an important food source for grazing fish) (Ogden and Lobel, 1978); and macroalgae (which as well as preventing settlement inhibit coral growth by competing for space) (Tanner, 1995). However, even a functional classification of benthic taxa can result in heterogeneity at the sub-pixel scale, and therefore, a composite spectral signal is the norm.

There are two approaches to deal with sub-pixel heterogeneity: (1) unmix the individual spectra from the composite signal (Adams *et al.*, 1986) or (2) incorporate the heterogeneity into a classification system based on hierarchical assemblages of taxa (Mumby and Harborne, 1999). The former approach requires a sound knowledge of substratum spectra and the latter approach is constrained by the spectral separability of meaningful ecological or functional classes.

3. OPTICAL PROPERTIES OF REEF BENTHOS

3.1. Macroalgae

The three major groups of reef macroalgae, the Chlorophyta, Phaeophyta and Rhodophyta (green, brown and red algae respectively), have distinctive complements of pigments which define the groups and result in a basic spectral dissimilarity between them (see Table 1, Plate 4) (Dawes, 1998). All contain chlorophyll-*a* (chl-*a*) but differ in the presence of other chlorophylls and accessory pigments. This is promising as regards the possibilities for spectrally discriminating some of the more abundant reef algae, such as *Halimeda* (a green alga), *Lobophora* (a brown alga) and the various forms of coralline red algae. However, within each division (i.e. phyla, Dawes, 1998) there may be considerable variation. For example, Caribbean species of red algae occur in colours ranging from pink and violet to brown-orange or blackish red (Littler *et al.*, 1989). Quantitative information on the pigment complexes present in reef macroalgae, which are significant in remote sensing seems currently to be lacking. This contrasts with detailed studies on zooxanthellae pigments, for example. However, studies on various other species of macroalgae suggest the patterns that can be expected.

The pigment composition of a single macroalgal species can vary as a result of environmental factors such as irradiation, nutrients and water flow. In nearly all algae studied, total pigments as well as the ratio of accessory pigments to chl-*a* increase at low levels of irradiance (Waaland *et al.*, 1974; Rosenberg and Ramus, 1982; Hannach, 1989). Further, algal pigments can shift in type to mirror the spectral nature of the light environment in a process referred to as ontogenetic chromatic acclimation (Dawes, 1998). This has been shown in several genera of green macroalgae with tropical representatives including *Cladophora*, *Valonia*, *Codium* and *Caulerpa* (Yokohoma *et al.*, 1977; Reichert and Dawes, 1986). Siphonxanthin, with an absorption peak at 540 nm (Table 1) is absent in shallow-water examples of *Caulerpa racemosa* but is produced when the same specimens are cultivated under low irradiation (Dawes, 1998). Pigment concentrations are also dependent on nutrient levels. In *Porphyra*, under high conditions of water motion, reducing nutrient levels by three-quarters reduces thallus absorbance to approximately one-third over 400–700 nm (Hannach, 1989).

In many cases the ratio of accessory pigments to chl-*a* appears consistent (Ramus, 1983). As a notable exception, limiting nitrogen reduces phycobiliproteins in some (possibly all) red algae (Lapointe, 1981) and conversely, phycobilins may be synthesized preferentially when nutrients are abundant (e.g. in *Porphyra* sp.) (Hannach, 1989).

The overall spectral appearance of an alga cannot necessarily be inferred from the pigments present. The morphology, thickness of thalli, and cellular architecture affect the relationship between pigment densities and absorption spectra (Ramus, 1978; Vogelmann and Björn, 1986; Hannach, 1989). Water flow and nutrient availability can form a complex relationship with growth and hence also with spectral absorption properties. Hannach (1989) has shown that water motion reduces thallus

Table 1 Photosynthetic pigments found within the divisions Chlorophyta, Phaeophyta, Rhodophyta and Pyrrhophyta. Data sources: Presence/absence, Davies (1976), Dawes (1998), Ambarsari (1997). Absorption peak solvents (solv.) and sources: a – acetone, e – ethanol (Davies, 1976), v – *in vivo* (Govindjee and Braun, 1974). ? – unknown (Dawes, 1998). • – indicates the pigment has been recorded in at least one species from the division.

Pigments	Absorption peaks (nm)	Solv.	Chloro. (green)	Phaeo. (brown)	Rhodo. (red)	Pyrrho. (zoox.)
Chlorophylls						
chl-*a*	435, 670–680	v	•	•	•	•
chl-*b*	480, 650	v	•	•	•	•
chl-*c*	645	v		•		•
Carotenoids						
α	423, 444, 473[a]	e			•	
β	427, 449, 475	e	•	•	•	•
Peridinin	475	e				•
Xanthophylls						
Zeaxanthin	428, 450, 478	e	•		•	
Neoxanthin	415, 438, 467	e	•			
Lutein	422, 445, 474	e	•		•	
Violaxanthin	417, 440, 469	e	•			
Fucoxanthin	426, 449, 465[b]	e		•		•
Diatoxanthin	425, 449, 475	e		•		•
Diadinoxanthin	424, 445, 474	e		•		•
Dinoxanthin	418, 442, 470	a		•		•
Siphonxanthin	540	?	•			
Phycobilins						
Phycocyanin	618	v			•	
Phycoerythrin	490, 546, 576	v			•	
Allophycocyanin	654	v			•	

[a] Third peak sometimes shifts to 500 nm *in vivo*.
[b] Absorption extends to 580 nm *in vivo*.

pigmentation in *Porphyra* via enhancement of growth, whereas Cousens (1982) found increased pigmentation in *Ascophyllum* at wave exposed sites, in a manner not related to growth.

It seems reasonable to assume that the sources of variation outlined above will be fairly ubiquitous and at least play some part in the variation in spectra of reef macroalgae. However, it may be that the variation due to irradiance levels (for example) is comparatively small over the range of depths at which remote sensing can be usefully applied. It is clear that intra-specific spectral variation is considerable in macroalgae, but currently there is insufficient information to assess the extent to which this compromises the spectral discrimination of species. Further studies are required.

3.2. Coral

The optical characteristics of corals are determined fundamentally by the pigments that they contain (Kennedy, 1979). Pigments occur in the zooxanthellae, the ectodermal and endodermal tissues of the polyp, and the skeletons of some species (Kawaguti, 1944; Fox, 1972; Dove *et al.*, 1995). At the macroscopic level, the physical distribution of the pigments combined with the morphology of a colony will affect the spectral signal received from it (Plate 4). This section will summarize what is known of the fundamental properties that contribute to the "spectral signal" of a coral and the sources of variation within and between species.

3.2.1. *Zooxanthellar pigments*

The pigments of the zooxanthellae are a significant component of the overall spectral appearance of a coral. The loss of zooxanthellar pigments or the expulsion of zooxanthellae from polyps during bleaching events (Brown, 1997) can lead to significant spectral changes that can be detected *in situ* (Holden and LeDrew, 1998b; Myers *et al.*, 1999) and possibly by remote sensing. The spectral changes that occur after bleaching, for example, as a result of coral mortality, have consequences for the important goal of monitoring reef "health" and will be discussed later (section 3.2.4)

Until fairly recently, all coral zooxanthellae were thought to be of the genus *Gymnodinium* of the class Dinophyceae (Chalker and Dunlap, 1981; Gil-Turnes and Corredor, 1981). Recent studies have shown that algal symbionts are more diverse, encompassing several orders. However, most of those characterized so far are dinoflagellates of the genus

Symbiodinium, a group related to *Gymnodinium* (the genus *Gymnodinium* is now confined mostly to free-living dinoflagellates) (Rowan, 1998). There are a variety of clades, some of which differ in their light tolerance (Rowan, 1998).

Most species of photosynthetic dinoflagellates contain approximately equal amounts of two classes of pigments; carotenoids and chlorophyll (Hoffman *et al.*, 1996) (see Table 1). The specific pigments present are dependent on species. For example, although peridinin is characteristic of dinoflagellates it is not found in all species (Jeffery *et al.*, 1975; Hoffman *et al.*, 1996). Studies on the identity of pigments in coral zooxanthellae report broadly similar results. Medium pressure liquid chromatographic extraction of zooxanthellar pigments of five Caribbean corals, *Meandrina meandrites*, *Agaricia agaricities*, *Porites astreoides*, *Montastraea cavernosa* and *Mycetophyllia aliciae* has shown a very consistent pattern of relative pigment concentrations (Gil-Turnes and Corredor, 1981). A complement of five pigments was found in the symbionts of all the corals, with the following approximate relative concentrations: chl-*a* (39%), peridinin (39%), chl-*c* (13%), dinoxanthin (7%), β-carotene (2%). High performance liquid chromatography (HPLC) performed on zooxanthellae from Australian *Acropora formosa* (Chalker and Dunlap, 1981) has shown the presence of structural isomers of chl-*a* and chl-*c* and several varieties of yellow and orange xanthophylls (of which dinoxanthin may be one). Pigments in the zooxanthellae of *Goniastrea aspera* were identified as chl-*a*, chl-c_2, peridinin, diadinoxanthin, diatoxanthin and β-carotene by Ambarsari *et al.* (1997). Differences in the specific yellow xanthophylls identified in these studies (i.e. dinoxanthin, diadinoxanthin and diatoxanthin) may be due to improved techniques in the more recent work, or variation between zooxanthellae varieties. The interconversion of diadinoxanthin and diatoxanthin (the xanthophyll cycle) is widely recognized to have a photoprotective role in algae containing chl-c_2 (Young and Frank, 1996). The existence of the same photoprotective system has been implicated in the zooxanthellae of seven species of coral by Brown *et al.* (1999a).

Gil-Turnes and Corredor (1981) found that there were no significant differences between the relative concentrations of pigments in each of their five coral species but the total pigment content per unit area of live coral differed significantly between species. These data are corroborated by studies of fluorescent signatures of seven Caribbean corals (Hardy *et al.*, 1992). Profiles of chlorophyll fluorescence spectra were very similar across these species but differed in amplitude, possibly indicating differing densities of the pigment complex.

There are several possible sources of the variation in total pigment. Several studies of corals in sun exposed and shaded sites indicate that the density of zooxanthellae (typically measured either as cells per polyp or cells per unit coral surface area) and zooxanthellar pigment density increase as a response to low light intensity (reviewed in Falkowski *et al.*, 1990; for specific examples see Titlyanov, 1981; Brown *et al.*, 1999b). In contrast, controlled field experiments and laboratory studies on irradiance effects are inconclusive regarding the effect on algal density, and overall indicate no change (reviewed in Stimson, 1997).

Brown *et al.* (1999b) conducted a detailed field study of the influence of various factors on the photophysiological parameters of corals, including irradiance, sea surface temperature, sea level, rainfall and nutrients. The conclusions were that sea surface temperature and irradiance had significant effects on algal density, and algal chlorophyll concentration was significantly affected by irradiance. Variation in sea surface temperature was reflected in symbiont density on a timescale of days (with algal density declining as temperature rose).

The overall spectral effect of variation in zooxanthellae density and pigment content is further complicated by seasonal variation in zooxanthellae cell volume. Brown *et al.* (1999b) reported an average 26% increase in algal cell volume from the seasonal minimum volume to the maximum volume, while algal numbers (per unit coral area) conversely decrease by 53% over this period.

One hypothesis that has been proposed to explain variation in zooxanthellae numbers is that autotrophic and heterotrophic functions are complementary, such that variation in predatory capacity would be reflected by inverse variation in photosynthetic capacity. There is, however, little support for this hypothesis (Gil-Turnes and Corredor, 1981), as increased nutrient availability leads to greater zooxanthellae biomass (Rowan, 1998; Brown *et al.*, 1999b).

Overall, the current evidence indicates that the relative composition of the more significant pigments in the zooxanthellae is fairly consistent across coral species but the actual concentration of this complement of pigments is highly variable. However, the recent discovery of high genetic diversity in zooxanthellae (Rowan, 1998) suggests that it may be prudent to reassess the pigment composition of the differing clades. Additionally, mechanisms such as the xanthophyll cycle can cause diurnal variation in accessory pigment concentrations. For example, data from Brown *et al.* (1999a) show the ratio of diatoxanthin to diadinoxanthin changing from 0.53 (at midday) to 0.05 (at night).

In summary, if the zooxanthellar pigment composition is considered to be fairly consistent (in line with the results of Gil-Turnes and Corredor, 1981), the spectral contribution of the symbionts to coral reflectance would consist of a predictable spectra but with unpredictable intensity, since zooxanthellar density, volume and total pigment content are variable. As the major groups of macroalgae contain different pigments to zooxanthellae (Table 1) this is promising for discrimination of corals from macroalgae. However, systematic variation in symbiont pigment composition between coral species has not been demonstrated; the variation that has been observed is typically due to environmental factors. This implies that other pigmentary components within the coral (specifically, in the host tissues) may be more useful for spectral discrimination between coral species.

3.2.2. *Coral tissue pigments*

The vivid colours for which reef-building corals are renowned are the result of host-based pigments which are often fluorescent (and hence referred to as fluorescent proteins or FPs) and which overlay the brown pigments of the zooxanthellae (Dove *et al.*, 2001). These pigments have been rather less clearly categorized than the pigments of the symbionts or the rarer skeletally fixed pigments (see section 3.2.3). Typically, they have been studied *in vivo* and *in situ* by means of their fluorescent properties. By considering several separate studies on these fluorescent pigments the broad patterns can be discerned. However, the contribution of non-fluorescent host-based pigments to coral spectra remains enigmatic. Matz *et al.* (1999) and Dove *et al.* (2001) have categorized coral FPs as being homologous (or partly homologous) to the green fluorescent protein (GFP) which was first categorized from the jellyfish *Aequorea victoria* (Shimomura *et al.*, 1962). In *Aequorea,* GFP is an accessory protein to the chemiluminescent pigment "aequorin". GFP absorbs aequorin's blue emission and re-emits it as green (Tsien, 1998). This can occur via a direct excitatory link that bypasses the actual radiation of blue light by aequorin (Morise *et al.*, 1974). The structure of GFP is relatively complex compared with other pigmentary compounds (such as carotenoids for example), consisting of a protein chain of 238 amino acid residues forming a hollow "basket" surrounding the chromophore (the actively fluorescent part). GFP-like proteins are found in a variety of coelenterates including hydrozoa and non-coral anthozoa (Cubitt *et al.*, 1995; Tsien, 1998). One of the characteristics of GFP-like proteins is that they typically exhibit a single excitation peak and emission peak (with some exceptions), and so the fluorescent properties of the variants are summarized by the location of these peaks (Tsien, 1998).

Dove *et al.* (1995) initially characterized the proteins, which are the principal agents of pink and blue tissue colouration of several Australian coral species from the Pocilloporidae, Acroporidae, Poritidae and

Faviidae. The pigments were denoted "pocilloporins" from the initial extraction from *Pocillopora damicornis*. Subsequently these proteins were shown to be physically similar to GFP and Dove *et al.* (2001) categorized the proteins into four classes dependent on the wavelength of their excitation peak, denoted *UV* (300–360 nm), *violet* (420–465 nm), *blue* (480–490 nm) and *green-yellow* (550–600 nm) respectively (Figure 1). Overall, six out of the ten species surveyed showed more than one FP and a mixture of three or more was not uncommon. Salih *et al.* (2000) have reported FPs with emission maxima consistent with those of the GFP-like pocilloporins of Dove *et al.* (2001) from over 120 Great Barrier Reef species, and in particular give the excitation and emission peaks of three examples (Figure 1). In a separate study Matz *et al.* (1999) extracted and described GFP homologues from six reef corals (these have also been incorporated into Figure 1).

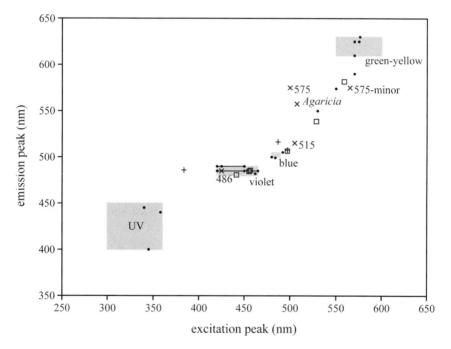

Figure 1 Coral host tissue fluorescent pigments (FPs). •, proteins isolated by Dove (2001), horizontal lines indicate a set of excitation peaks, grey rectangles indicate Dove's four categories: *UV*, *violet*, *blue* and *green-yellow*. ×, FP categories according to Mazel (1995, 1997a), name denotes the emission peak (*Agaricia* is based on three specimens of that species). □, coral GFP-like pigments isolated by Matz *et al.* (1999). +, example FPs based on data from Salih *et al.* (2000).

Mazel (1995, 1997a), Fux and Mazel (1999) and Fuchs (2001) have taken many *in situ* and *in vivo* measurements of coral fluorescence over at least 10 species from the Caribbean, but without extracting or characterizing the materials responsible. Their results suggest that four pigments are the source of most of the observed emissions, with emission peaks at 486 nm, 515 nm, 575 nm and 685 nm. A fifth pigment found in some morphs of *Agaricia* exhibited an emission peak at 557 nm. Chlorophyll is well documented as giving an emission peak at 685 nm (Dustan, 1985; Hardy *et al.*, 1992) and therefore accounts for this peak. The remaining pigments were labelled "486", "515" and "575". All had single excitation and emission peaks, in the manner of GFP-like proteins, with the exception of "575" which had a small secondary excitation peak, and therefore could represent two FPs with similar emission peaks.

The striking fact is that despite the categorization of FPs by Dove *et al.* (2001) and the postulated pigments of Mazel (1997a), pooling all the reported coral data indicates that coral fluorescent pigments vary in something more like a continuous spectrum as regards their emission and excitation peak (Figure 1). In particular, note that three pigments identified by Dove *et al.* (2001) do not fit into their own proposed categories and the only correspondence between the studies of Mazel (1997a) and Dove *et al.* (2001) is pigment "486" and category "violet". The different geographical locations of the two surveys may account for this variation.

Overall, the weight of evidence is that GFP-like homologues are largely responsible for coral colour across a wide range of species, but according to Figure 3 they form an almost continuous variety that resists categorization. The presence of other varieties of pigment should not be ruled out. Mazel (1995, 1997a) notes the similarity of the emission spectrum of "575" with that of phycoerythrin, a photosynthetic accessory pigment found in the red algae, and also suggests there is evidence that "486" and "515" may be biliproteins. Further, almost no work has been done to quantify non-fluorescent pigments in coral host tissue. Carotenoids have been demonstrated in the bodies of some skeletal corals, askeletal corals and other coelenterates (Fox and Wilkie, 1970; Zagalsky and Herring, 1977; Czeczuga, 1983) so it seems likely that pigments other than FPs may be present.

The function of pigments within the coral tissue has been the cause of some debate. One of the earliest suggestions (Kawaguti, 1944) was that pigments played a photoprotective role, shading the zooxanthellae from excessive sunlight that might otherwise lead to photobleaching of photo-synthetic pigments. Dove *et al.* (1995) initially rejected a photoprotective or photosynthesis-aiding role in favour of a possible immunological or chemical defence function, suggested by observations of a competitive advantage of some colour morphs (Rinkevich and Loya, 1983; Takabayashi and Hoegh-Guldberg, 1995). However, in a later paper, Dove et al. (2001) favoured a role for pocilloporin in "protecting the photosynthetic machinery of the symbiotic dinoflagellates of corals under high light conditions and enhancing the availability of photosynthetic light under shade conditions", based on the positions of the excitation and emission spectra of the characterized proteins. Salih et al. (2000) support the photoprotective role with the demonstration of the possibility of proteins acting in sequential fluorescence-excitance chains, which transfer energy to long non-photosynthetically active wavelengths. Further, the density of pigment chromatophores can be actively controlled by polyp expansion and contraction, under varying conditions of light (Salih et al., 2000).

The variation of pigments between and within species is the main question, as regards their spectral separability. Unfortunately very few systematic data exist for comparison of these kinds of variation, but observations reported in various papers indicate a high degree of variation between species, morphs and within polyps themselves. Colour morphs of coral species are well documented (e.g. Takabayashi and Hoegh-Guldberg, 1995). Salih et al. (2000) found 124 Great Barrier Reef species of corals across 16 families which displayed polymorphism for FPs. Although no figures are given for the number of species where such polymorphism is absent, it is clear that this is a common phenomenon. Dove et al. (2001) report that pocilloporins can be spatially inhomogenous within both colonies and polyps. A "green absorbing pocilloporin" is present only in the growing tip of *Acropora aspera* and within the tentacles of polyps of *Pavona decussata*, but not within the polyp body.

Overall, the variety of fluorescent pigments in coral host tissues is promising as regards a basis for spectral discrimination. However, studies to date imply a high degree of within-species variation and a complicated picture of pigment function and distribution, from which it may be difficult to establish systematic patterns. Additionally, fluorescence effects may be significant in the measurement of reflectance, as will be discussed in detail in section 4.2.

3.2.3. Coral skeletal pigments

The skeletons of true stony corals (Scleractinia) are characteristically white, whilst certain hydrocorallines and many members of the alcyonarian (octocorals) subclass are conspicuously coloured (Fox, 1972). The results of several investigations into the skeletally bound pigments of various non-scleractinian corals are summarized in Table 2. Note that the majority of

Table 2 Known "coral" skeletal pigments.

Species	Skeletal colour	Pigment	Group	Authority
ANTHOZOA OCTOCORALLIA				
Coenothecalia				
Heliopora coerulea	blue	biliverdin $1X_\alpha$	tetrapyrrole	Rüdiger *et al.*, 1968
Gorgonacea				
Adelogorgia phyllosclera	pink-purple	unknown	not carotenoid	Fox, 1972
Corallium rubrum	orange-red	unknown	not carotenoid	Fox, 1972
Eugoria ampla	yellow-orange	eugorgiaenoic acid	carotenoid	Fox *et al.*, 1969
Lophogorgia chilensis	salmon-pink	unknown	not carotenoid	Fox, 1972
Muricea californica	orange to brick red	unknown	not carotenoid	Fox, 1972
Stolonifera				
Tubipora musica	red	unknown	not carotenoid	Fox, 1972
HYDROZOA				
Stylasterina				
Allopora californica	purple	astaxanthin	carotenoid	Fox and Wilkie, 1970
Distichopora coccinea	vermilion	astaxanthin zeaxanthin (trace)	carotenoid carotenoid	Fox, 1972; Rønneberg *et al.*, 1979b
Distichopora nitida	yellow	astaxanthin zeaxanthin (1:1)	carotenoid carotenoid	Rønneberg *et al.*, 1979b
Distichopora violacea	vermilion	astaxanthin zeaxanthin (trace)	carotenoid carotenoid	Fox, 1972; Rønneberg *et al.*, 1979b
Stylaster elegans	pink and orange	astaxanthin zeaxanthin (1:1)	carotenoid carotenoid	Fox, 1972; Rønneberg *et al.*, 1979b
Stylaster roseus	purple	astaxanthin	carotenoid	Fox, 1972; Rønneberg *et al.*, 1979b
Stylaster sanguineus	pale pink	astaxanthin zeaxanthin (1:1)	carotenoid carotenoid	Fox, 1972; Rønneberg *et al.*, 1979b

skeletally pigmented corals are not significant as regards remote sensing: the hydrocorals *Distichopora* and *Stylaster* typically inhabit overhangs and caves (Veron, 1993). Some may be significant, *Heliopora* and *Tubipora* can form large colonies and *Heliopora* can be dominant on reef flats (Veron, 1993). However, the skeletal pigments of zooxanthellate corals are subjectively the least conspicuous spectral component in the living coral because zooxanthellar pigments disguise the more striking skeletal colouration (Veron, 1993).

The pigmentation of coral skeletons, where it occurs, appears to be the result of more diverse mechanisms than that of the pigmentation of polyp tissues. For example, although astaxanthin is a yellow-orange pigment, it alone is responsible for the purple colour of *Allopora californica* and the vermilion of the two *Distichopora* species. Similarly, a 1:1 mix of astaxanthin and zeaxanthin has been recovered from coral skeletons which exhibit colours ranging from yellow through pink and orange to purple (Table 2). It appears that such diverse colouration arises from the incorporation of carotenoids into a carotenoprotein within the calcium carbonate skeleton (Rønneberg *et al.*, 1979a). Variation in the protein structure and distribution of covalent bonds in the region of the bonded carotenoid(s) could presumably account for the observed spectral variation, but this remains untested.

In addition to pigment substances, pale yellow fluorescent chromolipid acids have been found bound to the microspicules of the gorgonian *Eugoria ampla* (Fox *et al.*, 1969). Boto and Isdale (1985) have found humic and fulvic acids in the skeletons of massive corals (especially *Porites* spp.), which impart a blue and yellow-green fluorescence, respectively. These acids are found only in inshore corals and their presence corresponds to adjacent river input, forming bands throughout the coral core which in the largest corals can record centuries of coastal rainfall (Isdale, 1984; Boto and Isdale, 1985).

3.2.4. *Pigmentary consequences of coral bleaching*

Coral bleaching due to environmental stress (see Brown, 1997, and Hoegh-Guldberg, 1999, for summary of theories) involves both the loss of pigments from the zooxanthellae and the expulsion of zooxanthellae from the coral (Kleppel *et al.*, 1989). The extent of bleaching may be due to the particular clades or species of *Symbiodinium* present (Rowan, 1998; Baker, 2001).

Kleppel *et al.* (1989) found that unbleached corals have 35, 17 and 20 times the levels of chl-*c*, peridinin and diadinoxanthin, respectively, than occurs in bleached corals. Chlorophyll-*b* was found in both bleached and unbleached corals, but as this is not a zooxanthellar pigment (Table 1) its presence was accounted for as the accidental incorporation of endolithic green algae into the analysis. For this reason the chl-a data were also considered unreliable (chl-a being present in both zooxanthellae and green algae; Table 1). This has important implications for remote sensing. The ubiquitous presence of chl-a in zooxanthellae and algae may limit the use of chl-a for distinguishing bleached coral or recently dead coral, due to the rapid colonization of dead coral by microalgae. It would seem that chl-c and peridinin may be preferred as indicators of bleaching over chl-a. However, Myers $et\ al.$ (1999) found that among corals with varying amounts of bleaching, densities per unit area of peridinin and chl-a were positively correlated with each other, indicating that chl-a could be at least as reliable an indicator of bleaching as peridinin. It is interesting to note that Kleppel $et\ al.$ (1989) estimate that 72% of the loss of chl-c from corals is via zooxanthellar pigment loss, rather than expulsion of zooxanthellae.

The impact of the pigmentary changes during bleaching on overall coral reflectance is discussed in section 5.2.

4. MEASUREMENT OF REFLECTANCE SPECTRA

Reference spectra of coral are typically taken $in\ situ$ on the reef (Miyazaki and Harashima, 1993; Mazel, 1997a, b; Holden and LeDrew, 1998b; Myers $et\ al.$, 1999; Clark $et\ al.$, 2000; Hochberg and Atkinson, 2000) or on sections of live coral displaced to a tank or flume (Miyazaki and Harashima, 1993; Hochberg and Atkinson, 2000). Occasionally reflectance spectra have been taken on $in\ vivo$ materials (Shibata, 1969). The resultant spectral libraries are an important resource that can find utility not only in remote sensing but in photosynthetic studies (Lesser $et\ al.$, 2000) and radiative transfer modelling (Lubin $et\ al.$, 2001).

Typically, the equipment used to measure reflectance $in\ situ$ consists of commercially available spectrophotometers adapted for aquatic use. Detailed descriptions can be found in Booth and Dustan (1979), Miyazaki and Harashima (1993) and Mazel (1997b). Equipment for $in\ vivo$ measurement of fluorescence and reflectance is described in Shibata (1969). Lassen $et\ al.$ (1992) and Kühl and Jørgensen (1992) describe a microprobe and coupled spectrometer which can be utilized to study the light environment of polyps at the microscale ($>100\,\mu$m) (Kühl $et\ al.$, 1995).

The following section will outline three problematic areas as regards current published reef benthos spectra. First, the equipment, methodology and definition of "reflectance" employed vary between differing studies, so there may be problems making cross-study comparisons. Secondly, there is evidence that in corals the potential contribution of fluorescence to measured reflectance needs much greater consideration. Finally, spectro-photometers have small probes typically used in close proximity to the subject, whereas a remotely sensed signal integrates over a substantially larger area (typically greater than the size of a whole colony). As such, *in situ* coral reflectance spectra are largely independent of the morphology of the colony, whereas a remotely sensed signal is not.

4.1. Reflectance

There is some inconsistency in both the coral reef and remote sensing literature as to what constitutes "reflectance" and the practice of estimat-ing it. Essentially, this inconsistency revolves around the terms *radiance* and *irradiance*, how reflectance should be derived from them, and what exactly is measured by a specific instrument.

Mobley (1994) provides a definition of unpolarized spectral radiance, L, as:

$$L(\vec{x}; t; \hat{\xi}; \lambda) \equiv \frac{\Delta Q}{\Delta t \Delta A \Delta \Omega \Delta \lambda} \; (\mathrm{W\,m^{-2}\,sr^{-1}\,nm^{-1}}) \qquad \text{[Eqn 1]}$$

and spectral downward plane irradiance, E_d as:

$$E_d(\vec{x}; t; \lambda) \equiv \frac{\Delta Q}{\Delta t \Delta A \Delta \lambda} \; (\mathrm{W\,m^{-2}\,nm^{-1}}) \qquad \text{[Eqn 2]}$$

where ΔQ, ΔA, $\Delta \Omega$, $\Delta \lambda$, Δt, are the radiant energy entering the sensor, sensor surface area, solid angle "seen" by any point on the sensor, the wavelength interval and time interval respectively. The terms in brackets indicate the dependence of the values on: $\vec{x}$, the point in space; t, the point in time; $\hat{\xi}$, the direction of observation and λ, the wavelength. Both radiance and irradiance measure the radiant energy incident on a surface per unit area, but in the case of radiance the energy reaching this surface is limited by a solid angular window centred around a particular direction, $\hat{\xi}$. In the conceptual limit, this angular window reduces to an infinitesimal size and radiance becomes the measurement of radiation in precisely the direction $\hat{\xi}$.

Irradiance, on the other hand, is essentially the integral of radiance over a hemisphere. Typically, if a flat surface is considered, irradiance is the sum of radiances weighted by the cosine of their angle of incidence:

$$E_d(\vec{x}; t; \lambda) = \int L(\vec{x}; t; \hat{\xi}; \lambda) \, | \cos \theta | \, d\Omega(\hat{\xi}) \qquad \text{[Eqn 3]}$$

The application of the term "irradiance" is generally restricted to the radiation incident on a horizontal surface (either from above or below), leading to the definition of spectral downward plane irradiance, E_d, as in Eqn 2 and similarly spectral upward plane irradiance, E_u (Mobley, 1994).

Reflectance, in general, is intuitively understood as the ratio of reflected radiant energy to the incident radiant energy on a material, dependent on wavelength. As either of these quantities can be measured as radiance or irradiance (dependent on the sensor used), this potentially gives several forms of reflectance that are not theoretically equivalent. Consider two definitions of reflectance from Mobley (1994), as follows.

Spectral irradiance reflectance, R, is defined as the ratio between the upwelling and downwelling plane irradiances:

$$R(z, \lambda) \equiv \frac{E_u(z, \lambda)}{E_d(z, \lambda)} \qquad \text{[Eqn 4]}$$

Spectral remote sensing reflectance, R_{rs}, is the ratio of upwelling radiance to downwelling plane irradiance:

$$R_{rs}(\theta, \phi; \lambda) \equiv \frac{L(z = a; \theta, \phi; \lambda)}{E_d(z = a; \lambda)} \; (\text{sr}^{-1}) \qquad \text{[Eqn 5]}$$

Here, z is the depth and $z = a$ in Eqn 5 indicates that by Mobley's definition R_{rs} is measured in the air just above the water surface. Also note that (θ, ϕ) is the directional component, equivalent to $\hat{\xi}$ in the preceding equations. The definition of R_{rs} is in accord with the use of the term "reflectance" in remote sensing (e.g. Mather, 1999). Note that R_{rs} is in units of inverse steradians whereas R is a dimensionless ratio. These two forms of reflectance have also been referred to as bi-hemispherical reflectance and hemispherical-conical reflectance, respectively (Kimes et al., 1980), indicating that in the latter the radiance measurement is taken by a sensor with a restricted (conical) field of view.

In practice, studies on coral have utilized differing equipment, techniques and assumptions to produce reflectance estimates that may or may not be comparable. Holden and LeDrew (1998b) do not explicitly state the form of reflectance they are using but they refer to "two measurements taken sequentially from the same instrument . . . a reference downwelling irradiance measurement . . . an upwelling radiance measurement". However, it is clear from their description of the instrument (a cosine receptor with a hemispherical field of view) that they have in fact taken two irradiance measurements and hence are calculating irradiance reflectance. Miyaki and Harashima (1993), on the other hand, describe a sensor that can be configured to read either radiance or irradiance, and thus their *in situ* reflectance spectra are in the form of remote sensing reflectance. The equipment used by Hochberg and Atkinson (2000) has a fairly small collection angle of ~ 0.098 sr that can be considered approximate to a radiance measurement. They estimated reflectance as the ratio between the radiance from the substratum (measured at a 45° angle) with the radiance of a Spectralon 99% diffuse reflectance target in the same position (i.e. on top of the substrate). This is common methodology which has been used elsewhere (Mazel, 1997b; Myers *et al.*, 1999; Lubin *et al.*, 2001). The Spectralon is a known Lambertian reflector, i.e. radiance reflected from it is constant in all directions. If the region of coral sampled is assumed to be equivalent to a flat Lambertian surface at the same orientation as the Spectralon, then this estimation of reflectance is also equivalent to irradiance reflectance (Eqn 4). If it is like a Lambertian surface at some other orientation, or not at all like a Lambertian surface, then this will affect the derived reflectance spectra in a way that is dependent on the form of the incident light field. Clark *et al.* (2000) provide an example of another methodological approach by taking their white reflectance reading at the surface while the coral measurement is submerged, explicitly revealing the effects of depth on the spectra.

Many of the factors mentioned above have been discussed in the context of terrestrial environments. Kimes *et al.* (1980) have shown that variation in solar zenith angle gives rise to a wide variability in calculated reflectance due to anisotropic irradiance, structure of the surface (canopy geometry) and type of reflectance measurement used. Gilabert and Melia (1993) give a further example of the effect of canopy structure and solar zenith angle on ground reflectance measurements. Models developed by Kriebel (1976) indicate a $\pm 1\%$ change in reflected radiation per degree of solar zenith angle or per 6% change in spectral atmospheric turbidity. According to Duggin (1980, 1982), when the upward and downward measurements are not simultaneous, atmospheric variations can give rise to an error of up to 10% on the calculated reflectance factors (see also Milton, 1982). These effects must surely be significant in aquatic environments, where refraction across a non-flat air–sea interface will further complicate the incident light field. In particular, the phenomenon of "wave focusing" can cause temporal variation in downwelling irradiance with short "flashes" (<25 ms) over four times the average intensity (Mobley, 1994). Wave-focusing effects are most pronounced at high solar elevation, slow wind speeds ($<5\,\text{ms}^{-1}$) and when the sky and water are clear; conditions which closely match those in which *in situ* reef spectra are commonly taken.

4.2. Contribution of fluorescence to reflectance

Induced coral fluorescence has some potential in reef remote sensing in its own right (see section 5) but the contribution of fluorescence to reflectance spectra in the ambient light field may be important when taking *in situ* measurements. Hochberg and Atkinson (2000) report a correspondence between peaks in coral reflectance spectra and the fluorescence peaks reported in Mazel (1997a), indicating that fluorescent features can be represented in reflectance. Quantitative studies on the contribution of fluorescence to the apparent reflectance of coral in the ambient light field are at relatively early stages (Mazel, 1997a; Fux and Mazel, 1999; Fuchs, 2001), but results indicate that fluorescence is a significant component that should be included in optical models and remote sensing interpretation.

Fluorescence may cause a depth-dependent effect on reflectance spectra measured *in situ*, and also to a somewhat lesser degree on remotely sensed reflectance. Due to the apparent wide variety of FPs and their variable distribution in corals, the upwelling light component due to fluorescence may vary unpredictably with depth. The fluorescent enhancement of reflectance depends on the spectral energy distribution of the incoming light (Fuchs, 2001). For example, a pigment with an excitation peak in a wavelength region for which light penetrates water well will continue to receive energy at depth. If that same pigment fluoresces in a region where light is heavily attenuated with depth, then it will cause a more prominent peak as depth increases. The significance of this phenomenon can be intuitively appreciated by considering the striking red or orange appearance of some corals at depths where these wavelengths are absent in the incident light (Limbaugh and North, 1956). Two separately published spectra of *Montastraea cavernosa*, from Mazel (1997b) and Myers *et al.* (1999), provide possible evidence of this proposition (Figure 2). They differ significantly in their spectral features, although reportedly measured by the same methodology (relative to a Spectralon on the substratum). Mazel's (1997b) example was taken at a depth of 4 m. No depth information is given by Myers *et al.* (1999), but there are superficial (although not exact) similarities between their reflectance spectra and a reported upwelling spectral signal from *M. cavernosa* at 18 m with a strong fluorescent component in evidence (Mazel, 1997a).

A simple model can be used to gain a quantitative estimate of the discrepancy between *in situ* reflectance measurements caused by the varying contributions of fluorescence as depth varies. Using the separated fluorescence and "true" reflectance data of various specimens of corals published in Fuchs (2001) and the excitation spectra of the pigments responsible from Mazel (1997a), the percentage discrepancy between

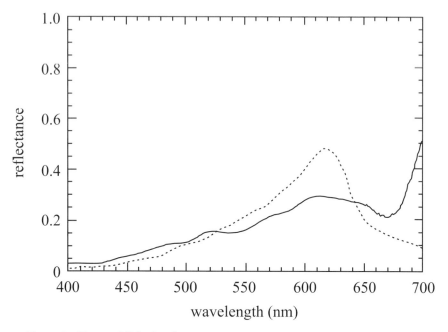

Figure 2   Two published reflectance spectra of *Montastraea cavernosa*, based on data from Mazel (1997b) (solid line) and Myers *et al.* (1999) (dotted line).

reflectance measurements at differing depths, as a function of wavelength, can be calculated. This estimation of discrepancy will be derived by developing a basic model for measured reflectance at a given depth and then applying it to data from Fuchs (2001) and Mazel (1997a).

By considering the contributions of fluorescence and reflectance to the exiting radiance, the measured irradiance reflectance of a coral (with a single type of fluorescent pigment) at a given wavelength λ and depth z can be modelled as:

$$R(z, \lambda) = (I_{\mathrm{FII}}(z)f(\lambda) + E_{\mathrm{d}}(z, \lambda)r(\lambda))/E_{\mathrm{d}}(z, \lambda) \qquad \text{[Eqn 6]}$$

The first part of this expression represents the fluorescent enhancement of reflectance. $I_{\mathrm{FII}}(z)$ denotes the "fluorescence inducing irradiance" that is the integral of the incident radiation at depth z multiplied by the fluorescence excitation spectrum. The value of $f(\lambda)$ for a given wavelength is the ratio of the incoming "fluorescence inducing irradiance" to the radiance that is actually subsequently fluoresced at wavelength λ. Assuming that $f(\lambda)$ is constant for a given emission wavelength λ is equivalent to the assumptions used in measuring excitation and emission spectra of pure substances: that the excitation spectrum can be character-ized by the strength of emission at a single wavelength as the excitation wavelength is varied, and conversely that the emission spectrum can be similarly characterized by excitation at a single wavelength (Mobley, 1994). The invariance of $f(\lambda)$ will be compromised for short emission wavelengths where there is some overlap between the emission spectrum and the excitation spectrum. In this analysis only wavelengths beyond this region will be considered. The second part of Eqn 6 represents the reflect-ance contribution to the exiting radiance from the coral, being the product of the downwelling irradiance at depth z, $E_d(z, \lambda)$, and the coral reflectance $r(\lambda)$. The fluorescence and reflectance components are summed and then divided by the downwelling irradiance, $E_d(z, \lambda)$, to model the "reflectance" $R(z, \lambda)$ as it would be measured at depth z by many of the *in situ* methodologies described earlier (Mazel 1997b; Holden and LeDrew, 1998b; Myers *et al.* 1999; Hochberg and Atkinson, 2000).

Fuchs (2001) describes the contribution of fluorescence to measured reflectance in a specimen of *Colpophyllia natans* under white light conditions. Fluorescence in this specimen of *C. natans* appeared to be dominated by the pigment type denoted "515" (Fuchs, 2001), for which the excitation spectrum is published in Mazel (1997a). The measured reflectance under white light, $R_W(\lambda)$, can be modelled as a special case of Eqn 6, where $E_W(\lambda)$ is the white light irradiance and W_{FII} is similarly the white light fluorescence inducing irradiance:

$$R_W(\lambda) = (W_{FII}f(\lambda) + E_W(\lambda)r(\lambda))/E_W(\lambda) \qquad \text{[Eqn 7]}$$

If for a given wavelength, λ, the proportion of the fluorescence con-tribution to $R_W(\lambda)$, relative to reflectance, is $p_f(\lambda)$, then Eqn 7 gives:

$$\frac{W_{FII}f(\lambda)}{E_W(\lambda)r(\lambda)} = \frac{p_f(\lambda)}{1 - p_f(\lambda)} \qquad \text{[Eqn 8]}$$

i.e. $$f(\lambda) = \frac{p_f(\lambda)}{1 - p_f(\lambda)} \times \frac{E_W(\lambda)r(\lambda)}{W_{FII}}$$

$$\text{[Eqn 9]}$$

In order to assess the effect of variable depth on measured reflectance, $R(z, \lambda)$ can be calculated at two depths, z_1 and z_2, and the percentage difference of the latter to the former calculated as:

$$D(\lambda) = 100 \times \frac{R(z_2, \lambda) - R(z_1, \lambda)}{R(z_1, \lambda)} \qquad \text{[Eqn 10]}$$

Substituting Eqns 6 and 9 into Eqn 10 reduces to:

$$D(\lambda) = 100 \times \left(\frac{I_{\text{FII}}(z_2)}{E_{\text{d}}(z_2, \lambda)} - \frac{I_{\text{FII}}(z_1)}{E_{\text{d}}(z_1, \lambda)} \right) \bigg/ \left(\frac{I_{\text{FII}}(z_1)}{E_{\text{d}}(z_1, \lambda)} + \frac{1 - p_{\text{f}}(\lambda)}{p_{\text{f}}(\lambda)} \frac{W_{\text{FII}}}{E_{\text{W}}(\lambda)} \right)$$

[Eqn 11]

Note that $r(\lambda)$ and $f(\lambda)$ have been eliminated. For this particular specimen of *C. natans*, $p_{\text{f}}(\lambda)$ is reported in Fuchs (2001). Sea surface irradiance $E_{\text{d}}(0, \lambda)$ under tropical clear sky conditions can be modelled by a radiative transfer program (e.g. SBDART, Ricchiazzi *et al.*, 1998) and then attenuated to depth z by an approximate exponential model (Mobley, 1994):

$$E_{\text{d}}(z, \lambda) = E_{\text{d}}(0, \lambda)e^{-zk(\lambda)}$$

[Eqn 12]

Appropriate values for attenuation coefficients $k(\lambda)$ are numerically twice those of the clearest natural waters (tabulated in Smith and Baker, 1981), as these are of the order of those reported on coral reefs (Maritorena and Guillocheau, 1996). $I_{\text{FII}}(z)$ may be found by summing the product of $E_{\text{d}}(z, \lambda)$ with the excitation spectrum of pigment 515 (published in Mazel, 1997a). Irradiance under white light, $E_{\text{W}}(\lambda)$, is uniform for all visible wavelengths and therefore W_{FII} has the same form as the excitation spectrum. Note that as Eqn 11 consists of four terms which are ratios between irradiance and "fluorescence inducing irradiance" it does not matter what relative scale the irradiance is measured in, as long as the same values are used to calculate $I_{\text{FII}}(z)$. In particular, it does not matter that we do not know the exact intensity of the white light used in Fuchs (2001) and can simply set $E_{\text{W}}(\lambda) = 1$ for all λ, and use these values to calculate W_{FII}. Similarly, the absolute scale of the excitation spectrum cancels out in Eqn 11 so it is sufficient to utilize a normalized excitation spectrum (e.g. with the peak value $= 1$, as in Mazel, 1997a).

The wavelength-dependent percentage differences between a reflectance measurement from the *Colpophyllia natans* specimen from Fuchs (2001) modelled at 2 m and subsequently at 6, 10 and 14 m are shown in Figure 3. Only values above 510 nm are shown, as below this the overlap between excitation and emission spectra (Mazel, 1997a) may affect the accuracy of the model. It can be seen that increased depth leads to a relative overestimate of reflectance over the wavelength range 510–580 nm, but even over this range the overestimation varies considerably

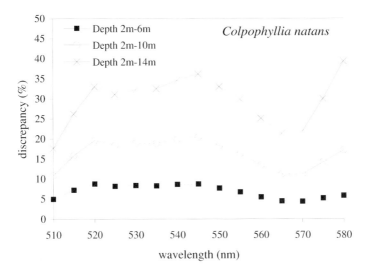

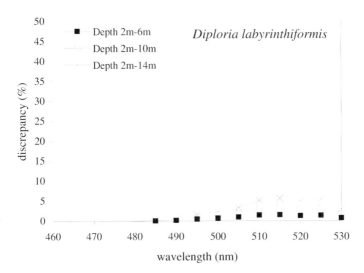

Figure 3 Modelled percentage discrepancy in reflectance measurements taken at differing depths on two coral specimens. Based on data from Mazel (1977a) and Fuchs (2001). Discrepancy is percentage error of the deeper measurement (at 6, 10 and 14 m) relative to the shallower measurement (2 m).

with wavelength. For a depth difference of 8 m (measurements at 2 m and 10 m) the deeper measurement is at most approximately 20% in discrepancy to the shallower one. This specimen of *C. natans* exhibited a comparatively high contribution of fluorescence to apparent reflectance (70% at 500 nm) (Fuchs, 2001), so the analysis was repeated with data from a specimen of *Diploria labyrinthiformis* in which pigment 486 dominated (excitation spectrum published in Mazel, 1997a) and the fluorescent component of reflectance was at most ~20% (Figure 5). The modelled discrepancy between reflectance measurements at differing depths is substantially less than for the *Colpophyllia natans* data, but still reaches 6% when the depth difference is 12 m (measurements at 2 m and 14 m). A 20% difference in reflectance values is of the order of the differences found between some coral species. For example, 95% confidence intervals of coral reflectance spectra by species (Andréfouët *et al.*, 2001) indicate that an individual *Acropora* may not differ from an individual *Pocillopora* by >30% with respect to reflectance values between 500 nm and 600 nm. The reflectance spectra of some *Porites* and *Pavona* are even more similar, potentially differing by <10% over most of the visible reflectance range (Andréfouët *et al.*, 2001). Many other contributory factors will vary with depth, including, particularly, the amount of fluorescent pigments present (Salih *et al.*, 2001). However, the indications are that the effect of fluorescence alone may be a serious contribution to variation in coral reflectance spectra measured *in situ*, and this may have a significant impact on the utility of these spectra for identifying criteria for species discrimination. As regards remotely-sensed reflectance, the wavelength-dependent attenuation of light in water alone causes a significant depth-dependent effect and benthic fluorescence will add another layer of complexity to this relationship. However, the effect of fluorescence on remotely-sensed reflectance may be somewhat smaller than that estimated for *in situ* reflectance measurements because of the attenuation of the fluoresced light as it passes upwards through the water column.

The simplest implication of fluorescence effects on measured reflectance is that the depth at which *in situ* reflectance measurements are recorded must always be noted. At worst, it indicates that for some corals there may be a complex relationship between depth and the spectral quality of the light which emanates from them (the sum of fluoresced and reflected light). A spectral library of species reflectance at depth may be required to analyse this situation, or the development of models that predict the fluorescent component under given incident radiation (Fux and Mazel, 1999; Fuchs, 2001). Given that polyps can actively control FP densities in response to ambient light (Salih *et al.*, 2000) further work on the contribution of fluorescence to reflectance is needed.

4.3 Morphology

Corals exhibit a wide variety of growth forms dependent on species (Plate 4), light levels (i.e. depth) and water flow (Barnes, 1973; Veron, 1993; Helmuth et al., 1997). While differing species have the potential to take on a certain set of forms, it seems that within-species variation is usually due to phenotypic plasticity alone (Muko et al., 2000).

Most in situ reflectance spectra of corals are taken so close that the morphology of the colony is relatively unimportant. However, when considering the relatively coarse pixels obtained in remote sensing (usually $>16\,m^2$), the spectral signal will be a function of this basic reflectance and the morphology of the coral. A horizontal plate-like form will clearly give a different spectral signal when viewed from above than a branching colony, even if the material from which they are composed is spectrally identical. In the latter case, divergent reflection will tend to darken the image. Even the difference between a lobate and massive form may contribute significantly to the variance in reflectance of a given species across different colonies. Similar arguments may be applied to macroalgae, where an encrusting form will interact with the incident radiation differently to a branching or leafy form. To date it seems that no work has been done to quantify the extent of this effect.

Additionally, the problem of anisotropic irradiance (as discussed in a previous section) will be more conspicuous with some morphologies than others. Analogous to the heightened effect in terrestrial environments of canopies with vertical elements (Milton, 1982), certain morphologies such as the vertical plate-like form of Agaricia tenifolia (Helmuth et al., 1997) will interfere with anisotropic irradiance to a greater extent. Even subtle morphological variations, such as branch spacing and height in Agaricia (Helmuth et al., 1997) may result in complex interactions with the incident light. As mentioned previously, zooxanthellar density and fluorescent pigments are also affected by the light environment, so a very complex picture of coral spectral variation with depth is built up (see also Jaubert, 1981). Patterns of coral species' morphological variation can be predicted on reefs (Helmuth et al., 1997), but given the complexity and number of interacting factors it would be a challenge to translate this to a prediction of spectral variation.

5. SEPARABILITY OF SPECTRA

Remote sensing studies and analyses of spectra taken in situ have had some success in statistically demonstrating the spectral separability of benthic classes (Holden and LeDrew, 1998b, c; Clark et al., 2000; Hochberg and Atkinson, 2000; Andréfouët et al., 2001; Mumby et al., 2001). Others qualitatively suggest that separability is possible (Myers *et al.*, 1999) and some use spectral features to quantify abundance (Rundquist *et al,*. 1996; Mumby *et al.*, 1997a). A summary of results is presented in Table 3. Although recent studies imply that Landsat TM can detect coarse community structure in a reef (e.g. discriminating sand from the living reef "foreground", see Andréfouët *et al.*, 2001; and also Dustan *et al.*, 2001; Phinney *et al.*, 2001) the emphasis here is on direct identification of benthic classes using hyperspectral information. If discriminant criteria can be determined hyperspectrally, the application to sensors with fewer and wider bands can be inferred (e.g. to answer questions such as: is it feasible to distinguish the spectra of massive corals and brown macroalgae using the broad bands on the Landsat Thematic Mapper sensor?). For a summary of reef mapping using Landsat TM or SPOT see Green *et al.* (1996) and Holden and LeDrew (1998a).

Two distinct approaches can be taken in analysing spectral data. Firstly, general separability can be assessed by multivariate techniques such as principal components analysis (Holden and LeDrew, 1998b, c), linear discriminant functions (Hochberg and Atkinson, 2000), or classification (Mumby *et al.*, 2001). Even if the discrimination or classification is achieved successfully, without further analysis these techniques do not identify which spectral features enable discrimination. To create generic and "transparent" means of analysing spectra, it is necessary to determine the location and nature of spectral features by which discrimination is possible. This is typically done by examining reflectance spectra and their derivatives (e.g. Holden and LeDrew, 1998b, c; Hochberg and Atkinson, 2000; Mumby *et al.*, 2001). Smoothing and derivative analysis (Tsai and Philpot, 1998) effectively remove a level of information from spectra and present the remainder in a way that simplifies the identification of certain types of features. First derivatives allow the shape of spectra to be compared independently of any relative shifts in overall reflectance, thus emphasizing the slope of the original spectra between wavelengths. Second derivatives go further, quantifying the relative curvature at that point on the reflectance spectra. A region of convexity (viewed from above the plotted reflectance line) will have a negative second derivative, whereas concavity will give a positive value. With fourth derivatives it becomes more difficult to intuitively grasp what is actually being measured, but improved spectral resolution and signal-to-noise ratios have been demonstrated by this method (Butler and Hopkins, 1970a, b).

By reviewing the discriminant criteria demonstrated in various studies (Table 3) it might be hoped that some general patterns would be evident to guide the selection of wavelength bands and discriminant criteria for future remote sensing studies. In practice, although there is some consensus, many of the findings are unique to particular studies and some are

Table 3 Demonstrated and suggested methods for spectrally distinguishing reef benthos.

Category	Wavelengths (nm)	Treatment	Method	Conditions	Study
DEMONSTRATED DISCRIMINATION					
Live *Porites* Dead *Porites*	506–565	1st derivative	Relative values	Airborne sensor	Mumby *et al.*, 2001
Live *Porites* Recently dead *Porites* Dead *Pocillopora* Red coralline algae *Halimeda* Sand	413, 450, 506, 550, 565, 575, 600, 626, 650, 675 (widths ca. 5 nm)	Depth-corrected spectra	Unsupervised classification		(note: wavelength bands from unpublished data)
Coral (3 spp.) Algae (5 spp.) Sand/rubble (3 types)	400, 427, 452, 462, 494, 526, 596 (width 0.5 nm) 484, 545, 567, 584 (width 5 nm)	4th derivative	Linear discriminant function	*In situ* and airborne sensor	Hochberg and Atkinson, 2000
Healthy *Acropora*	500–590 also 640 and 676 (width 4 nm)	1st derivative	Relative values	*In situ*	Holden and LeDrew, 1998b
Bleached *Acropora* + dead coral debris	556 and 577 (width 5.7 nm)	2nd derivative			
Live *Porites* Dead *Porites*	550	2nd derivative	Relative values	*In situ*	Clark *et al.*, 2000
Dead *Porites* >6 months and <6 months	536	1st derivative			
All three above	596	1st derivative			
Bleached coral Algae Dead coral debris Healthy coral	654–674 (sand) 582–686 (algae) 506–566 (bleached coral)	1st derivative	Relative values	*In situ*	Holden and LeDrew, 1998c
SUGGESTED DISCRIMINATION					
Bleached coral Non-bleached coral	450–550	None	Relative values	*In situ*	Myers *et al.*, 1999
Corals Macroalgae	500–530	None	Presence of peaks (coral)		
QUANTIFICATION					
Algal chlorophyll (microalgae)	690	1st derivative		Sensor over tank	Rundquist *et al.*, 1996
Seagrass	640 and 630	Depth invariant ratio	Regression equation	Airborne sensor	Mumby *et al.*, 1997a

inconsistent or even contradictory. However, by relating the findings to what is known of the pigments some confidence can be attained as to which wavelength regions and criteria are valid.

5.1 Chlorophyll

Of all the pigments, chlorophyll has been most frequently used in aquatic remote sensing. Rundquist *et al.* (1996) have indicated that the first derivative of reflectance spectra at 690 nm is a good indicator of microalgal chlorophyll in suspension, which corresponds well with the downslope of the absorption peak in chl-*a* at around 675 nm (Govindjee and Braun, 1974; Bidigare *et al.*, 1990) (see Table 1). The same absorption peak (i.e. dip in reflectance) is evident in the spectra of seagrass (Armstrong, 1993), most macroalgae and corals, and even to some extent, sand (Clark *et al.*, 2000; Hochberg and Atkinson, 2000; Andréfouët *et al.*, 2001; Lubin *et al.*, 2001). Holden and LeDrew (1998c) suggest that sand may be distinguished from other reef benthos by virtue of a positive slope in reflectance between 654 and 674 nm. It seems that this spectral feature may be based on the absence of a dip in reflectance due to chl-*a* which is present in the spectra of most other classes of benthos. However, as some published spectra of sand also exhibit this feature (Lubin *et al.*, 2001), it may be that this measure is only appropriate for the "purest" sands with no algal component, or algae in the water column above.

Mumby *et al.* (1997a) used the ratio of two reflectance bands, 540 and 630 nm, to quantify seagrass standing crop over sand. These two wavelengths highlight differences in the spectra of sand and seagrass, 630 nm being close to the chl-*a* absorption peak at ~680 nm and therefore of comparatively low reflectance in seagrass. It may seem that the technique could be further improved by measuring at the peak (680 nm), but this would be mitigated by the relatively poor transmission of light through water at that wavelength (Smith and Baker, 1981). Chl-*a*, -*b* and -*c* also have significant absorption peaks over 400–460 nm (Govindjee and Braun, 1974; Bidigare *et al.*, 1990), but these seem less noticeable in published spectra, occurring in a wavelength region where most reef benthos have relatively low reflectance (Andréfouët *et al.*, 2001; Lubin *et al.*, 2001).

5.2 Corals and bleaching responses

Non-bleached and bleached (or similarly live and dead) corals are variously reported to be separable by consideration of the first or second derivative of reflectance at locations between 500 and 600 nm (Holden and LeDrew, 1998b, c; Clark *et al.*, 2000). There are no features in the chlorophyll absorption spectra either in solvent (Govindjee and Braun, 1974; Jeffery *et al.*, 1997b) or *in vivo* (Bidigare, 1990) that indicate that reduced chlorophyll is the source of this separability (with the possible exception of a slight peak at ca. 580 nm from chl-*c*). Hardy *et al.* (1992) have published the absorption spectra of temperature-stressed and unstressed *Solenastrea bournoni*, and in both cases the spectra are fairly flat over this range. Instead, peridinin seems the most likely cause of spectral separation in this wavelength region. Peridinin is one of the most prominent zooxanthellar pigments (being approximately 39% of the total zooxanthellar pigment complement) (Gil-Turnes and Corredor, 1981) and exhibits an absorption peak at ca. 475 nm, the downward slope of which extends over 500–550 nm *in vitro* (Jeffery *et al.*, 1997b). Peridinin will therefore contribute an upward slope in reflectance between 500 and 550 nm, a decrease in peridinin will reduce the gradient over this region, which will be reflected by a decrease in the first derivative. Reflectance spectra of bleached and unbleached coral published in Holden and LeDrew (1998b) do correspond with this prediction, unbleached corals having a greater overall increase in reflectance between 500 and 580 nm. However, the corresponding first derivatives published in the same paper seem inconsistent with the reflectance spectra, and consequently the discriminatory criteria suggested, that bleached corals exhibit a higher first derivative over 500–590 nm, is exactly the opposite to what would be expected. Clark *et al.* (2000) have utilized the first derivative of reflectance at 536 nm and 596 nm to discriminate between live, recently dead and old dead *Porites*. In this case the suggested discriminatory criteria are consistent with the proposed effect of reduced peridinin with bleaching: live *Porites* have the greatest first derivative at 596 nm; recently dead *Porites* are greater in the first derivative at 536 nm than old dead *Porites*.

Separation by second derivatives requires some concavity or convexity to be present in the reflectance spectra. Hochberg and Atkinson (2000) report that coral exhibits a reflectance peak near 570 nm and this feature can be seen in some other published coral reflectance spectra (Andréfouët *et al.*, 2001; Lubin *et al.*, 2001). Fluorescence at 575 nm has been measured in many corals by Mazel (1995, 1997a) and could be the cause of a "reflectance" peak at 575 nm. Holden and LeDrew (1998b) have suggested separation of bleached coral from non-bleached coral by second derivatives at 556 and 577 nm, the latter of these figures is strikingly coincident with the observed fluorescence peak at 575 nm. The sign of the expected values at 577 nm reported by Holden and LeDrew (1998b) indicates a peak in non-bleached coral and a dip in bleached coral, as might be expected. The actual cause of this reflectance peak in terms of pigments is not clear. Although phycoerythrin exhibits fluorescence at 575 nm and this lends circumstantial evidence to its presence (Mazel,

1997a), its occurrence in corals has not been demonstrated. The GFP-like fluorescent proteins (FPs) of coral host tissue could be responsible for a peak at ca. 570 nm, as several FPs have been characterized with emission peaks in this region (Figure 3) (Matz *et al.*, 1999; Dove *et al.*, 2001).

5.3 Algae

Discrimination between coral and the various types of algae is less easy to assess from published data, as most studies have defined a single group "algae" (Holden and LeDrew, 1998b, c; Hochberg and Atkinson, 2000), restricting a possible pigment-based assessment of separability. For example, Holden and LeDrew (1998c) cite a positive slope between 582 and 686 nm as indicative of "algal covered surfaces", but a qualitative consideration of the published reflectance spectra of various algal species (Andréfouët *et al.*, 2001; Lubin *et al.*, 2001) indicates that some variability and probable inconsistency with this measure is likely, as negative slopes in this region are evident in many cases. One noticeable feature is a prominent dip in the reflectance spectra of coralline red algae at around 570 nm (Lubin *et al.*, 2001), which is coincident with the phycoerythrin absorption spectrum (Table 1). This dip occurs where most corals have a peak, and is arguably also discernible in published reflectance spectra of turf algae (where some red algae might be expected) but not in the brown *Padina, Sargassum* or *Turbinaria* (Andréfouët *et al.*, 2001).

5.4 Fluorescent signatures

Fluorescent signatures have been used to characterize phytoplankton populations and macroalgae (Yentsch and Yentsch, 1979; Topinka *et al.*, 1990), and to detect algal blooms (Mumola *et al.*, 1975). Fluorescence can be induced in specific accessory pigments by use of a laser and the results may have greater discriminatory power than can be attained from passive ambient reflectance (Yentsch and Yentsch, 1979). In practice, active laser systems can be mounted on a ship or aircraft (Hardy *et al.*, 1992). One difficulty, however, is distinguishing between benthic and pelagic sources of fluorescence.

Fluorescence from coral reef organisms has been detected as an incidental signal while making aircraft measurements of laser-induced phytoplankton fluorescence (Mazel, 1990), and laboratory studies indicate that this technique could be used to monitor coral bleaching (Hardy *et al.*, 1992). If sufficiently detailed information were available on pigments of reef benthos, and their fluorescent characteristics, this could be a promising approach for more detailed mapping. As induced-fluorescence remote sensing relies on specifically targeting the excitatory wavelengths of fluorescent pigments with irradiance (Yentsch and Yentsch, 1979; Topinka *et al.*, 1990) a far more detailed *a priori* knowledge of pigments is required than in passive remote sensing. It appears that fluorescent pigments of corals are diverse (Figure 1) and therefore could form a good basis for species separation. However, the pattern of this diversity across species is currently not well known. If the within-species diversity is nearly as high as that found between species, the potential for separation may be limited.

5.5 Conclusions on spectral separability

Overall, there is a great deal of uncertainty and inconsistency in the reported spectral features that can be employed to discriminate reef benthos. Although Clark *et al.* (2000) note the similarity of their second derivative discrimination between live and dead coral to that of Holden and LeDrew (1998b, c) and identify the same wavelength region for the application of first derivatives to separate live and dead coral, the expected relative values they cite are contradictory. Clark *et al.* (2000) suggest lower first derivatives of reflectance in the 536 nm and 596 nm region as being indicative of coral mortality, whereas Holden and LeDrew (1998b) suggest that over the region 500–590 nm the first derivative of bleached coral reflectance is consistently higher than that of healthy coral.

In general, analyses of reflectance spectra will probably continue to give ambiguous results in the absence of a causal spectral context. Applying derivative analysis and multivariate techniques to spectral libraries on an *ad hoc* basis can produce discriminative criteria, but without knowing the underlying reasons for the results, the general applicability will be limited. For example, by determining that the suggested discrimination of sand by a positive slope in reflectance between 654 and 674 nm (Holden and LeDrew, 1998c) is due to the absence of chl-*a*, the weakness of this discriminatory criterion is revealed: if chl-*a* is present in the sand or water column, then the discrimination may fail. Developing robust discrimination techniques for various coral and algal species will require a similar but more complex understanding of the contributions of pigment absorbance and fluorescence to reflectance.

As a final word, there are some caveats required for a pigment-based view of reflectance spectra. Firstly, caution should be exercised as to the basic accuracy of estimated pigment quantities. High performance liquid chromatography (HPLC) is one of the most commonly used techniques for pigment extraction. Latasa *et al.* (1996) have shown that HPLC analysis performed at several laboratories on mixed standards of algal pigments disagreed significantly, with concentration estimates differing by >20% in some cases. Spectrophotometric methods can be equally error-prone, chlorophyll can be underestimated by up to 9% dependent on the type of spectrophotometer used, due to the fluorescence of chlorophyll contaminating the reading (Latasa *et al.*, 1996). At least one other spectrophotometer gives systematically low readings for chlorophyll for another reason, as yet undetermined (Latasa *et al.*, 1999). Many of these problems could be alleviated by calibration with high-purity pigment standards, but these are not commercially available (Latasa *et al.*, 1999).

Secondly, published absorption spectra of pigments vary depending on the solvent used to extract them (Davies, 1976) and may not reflect their appearance *in vivo* where they may be associated with proteins which affect their spectra (e.g. coral skeletal pigments). Increased packaging of light absorbing material within particles reduces the absorbance of light by those particles (Bisset *et al.*, 1997) and this "packaging effect" has been shown to affect the light absorption by phytoplankton (Nelson *et al.*, 1993). Whether this effect is applicable to reef benthos (e.g. zooxanthellae) is undetermined, but it may complicate the relationship between pigment densities and spectral absorption.

ACKNOWLEDGEMENTS

J.D.H. is funded by Newcastle University, P.J.M. is a Royal Society Research Fellow.

REFERENCES

Adams, J. B., Smith, M. O. and Johnson, P. E. (1986). Spectral mixture modelling: a new analysis of rock and soil types at the Viking Lander I site. *Journal of Geophysical Research* **91**, 8098–8112.

Ambarsari, I., Brown, B. E., Barlow, R. G., Britton, G. and Cummings, D. (1997). Fluctuations in algal chlorophyll and carotenoid pigments during solar bleaching in the coral *Goniastrea aspera* at Phuket, Thailand. *Marine Ecology Progress Series* **159**, 303–307.

Andréfouët, S., Muller-Karger, S. E., Hochberg, E. J., Hu, C. and Carder, K. L. (2001). Change detection in shallow reef environments using Landsat 7 ETM+ data. *Remote Sensing of Environment* **79**, 1–13.

Armstrong, R. A. (1993). Remote sensing of submerged vegetation canopies for biomass estimation. *International Journal of Remote Sensing* **14**, 10–16.

Baker, A. C. (2001). Reef corals bleach to survive change. *Nature* **411**, 765–766.

Barnes, D. J. (1973). Growth in colonial scleractinians. *Bulletin of Marine Science* **23**, 280–298.

Bidigare, R. R., Ondrusek, M., Morrow, J. and Keifer, D. (1990). *In vivo* absorption properties of algal pigments. *Proceedings – SPIE the International Society for Optical Engineering, Ocean Optics X* **1302**, 290–302.

Bisset, W. P., Patch, J. S., Carder, K. L. and Lee, Z. P. (1997). Pigment packaging and Chl *a*-specific absorption in high-light ocean waters. *Limnology and Oceanography* **42**, 961–968.

Booth, C. R. and Dustan, P. (1979). Diver-Operable Multiwavelength Radiometer. *Proceedings of the Society of Photo-Optical Instrumentation Engineers* **196**, 33–39.

Boto, K. and Isdale, P. (1985). Fluorescent bands in massive corals result from terrestrial fulvic acid inputs to nearshore zone. *Nature* **315**, 396–397.

Brown, B. E. (1997). Coral bleaching: causes and consequences. *Coral Reefs* **16**, S129–S138.

Brown, B. E., Ambarsari, I., Warner, M. E., Fitt, W. K., Dunne, R. P., Gibb, S. W. and Cummings, D. G. (1999a). Diurnal changes in photochemical efficiency and xanthophyll concentrations in shallow water reef corals: evidence for photo-inhibition and photoprotection. *Coral Reefs* **18**, 99–105.

Brown, B. E., Dunne, R. P., Ambarsari, I., Le Tessier, M. D. A. and Satapoomin, U. (1999b). Seasonal fluctuations in environmental factors and variations in symbiotic algae and chlorophyll pigments in four Indo-Pacific coral species. *Marine Ecology Progress Series* **191**, 53–69.

Butler, R. W. and Hopkins, D. W. (1970a). Higher derivative analysis of complex absorption spectra. *Photochemistry and Photobiology* **12**, 439–450.

Butler, R. W. and Hopkins, D. W. (1970b). An analysis of fourth derivative spectra. *Photochemistry and Photobiology* **12**, 451–456.

Chalker, B. E. and Dunlap, W. C. (1981). Extraction and quantification of endosymbiotic algal pigments from reef-building corals. *In* "Proceedings of the Fourth International Coral Reef Symposium" (E. D. Gomez *et al.*, eds), pp. 45–49, University of the Philippines, Manila, Philippines.

Clark, C. D., Mumby, P. J., Chisholm, J., Jaubert, J. and Andréfouët, S. (2000). Spectral discrimination of coral mortality states following a severe bleaching event. *International Journal of Remote Sensing* **21**, 2321–2327.

Cousens, R. (1982). The effect of exposure to wave action on the morphology and pigmentation of *Ascophyllum nodosum* (L) Le Jolis in southeastern Canada. *Botanica Marina* **25**, 191–195.

Cubitt, A. B., Heim, R., Adams, S. R., Boyd, A. E., Gross, L. A. and Tsien, R. Y. (1995). Understanding, improving and using green fluorescent proteins. *Trends in Biochemical Sciences* **20**, 448–455.

Czeczuga, B. (1983). Investigations of carotenoprotein complexes in animals – VI. *Anemonia sulcata*, the representative of askeletal corals. *Comparative Biochemistry and Physiology* B **75**, 181–183.

Davies, B. H. (1976). Carotenoids. *In* "Chemistry and Biochemistry of Plant Pigments" (T. W. Goodwin, ed.), pp. 38–165. Academic Press, London.

Dawes, C. J. (1998). "Marine Botany, 2nd edn". John Wiley and Sons, New York.

de Vries, D. H. (1994). Imaging spectroscopy: CASI operations in Australia during summer 1992/93. *In* "Seventh Australasian Remote Sensing Conference Proceedings", pp. 136–140, ARSC, Melbourne, Australia.

Dove, S. G., Takabayashi, M. and Hoegh-Guldberg, O. (1995). Isolation and partial characterization of the pink and blue pigments of Pocilloporid and Acroporid corals. *Biological Bulletin* **189**, 288–297.

Dove, S. G., Hoegh-Guldberg, O. and Ranganathan, S. (2001). Major colour patterns of reef-building corals are due to a family of GFP-like proteins. *Coral Reefs* **19**, 197–204.

Duggin, M. J. (1980). The field measurement of reflectance factors. *Photogrammetric Engineering and Remote Sensing* **46**, 643–647.

Duggin, M. J. (1982). The need to use two radiometers simultaneously to make reflectance measurements in field conditions. *Photogrammetric Engineering and Remote Sensing* **48**, 142–144.

Dustan, P. (1985). Studies on the bio-optics of coral reefs. *In* "The Ecology of Coral Reefs: Symposium Series for Undersea Research, Vol. 3" (M. L. Reaka, ed.), pp. 189–198, NOAA, Rockville, MD.

Dustan, P., Dobson, E. and Nelson, G. (2001). Landsat thematic mapper: detection of shifts in community composition of coral reefs. *Conservation Biology* **15**, 892–902.

Falkowski, P. G., Jokiel, P. L. and Kinzie, R. A. (1990). Irradiance and corals. *In* "Ecosystems of the World: Coral Reefs, Vol. 25" (Z. Dubinsky, ed.), pp. 89–107. Elsevier Science, New York.

Fox, D. L. (1972). Pigmented calcerous skeletons of some corals. *Comparative Biochemistry and Physiology B* **43**, 919–927.

Fox, D. L. and Wilkie, D. W. (1970). Somatic and skeletally fixed carotenoids of the purple hydrocoral, *Allopora californica. Comparative Biochemistry and Physiology* **36**, 49–60.

Fox, D. L., Smith, V. E., Grigg, R. W. and Macleod, W. D. (1969). Some structural and chemical studies of the micro spicules in the fan-coral *Eugorgia ampla* Verrill. *Biochemical Physiology* **28**, 1103–1114.

Fuchs, E. (2001). Separating the fluorescence and reflectance components of coral spectra. *Applied Optics* **40**, 3614–3621.

Fux, E. and Mazel, C. (1999). Unmixing coral fluorescence emission spectra and predicting new spectra under different excitation conditions. *Applied Optics* **38**, 486–494.

Gilabert, M. A. and Melia, J. (1993). Solar angle and sky light effects on ground reflectance measurements in a citrus canopy. *Remote Sensing of Environment* **45**, 281–293.

Gil-Turnes, S. and Corredor, J. (1981). Studies of the photosynthetic pigments of zooxanthellae in Carribbean hermatypic corals. *In* "Proceedings of the Fourth International Coral Reef Symposium" (E. D. Gomez *et al.*, eds), pp. 51–54, University of the Philippines, Manila, Philippines.

Govindjee, and Braun, B. Z. (1974). Light absorption, emission and photosynthesis. *In* "Algal Physiology and Biochemistry" (W. D. P. Stewart, ed.), pp. 346–390. Blackwell, Oxford.

Green, E. P., Mumby, P. J., Edwards, A. J. and Clark, C. D. (1996). A review of remote sensing for the assessment and management of tropical coastal resources. *Coastal Management* **24**, 1–40.

Green, E. P., Mumby, P. J., Edwards, A. J. and Clark, C. D. (2000). "Remote Sensing Handbook for Tropical Coastal Management". UNESCO, Paris.

Hannach, G. (1989). Spectral light absorption by intact blades of *Porphyra abbottae* (Rhodophyta): effects of environmental factors in culture. *Journal of Phycology* **25**, 522–529.

Hardy, J. T., Hoge, F. E., Yungel, J. K. and Dodge, R. E. (1992). Remote detection of coral bleaching using pulsed-laser fluorescence spectroscopy. *Marine Ecology Progress Series* **88**, 247–255.

Helmuth, B. S. T., Sebens, K. P. and Daniel, T. L. (1997). Morphological variation in coral aggregations: branch spacing and mass flux to coral tissues. *Journal of Experimental Marine Biology and Ecology* **209**, 233–259.

Hochberg, E. F. and Atkinson, M. J. (2000). Spectral discrimination of coral reef benthic communities. *Coral Reefs* **19**, 164–171.

Hoegh-Guldberg, O. (1999). Climate change, coral bleaching and the future of the world's coral reefs. *Marine and Freshwater Research* **50**, 839–866.

Hoffman, E., Wrench, P. M., Sharples, F. P., Hiller, R. G., Welte, W. and Diederichs, K. (1996). Structural basis of light harvesting by carotenoids: peridinin-chlorophyll-protein from *Amphidium carterae*. *Science* **272**, 1788–1791.

Holden, H. M. and LeDrew, E. F. (1998a). The scientific issues surrounding remote detection of submerged coral ecosystems. *Progress in Physical Geography* **22**, 190–221.

Holden, H. M. and LeDrew, E. F. (1998b). Spectral discrimination of healthy and non-healthy corals based on cluster analysis, principal components analysis, and derivative spectroscopy. *Remote Sensing of Envronment* **65**, 217–224.

Holden, H. M. and LeDrew, E. F. (1998c). Hyperspectral identification of coral reef features in Fiji and Indonesia. *In* "Proceedings of the Fifth International Conference on Remote Sensing for Marine and Coastal Environments", pp. 78–84. Environmental Research Institute of Michigan, Ann Arbor.

Isdale, P. (1984). Fluorescent bands in massive corals record centuries of coastal rainfall. *Nature* **310**, 578–579.

Jaubert, J. (1981). Variations of the shape and of the chlorophyll concentration of the scleractinian coral *Synaraea convexa* Verrill: two complementary processes to adapt to light variations. *In* "Proceedings of the Fourth International Coral Reef Symposium" (E. D. Gomez *et al.*, eds), pp. 53–58. University of the Philippines, Manila, Philippines.

Jeffery, S. W., Sielicki, M. and Haxo, F. T. (1975). Chloroplast pigment patterns in dinoflagellates. *Journal of Phycology* **11**, 374–384.

Jeffery, S. W., Mantoura, R. F. C. and Wright, S. W. (editors) (1997a). "Phytoplankton Pigments in Oceanography: Guidelines to Modern Methods". UNESCO, Paris.

Jeffery, S. W., Mantoura, R. F. C. and Bjørnland, T. (1997b). Data for the identification of 47 key phytoplankton pigments. *In* "Phytoplankton Pigments in Oceanography: Guidelines to Modern Methods" (S. W. Jeffery, R. F. C. Mantoura and S. W. Wright, eds), pp. 449–559. UNESCO, Paris.

Kawaguti, S. (1944). On the physiology of reef corals. VI. Study on the pigments. *Palao Tropical Biological Station Studies* **2**, 617–673.

Kennedy, G. Y. (1979). Pigments of marine invertebrates. *Advances in Marine Biology* **16**, 309–381.

Kimes, D. S., Smith, J. A. and Ranson, K. J. (1980). Vegetation reflectance measurements as a function of solar zenith angle. *Photogrammetric Engineering and Remote Sensing* **46**, 1563–1573.

Kleppel, G. S., Dodge, R. E. and Reese, C. J. (1989). Changes in pigmentation associated with the bleaching of stony corals. *Limnology and Oceanography* **37**, 1331–1335.

Kriebel, K. T. (1976). On the variability of the reflected radiation field due to differing distributions of the irradiation. *Remote Sensing of Environment* **4**, 257–264.

Kühl, M. and Jørgensen, B. B. (1992). Spectral light measurements in microbenthic phototropic communities with a fibre-optic microprobe coupled to a sensitive diode array detector. *Limnology and Oceanography* **37**, 1813–1823.

Kühl, M., Cohen, Y., Dalsgaard, T., Jørgensen, B. B. and Revsbech, N. P. (1995). Microenvironment and photosynthesis of zooxanthellae in scleractinian corals studied with microsensors for O_2, pH and light. *Marine Ecology Progress Series* **117**, 159–172.

Lapointe, B. E. (1981). The effects of light on nitrogen and growth, pigment content, and biochemical composition of *Gracilaria foliifera* var. *augustissima* (Gigartinales, Rhodophyta). *Journal of Phycology* **17**, 90–95.

Lassen, C., Ploug, H. and Jørgensen, B. B. (1992). A fibre-optic scalar irradiance microsensor: application for spectral light measurements in sediments. *FEMS Microbiology Ecology* **86**, 247–254.

Latasa, M., Bidigare, R. R., Ondrusek, M. E. and Kennicutt, M. C. (1996). HPLC analysis of algal pigments: a comparison exercise among laboratories and recommendations for improved analytical performance. *Marine Chemistry* **51**, 315–324.

Latasa, M., Bidigare, R. R., Ondrusek, M. E. and Kennicutt, M. C. (1999). On the measurement of pigment concentrations by monochromator and diode-array spectrophotometers. *Marine Chemistry* **66**, 253–254.

Lesser, M. P., Mazel, C. H., Phinney, D. and Yentsch, C. S. (2000). Light absorption and utilization by colonies of the congeneric hermatypic corals *Montastraea faveolata* and *Montastraea cavernosa*. *Coral Reefs* **45**, 76–86.

Limbaugh, C. and North, W. J. (1956). Fluorescent, benthic, Pacific coast coelenterates. *Nature* **178**, 497–498.

Littler, D. S., Littler, M. M., Bucher, K. E. and Norris, J. N. (1989). "Marine Plants of the Caribbean". Smithsonian Institute, Washington.

Lubin, D., Li, W., Dustan, P., Mazel, C. H. and Stamnes, K. (2001). Spectral signatures of coral reefs: features from space. *Remote Sensing of Environment* **75**, 127–137.

Maritorena, S. and Guillocheau, N. (1996). Optical properties of water and spectral light absorption by living and non-living particles and by yellow substances in coral reef waters of French Polynesia. *Marine Ecology Progress Series* **131**, 245–255.

Mather, P. M. (1999). "Computer Processing of Remotely-Sensed Images, 2nd edn". John Wiley and Sons, Chichester.

Matz, M. V., Fradkov, A. F., Labas, Y. A., Savitsky, A. P., Zaraisky, A. G., Markelov, M. L. and Lukyanov, S. A. (1999). Fluorescent proteins from non-bioluminescent Anthozoa species. *Nature Biotechnology* **17**, 969–973.

Mazel, C. H. (1990). Spectral transformation of downwelling radiation by auto-fluorescent organisms in the sea. *Proceedings – SPIE the International Society for Optical Engineering, Ocean Optics X* **1302**, 320–327.

Mazel, C. H. (1995). Spectral measurements of the fluorescence emission in Caribbean cnidarians. *Marine Ecology Progress Series* **120**, 185–191.

Mazel, C. H. (1997a). Coral fluorescence characteristics: excitation–emission spectra, fluorescence efficiencies, and contribution to apparent reflectance. *Proceedings – SPIE the International Society for Optical Engineering* **2963**, 240–245.

Mazel, C. H. (1997b). Diver-operated instrument for *in situ* measurement of spectral fluorescence and reflectance of benthic marine organisms and substrates. *Optical Engineering* **36**, 2612–2617.

Milton, E. J. (1982). Field measurement of reflectance factors: a further note. *Photogrammetric Engineering and Remote Sensing* **48**, 1474–1476.

Miyazaki, T. and Harashima, A. (1993). Measuring the spectral signatures of coral reefs. *In* "13th Annual International Geoscience and Remote Sensing Symposium, Vol. 2", pp. 693–695. Institute of Electrical and Electronic Engineers, Tokyo.

Mobley, C. D. (1994). "Light and Water". Academic Press, San Diego.

Morise, H., Shimomura, O., Johnson, F. H. and Winant, J. (1974). Intermolecular energy transfer in the bioluminescent system of *Aequorea. Biochemistry* **13**, 2656–2662.

Morse, D. E., Hooker, N., More, A. N. C. and Jensen, R. A. (1988). Control of larval metamorphosis and recruitment in sympatric agariciid corals. *Journal of Experimental Marine Biology and Ecology* **116**, 193–217.

Muko, S., Kawasaki, K., Sakai, K., Takusa, F. and Shigesada, N. (2000). Morphological plasticity in the coral *Porites sillimaniani* and its adaptive significance. *Bulletin of Marine Science* **66**, 225–239.

Mumby, P. J. and Harborne, A. R. (1999). Development of a systematic classification scheme of marine habitats to facilitate regional management and mapping of Caribbean coral reefs. *Biological Conservation* **88**, 155–163.

Mumby, P. J., Green, E. P., Edwards, A. J. and Clark, C. D. (1997a). Measurement of seagrass standing crop using satellite and digital airborne remote sensing. *Marine Ecology Progress Series* **159**, 51–60.

Mumby, P. J., Green, E. P., Edwards, A. J. and Clark, C. D. (1997b). Coral reef habitat-mapping: how much detail can remote sensing provide? *Marine Biology* **130**, 193–202.

Mumby, P. J., Chisholm, J. R. M., Clark, C. D., Hedley, J. D. and Jaubert, J. (2001). A bird's-eye view of the health of coral reefs. *Nature* **413**, 36.

Mumola, P. B., Jarrett, O. and Brown, C. (1975). Multiwavelength LIDAR for remote sensing of chlorophyll in algae and phytoplankton. *In* "The Use of Lasers for Hydrographic Studies. NASA SP-375" (H. H. Kim and P. T. Ryan, eds), p. 207. National Aeronautics and Space Administration, Washington.

Myers, M. R., Hardy, J. T., Mazel, C. H. and Dustan, P. (1999). Optical spectra and pigmentation of Caribbean reef corals and macroalgae. *Coral Reefs* **18**, 179–186.

Nelson, N. B., Prézelin, B. B. and Bidigare, R. R. (1993). Phytoplankton light absorption and the package effect in California coastal waters. *Marine Ecology Progress Series* **94**, 217–227.

Ogden, J. C. and Lobel, P. S. (1978). The role of herbivorous fishes and urchins in coral reef communities. *Environmental Biology of Fishes* **3**, 49–63.

Phinney, J. T., Muller-Karger, F., Dustan, P. and Sobel, J. (2001). Using remote sensing to reassess the mass mortality of *Diadema antillarum* 1983–1984. *Conservation Biology* **15**, 885–891.

Ramus, J. (1978). Seaweed anatomy and photosynthetic performance: the ecological significance of light guides, heterogenous absorption and multiple scatter. *Journal of Phycology* **14**, 352–362.

Ramus, J. (1983). A physiological test of the theory of complementary chromatic adaptation. II. Brown, green and red seaweeds. *Journal of Phycology* **19**, 173–178.

Reichert, R. and Dawes, C. J. (1986). Acclimation of the green alga *Caulerpa racemosa* var. *unifera* to light. *Botanica Marina* **24**, 533–537.

Ricchiazzi, P., Yang, S., Gautier, C. and Sowle, D. (1998). SBDART: a research and teaching software tool for plane parallel radiative transfer in the Earth's atmosphere. *Bulletin of the American Meteorological Society* **79**, 2101–2114.

Rosenberg, G. and Ramus, J. (1982). Ecological growth strategies in the seaweeds *Gracilaria foliifera* (Rhodophyceae) and *Ulva* sp. (Chlorophyceae): photosynthesis and antenna composition. *Marine Ecology Progress Series* **8**, 233–241.

Rinkevich, B. and Loya, Y. (1983). Short term fate of photosynthetic products in a hermatypic coral. *Journal of Experimental Marine Biology and Ecology* **73**, 175–184.

Rønneberg, H., Borch G., Fox, D. L. and Liaaen-Jensen, S. (1979a). Alloporin, a new carotenoprotein. *Comparative Biochemistry and Physiology B* **62**, 309–312.

Rønneberg, H., Fox, D. L. and Liaaen-Jensen, S. (1979b). Animal carotenoids – carotenoproteins from hydrocorals. *Comparative Biochemistry and Physiology* B **64**, 407–408.

Rowan, R. (1998). Diversity and ecology of zooxanthellae on coral reefs. *Journal of Phycology* **34**, 407–417.

Rüdiger, W., Klose, W., Tursch, B., Houvenaghel-Crevecoeur, N. and Budzikiewicz, H. (1968). Isolierung von biliverdin $1X_\alpha$ aus der blaue koralle *Heliopora caerulea* Pall. *Liebig's Annalen der Chemie* **713**, 209–211.

Rundquist, D. C., Han, L., Schalles, J. F. and Peake, J. S. (1996). Remote measurement of algal chlorophyll in surface waters: the case for the first derivative of reflectance near 690 nm. *Photogrammetric Engineering and Remote Sensing* **62**, 195–200.

Salih, A., Larkum, A., Cox, G., Kühl, M. and Hoegh-Guldberg, O. (2000). Fluorescent pigments in corals are photoprotective. *Nature* **408**, 850–853.

Shibata, K. (1969). Pigments and a UV-absorbing substance in corals and a blue-green alga living in the Great Barrier Reef. *Plant and Cell Physiology* **10**, 325–335.

Shimomura, O., Johnson, F. H. and Saiga, Y. (1962). Extraction, purification, and properties of aequorin, a bioluminescent hydromedusan, *Aequorea*. *Journal of Cellular and Comparative Physiology* **77**, 305–312.

Smith, R. C. and Baker, K. S. (1981). Optical properties of the clearest natural waters (200–800 nm). *Applied Optics* **20**, 177–184.

Stimson, J. (1997). The annual cycle of density of zooxanthellae in the tissues of field and laboratory-held *Pocillopora damicornis*. *Marine Ecology Progress Series* **23**, 153–164.

Takabayashi, M. and Hoegh-Guldberg, O. (1995). Ecological and physiological differences between two colour morphs of the coral *Pocillopora damicornis*. *Marine Biology* **123**, 705–714.

Tanner, J. E. (1995). Competition between scleractinian corals and macroalgae: an experimental investigation of coral growth, survival and reproduction. *Journal of Experimental Marine Biology and Ecology* **190**, 151–168.

Titlyanov, E. A. (1981). Adaptation of reef-building corals to low light intensity. *In* "Proceedings of the Fourth International Coral Reef Symposium, Vol. 2" (E. D. Gomez *et al.*, eds), pp. 39–43. University of the Philippines, Manila, Philippines.

Topinka, J. A., Korjeff-Bellows, W. and Yentsch, C. S. (1990). Characterization of marine macroalgae by fluorescence signatures. *International Journal of Remote Sensing* **11**, 2329–2335.

Tsai, F. and Philpot, W. (1998). Derivative analysis of hyperspectral data. *Remote Sensing of Environment* **66**, 41–51.

Tsien, R. Y. (1998). The green fluorescent protein. *Annual Review of Biochemistry* **67**, 509–544.

Veron, J. E. N. (1993). "Corals of Australia and the Indo-Pacific". University of Hawaii Press, Honolulu.

Vogelmann, T. C. and Björn, L. O. (1986). Plants as light traps. *Physiologia Plantarum* **68**, 704–708.

Waaland, J. R., Waaland, S. D. and Bates, G. (1974). Chloroplast structure and pigment composition in the red alga *Griffithsia pacifica*: regulation by light intensity. *Journal of Phycology* **10**, 193–199.

Wilkinson, C. R. (1999). Global and local threats to coral reef functioning and existence: review and predictions. *Marine and Freshwater Research* **50**, 867–878.

Yentsch, C. S. and Yentsch, C. M. (1979). Fluorescence spectral signatures: the characterization of phytoplankton populations by the use of excitation and emission spectra. *Journal of Marine Research* **37**, 471–483.

Yokohoma, Y., Kageyama, A., Ikawa, T. and Shimura., S. (1977). A carotenoid characteristic of chlorophycean seaweeds living in deep coastal waters. *Botanica Marina* **20**, 433–436.

Young, A. J. and Frank, H. A. (1996). Energy transfer reactions involving carotenoids: quenching of chlorophyll fluorescence. *Photochemistry and Photobiology* **36**, 3-15.

Zagalsky, P. F. and Herring, P. J. (1977). Studies on the blue astaxanthin-proteins of *Velella velella* (Coelenterata: Chondrophora). *Philosophical Transactions of the Royal Society of London* B **279**, 289–326.

Taxonomic Index

Note: page numbers in *italics* indicate tables and diagrams

Subject Index

Note: **bold** page numbers indicate major entries; *italics* are used for illustrations, maps and diagrams